338
Topics in Current Chemistry

For further volumes:
http://www.springer.com/series/128

Aims and Scope

The series Topics in Current Chemistry presents critical reviews of the present and future trends in modern chemical research. The scope of coverage includes all areas of chemical science including the interfaces with related disciplines such as biology, medicine and materials science.

The goal of each thematic volume is to give the non-specialist reader, whether at the university or in industry, a comprehensive overview of an area where new insights are emerging that are of interest to larger scientific audience.

Thus each review within the volume critically surveys one aspect of that topic and places it within the context of the volume as a whole. The most significant developments of the last 5 to 10 years should be presented. A description of the laboratory procedures involved is often useful to the reader. The coverage should not be exhaustive in data, but should rather be conceptual, concentrating on the methodological thinking that will allow the non-specialist reader to understand the information presented.

Discussion of possible future research directions in the area is welcome.

Review articles for the individual volumes are invited by the volume editors.

Readership: research chemists at universities or in industry, graduate students.

Lars T. Kuhn
Editor

Hyperpolarization Methods in NMR Spectroscopy

With contributions by

Ü. Akbey · A. Alia · D.A. Barskiy · B.E. Bode · L. Buljubasich · S.B. Duckett · W.T. Franks · M.B. Franzoni · M. Goez · K.B.S.S. Gupta · G. Jeschke · I.V. Koptyug · K.V. Kovtunov · L.T. Kuhn · S. Lange · A. Linden · J. Matysik · R.E. Mewis · K. Münnemann · M. Orwick-Rydmark · H. Oschkinat · I.V. Skovpin · S.S. Thamarath · V.V. Zhivonitko

Editor
Lars T. Kuhn
Europ. Neuroscience Institute
Göttingen
Göttingen, Germany

ISSN 0340-1022 ISSN 1436-5049 (electronic)
ISBN 978-3-642-39727-1 ISBN 978-3-642-39728-8 (eBook)
DOI 10.1007/978-3-642-39728-8
Springer Heidelberg New York Dordrecht London

Library of Congress Control Number: 2013948609

Printed on acid-free paper

Springer is part of Springer Science+Business Media (www.springer.com)

Preface

Nuclear magnetic resonance (NMR) spectroscopy is arguably one of the most potent experimental techniques for probing molecular structure at ultra-high resolution both in liquids and solids. Despite its pivotal role in chemical analysis both in biological and non-biological contexts, the technique itself suffers inherently from low sensitivity, owing to the extremely small energy separation existing between nuclear spin states in the presence of the static magnetic field of an NMR spectrometer. Thus, a range of different measures aimed at increasing nuclear polarization has been established ever since the introduction of NMR as an analytical technique, including, among others, the use of sensitivity-enhancing pulse sequences, higher magnetic fields, cryogenically cooled probes, and particularly in the context of the present contribution, the application of methods to increase nuclear sensitivity in a chemical or physical manner. In particular, the latter approach has yielded highly satisfactory results, making it an extremely attractive alternative to some of the more costly setup modifications described before.

Here, key players working in the area of nuclear hyperpolarization methods in NMR spectroscopy have gathered to present the state-of-the-art and to report on cutting-edge developments accomplished in the field during the last 10–15 years or so. The selection of contributions highlights, in particular, nuclear hyperpolarization methods based on two chemical, i.e., *para*-Hydrogen Induced Polarization (PHIP) and *photo*-Chemically Induced Dynamic Nuclear Polarization (CIDNP), and a physical phenomenon, i.e., dynamic nuclear polarization (DNP), which can be exploited to perturb nuclear spin state distributions in situ, i.e., inside the NMR spectrometer during an ongoing chemical or physical process. Interestingly, all of these methods have experienced renewed interest in recent years, although their underlying chemical or physical principles have been known for a relatively long time. In addition, continuous efforts to improve methodologically the different techniques have transformed their application from purely a means to increase spectral sensitivity to a valuable tool for addressing problems both in mechanistic organic chemistry and, structural biology. The emerging importance of magnetic resonance imaging/tomography (MRI/T) techniques has fuelled the interest in hyperpolarizing nuclear spins using in situ NMR methods, e.g., DNP or PHIP, even further. Nevertheless, the present collection of texts does not claim to provide a complete picture of the entire field of nuclear hyperpolarization as this would have surely exceeded the intended boundaries of this book, both in size and scope. For example, the area of hyperpolarized noble gases has, on purpose, been omitted entirely. In principle, this volume of "Topics in Current Chemistry" is believed to be suitable both for scientists working in the field of nuclear hyperpolarization and for researchers expecting to get a first glimpse of the subject. In addition, spectroscopists entirely unfamiliar with the methods presented here might use this collection of texts as a source of thought-provoking impulses to add, if desired, sensitivity-enhancing features to their particular experiments.

Last but not least, I would like to thank sincerely all contributors for their efforts and patience to make this highly stimulating and, in my humble opinion, worthwhile endeavor possible. On a more personal note, deep gratitude is owed to Joe Bargon, a good friend and scientific mentor, who first kindled the editor's interest in this fascinating area of nuclear magnetic resonance research.

Göttingen, Germany
June 2013

Lars T. Kuhn

Contents

Top Curr Chem (2013) 338: 1–32
DOI: 10.1007/128_2012_348

Published online: 22 September 2012

Elucidating Organic Reaction Mechanisms Using Photo-CIDNP Spectroscopy

Martin Goez

Abstract CIDNP (chemically induced dynamic nuclear polarization) arises in radical pairs or biradicals but is detected in the diamagnetic reaction products. Hence, it can be used not only to identify and characterize both types of species but also to establish the pathways connecting precursors, paramagnetic intermediates and products, and to employ the polarizations as labels to individual nuclei. Recent applications of CIDNP to elucidate the mechanisms of photochemical reactions are reviewed, which illustrate all these facets.

Keywords Biradicals · CIDNP-Spectroscopy · Photochemistry · Radicals · Reaction mechanisms

Contents

M. Goez (✉)
Institut für Chemie, Martin-Luther-Universität Halle-Wittenberg, Kurt-Mothes-Str. 2, 06120 Halle/Saale, Germany
e-mail: martin.goez@chemie.uni-halle.de

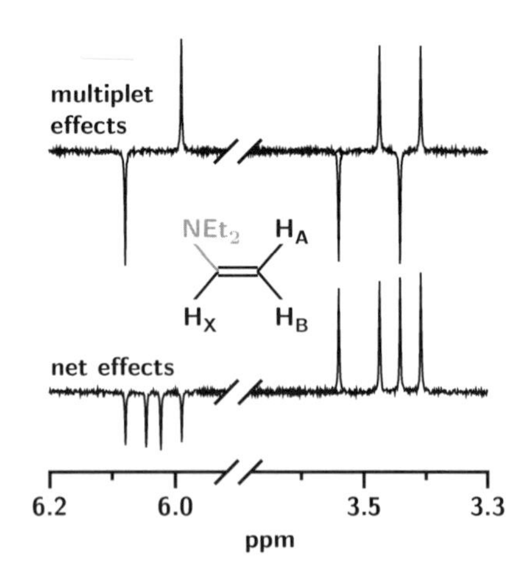

Fig. 1 CIDNP net (*bottom*) and multiplet effects (*top*) in the photoreaction of excited xanthone with triethylamine in acetonitrile. Shown are spectra of the product *N*, *N*-diethylvinylamine (for the formula, see *inset*); left half, signals of H_X; right half, signals of H_A and H_B. For further explanation, see text

1 The CIDNP Effect

In a nutshell, CIDNP ("chemically induced dynamic nuclear polarizations"; for earlier reviews, see, e.g., [1–5]) means that transient anomalous NMR line intensities are observed in the products of chemical reactions that are carried out in a magnetic field, usually that of the spectrometer, and involve radical intermediates.

The spectra of Fig. 1 illustrate that. The obviously anomalous signal patterns in the two traces only persist for a time on the order of T_1; after relaxation, the normal NMR signal intensities would result in principle but are essentially unobservable in the example because they are significantly weaker and the product is only metastable.

Nuclear spin polarizations are populations of the nuclear spin states that deviate from the equilibrium populations. Only the line intensities are thus affected, not the resonance frequencies. There are two important limiting cases that can easily be isolated experimentally. The bottom trace of Fig. 1 shows so-called net effects. Each line of a multiplet is scaled with the same positive or negative factor, so the appearance of each multiplet, apart from its overall scaling factor relative to other signals in the spectrum, is as normal; in the product operator formalism, net effects are longitudinal one-spin order, such as I_{1z} or I_{2z}. Quite different from these are the spectral habits of multiplet effects, as displayed in the top trace of the figure. For these the intensities within multiplets exhibit up/down patterns, which are due to longitudinal two-spin order, e.g., $2I_{1z}I_{2z}$. In general, CIDNP causes anomalies of all populations, and thus of all line intensities, but any population distribution can be described as a superposition of one-spin order, two-spin order, and higher spin order, so net and multiplet effects are the first, and most important terms in that expansion. As is well known, a pulse of flip angle ν transforms n-spin order into an observable signal of amplitude proportional to $\sin \nu \cos^{n-1} \nu$. Separating net and multiplet effects by the flip-angle dependence is thus a straightforward protocol [6]. As a special case, for $n > 1$ this expression is zero for $\nu = 90°$, so observation with $\pi/2$ pulses suppresses all multiplet and higher effects, and leaves only net effects.

CIDNP was discovered in 1967 by Bargon, Fischer, and Johnsen [7], and independently by Ward and Lawler [8]; in 1969, Closs [9], and – again independently – Kaptein and Oosterhoff [10], provided its now firmly established theoretical framework. An initial misinterpretation as an electron-nuclear Overhauser effect is responsible for the fragment "dynamic nuclear polarizations" of its name, which found so much favor in the scientific community that it was never changed, even though the polarizations are now known to arise through a completely different mechanism (see Sect. 2.1).

The other half of the name, "chemically induced," is correct, with the restriction that only radical and biradical intermediates cause CIDNP. In the very early days, those intermediates were produced by heating the samples but nowadays the scale has tipped almost completely in favor of photochemical radical generation, partly because this makes it possible to switch the chemical reaction on and off easily and rapidly, partly because relatively cheap lasers can be used to carry out time-resolved CIDNP experiments on a submicrosecond timescale, and partly because photochemistry offers a richer set of problems that can be tackled by CIDNP.

With its 50th birthday looming, CIDNP has certainly reached maturity. Some of its applications, e.g., probing the solvent accessibility of amino acid residues in proteins [11], have been taken up into the standard experimental repertoire by other disciplines. Chemical applications, such as using CIDNP as a tool to investigate complex reaction mechanisms, recently seem to have moved out of the focus of interest a little. Chemists possibly shun the method because its theoretical background and mathematical treatment involve some concepts less familiar to them while – no provocation intended – physicists, who probably have fewer problems with the theory, might not always recognize what a chemically relevant problem is. However, in the eyes of the present reviewer, the possibility to elucidate complex organic reaction mechanisms is one of the greatest assets of CIDNP.

In Sect. 2 the necessary theoretical background is discussed; Sect. 3 describes hardware and pulse sequences of CIDNP experiments. Both sections have deliberately been kept very brief because this review focuses on using CIDNP for mechanistic investigations. For more detailed treatments of theory and instrumentation, the reader is referred to another recent review on CIDNP [3]. Section 4 puts the specific questions that CIDNP can help address into the context of elucidating reaction mechanisms, illustrating them with case studies, and compiling relevant examples from the recent literature.

2 CIDNP Theory

2.1 Radical Pair Mechanism

Radical *pairs* are the key intermediates for CIDNP and related phenomena, e.g., magnetic field effects. As any survey of textbooks on mechanistic organic chemistry will convincingly demonstrate, chemists are trained to think of free radicals as

Chart 1 Radical pair mechanism. For further explanation, see text

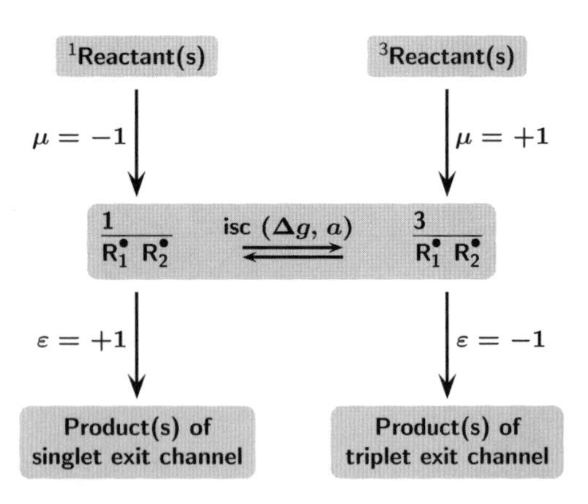

common intermediates and would regard radical pairs as a rare variant. However, the opposite is true. When radicals are formed from nonradical precursors (singlets as in almost all thermal reactions, or triplets as in many photochemical reactions) they are formed in pairs, and when radicals disappear to give nonradical products (singlets in the majority of cases) they disappear in pairs.

One fundamental principle of chemistry is spin conservation during elementary reaction steps. As its immediate consequence, a triplet entity must give a triplet product and a singlet entity a singlet product. Hence, if a gross reaction affords a product of a multiplicity that differs from that of the reactants, intersystem crossing must have taken place as an intervening physical process. The radical pair mechanism [9, 10] is one of the simplest conceivable instances of that; it comprises only the reactants, one intermediate (a radical pair), which undergoes intersystem crossing, and the products (see Chart 1).

Generation of the radical pair is necessarily the first step. Spin conservation demands that the pair is born with the electron-spin multiplicity of its precursors, which implies that the electron spins of its two radicals, $R_1^\bullet$ and $R_2^\bullet$, must be correlated accordingly. To indicate that correlation, an overbar is used, with the multiplicity given as a superscript. Because there are two possible precursor multiplicities, the diagram has two entry channels and comprises both the singlet and the triplet radical pair.

Similar considerations apply to the exit channels leading from the pair to the products. There must be two of them because a singlet pair can only yield a singlet product and a triplet pair a triplet product. In the majority of cases, however, a triplet "product" initially means a nonproduct because the radical pair very often is so much lower in energy than a triplet molecule that formation of the latter is precluded. Under these circumstances, the radicals diffuse apart permanently, become free (i.e., no longer spin-correlated) radicals, and react individually with other molecules in the sample. In contrast, a reaction of the two radicals of a singlet pair with each other is normally fully allowed thermodynamically. The terms "escape products" and "cage products," which are frequently used as synonyms for products through the triplet and singlet exit channels, reflect that behavior but are not generally applicable, because formation of true triplet products can also occur in the cage, i.e., by a geminate reaction of the two radicals, e.g., in the deactivation of photogenerated radical ion pairs by reverse electron transfer [12, 13].

Without intersystem crossing, the diagram of Chart 1 could thus only be traversed vertically, via either of the two pathways, singlet reactants → singlet products or triplet reactants → triplet products. However, radical pairs can undergo intersystem crossing, which provides a leakage between the two vertical pathways, so also allows a diagonal route through the diagram, e.g., triplet reactants → singlet products.

The mechanism of intersystem crossing peculiar to radical pairs will be explained in Sect. 2.2. It is driven by Zeeman interactions (g values of the radicals) and hyperfine interactions (hyperfine coupling constants a) in the pairs. Through the latter interactions, the intersystem crossing rate further depends on the nuclear spin states of the radicals $R_1^\bullet$ and $R_2^\bullet$. For certain nuclear spin states it is faster, for others slower. Hence, after radical pairs have been produced through, e.g., the triplet entry channel, those pairs containing the "faster" nuclear spins will leak more to the singlet side than those containing the "slower" ones. Focusing on the nuclear spins rather than on the molecules, the "faster" nuclear spins are seen to be enriched in the singlet pairs and depleted in the triplet pairs; the opposite holds for the "slower" nuclear spins. These deviations from the initial nuclear spin-state populations ultimately show up in the respective products, where they can be detected by NMR.

Regarding two nuclear spin states of a product that are connected by an NMR transition, an overpopulation of the lower of the two will produce an enhanced absorption signal, and an overpopulation of the higher one an emission signal. Because the overpopulations are relative to the Boltzmann populations, which are almost equal for all nuclear spin states, even at the highest available fields, even a small disparity of the intersystem crossing rates can produce a noticeable CIDNP effect.

It must be stressed that the nuclear spins are never flipped in the radical pair mechanism; they are only sorted, and that sorting is bidirectional. Thus for the nuclear spin polarizations it is unimportant whether the diagram of Chart 1 is traversed from singlet reactants to triplet products or from triplet reactants to singlet products – the "faster" nuclear spins will always be enriched in the product corner opposite to the reactant corner. By the same reasoning it can be seen that there is also no qualitative difference between the routes singlet reactants → singlet products and triplet reactants → triplet products, but that diagonal and vertical routes give rise to exactly opposite polarizations. These features are nicely modeled by assigning a variable μ to the entry channel that takes the value +1 for triplet and -1 for singlet, and a variable ε to the exit channel with the opposite assignment of these numbers to the multiplicities. Multiplying μ and ε yields +1 for any diagonal route through the diagram and -1 for any vertical route. Together with the signs of the magnetic parameters, this forms the basis of Kaptein's rules [14] for the CIDNP signs (see below).

As a fairly obvious first consequence of the mechanism, the sorting is undone for any product that is formed with the same probability through both exit channels.

A second, perhaps slightly less obvious consequence is that no polarizations can arise if singlet and triplet radical pairs are formed with the same probability, because then the leakages from the singlet pairs to the triplet pairs and in the opposite direction exactly cancel.

In summary, there are three indispensable prerequisites for CIDNP by the radical pair mechanism:

1. Generation of radical pairs with a predominance of one electron-spin multiplicity
2. Nuclear-spin selective intersystem crossing of the pairs
3. A disbalance of reaction rates for the two exit channels

The life of a radical pair is short, a few nanoseconds at most in homogeneous phase. Within that time, all the spin sorting and the recombination of the two radicals of the pair with each other occur; in other words, CIDNP generation is completed on that timescale. Free radicals, however, usually live much longer. The spin polarizations they have inherited from the radical pair are lost to some extent by relaxation in these paramagnetic intermediates, but the part that is not lost will eventually turn up in the subsequent products with the rate of free radical decay (i.e., on a much slower timescale than the polarizations in the products of a direct radical-pair collapse).

2.2 Intersystem Crossing of Radical Pairs

Given a two-state system with energy difference ΔE of its states and a mixing interaction of strength M, the time-dependent probability $p_2(t)$ of finding the system in state 2 when it initially was completely in state 1 is, according to time-dependent perturbation theory [15]

$$p_2(t) = \frac{1}{2} \frac{(2M)^2}{(\Delta E)^2 + (2M)^2} \left[1 - \cos \sqrt{(\Delta E)^2 + (2M)^2} t \right], \qquad (1)$$

where energies are specified in angular frequencies. In the limiting case of degenerate states ($\Delta E = 0$), the system is seen periodically to exist in the pure state 2 regardless of how weak the mixing interaction might be.

Unfortunately, the other limiting situation, $|\Delta E| \gg |M|$, for which the probability of finding the system in state 2 cannot exceed the small value $(2M/\Delta E)^2$, seems to be realized in our case. Magnetic interactions M of radicals, i.e., Zeeman and hyperfine interactions, are at maximum comparable to kT. This contrasts unfavorably with the singlet–triplet splittings ΔE in organic molecules, which are at least several tens of kJ/mol. How can the tiny magnetic interactions thus effect intersystem crossing to any measurable degree?

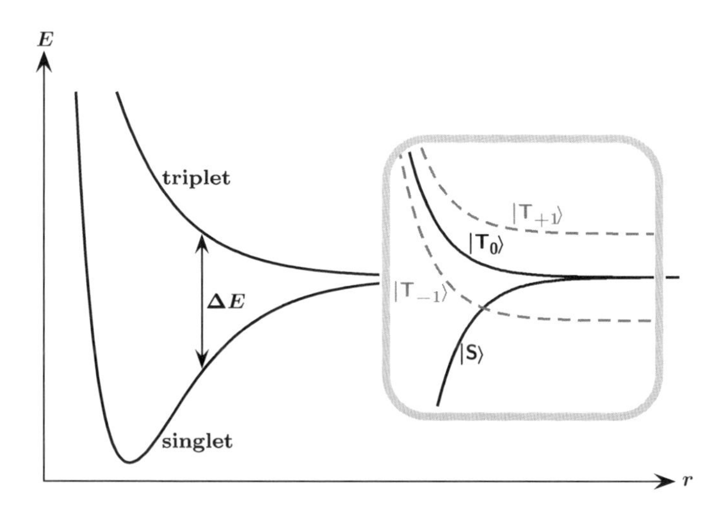

Fig. 2 Potential energy E of a radical pair of different multiplicity as function of the interradical separation r. ΔE is the singlet-triplet splitting. The *inset* shows an enlarged view of the additional splitting of the triplet sublevels in a magnetic field. For further explanation, see text

The answer lies in the fact that ΔE is distance dependent and a radical pair is a dynamic system. In homogeneous phase the radicals of a pair are not constrained to keep together, so immediately after their generation diffusion sets in and separates them. The mixing interaction M is not influenced by a separation but the energy difference ΔE is a strong function of the interradical distance [16] because its origin is the electrostatic interaction between the electrons.

As Fig. 2 shows, the energies of singlet and triplet converge at sufficient interradical separations (i.e., larger than a few molecular diameters), but the magnetic field of an NMR spectrometer splits off the triplet sublevels $|T_{+1}\rangle$ and $|T_{-1}\rangle$, so only $|S\rangle$ and $|T_0\rangle$ become degenerate. Hence, despite the smallness of the magnetic interactions, intersystem crossing between $|S\rangle$ and $|T_0\rangle$ is feasible for all interradical distance exceeding a critical distance. (As Fig. 2 indicates, it would also be possible between $|S\rangle$ and $|T_{-1}\rangle$ at a specific separation, but this comparatively rare case, which is only realized for biradicals or micellar systems and gives rise to different CIDNP phenomena [17–19], is outside the scope of this review).

The described distance dependence of ΔE has an interesting consequence for the eigenfunctions of a radical pair, which is summed up in Chart 2. $|S\rangle$ and $|T_0\rangle$ are only eigenfunctions when the two radicals are near each other; when $R_1^\bullet$ and $R_2^\bullet$ are well separated, their electron spins become independent, so the eigenfunctions are two doublets, $|D_1\rangle$ and $|D_2\rangle$, instead. Vector models [20] provide an intuitive, although quantum mechanically shaky, illustration of that concept. When the two radicals are in contact, the two vectors symbolizing the two electron spins have to be drawn in a fixed orientation relative to each other, antiparallel for $|S\rangle$ (vanishing total spin) and symmetric to, as well as in a plane perpendicular to, the xy plane for $|T_0\rangle$ (nonzero total spin but zero component along the quantization axis z. At large interradical

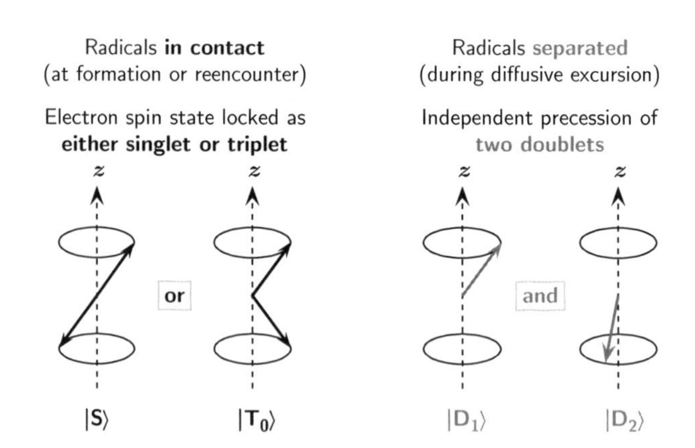

Chart 2 Vector model of a radical pair in different circumstances. For further explanation, see text

separations the two electron spins are decoupled, so each acquires a life of its own and processes independently from the other under the influence of Zeeman and hyperfine interactions; hence two individual vectors have to be drawn in the model. Even in the infrequent case of two chemically equivalent radicals, the probability that the populations of all corresponding nuclear spin states in $R_1^\bullet$ and $R_2^\bullet$ are identical is extremely small, so the precession frequencies will practically always be different. The effect of such a differential precession for some time is illustrated in the figure. The two independent arrows get out of phase. It is immediately perceived that at regular intervals their relative orientation will be that of $|S\rangle$ and, exactly halfway between those points of time, that of $|T_0\rangle$. The system thus oscillates between these two pure states, and at any time its state is describable as a superposition of $|S\rangle$ and $|T_0\rangle$ with time-dependent coefficients.

Because singlet and triplet are not eigenfunctions under these circumstances, they do not correspond to states that can be observed. However, the probability that the two radicals of the pair reencounter at some later time is substantial, and these reencounters provide the last ingredient to this mechanism: As the radicals approach each other, the exchange interaction rises steeply, and forces the pair into one of its eigenstates at contact, $|S\rangle$ or $|T_0\rangle$, with a probability given by the square of the respective coefficient in the superposition state. The ensuing multiplicity-dependent chemical reactions of the pair in contact serve to distinguish whether or not (or, in an ensemble of radical pairs, to what degree) intersystem crossing has taken place during the diffusive excursion of the radicals.

2.3 *CIDNP Intensities and Signs*

In the high field of an NMR magnet, the electron and nuclear spin states of a radical pair can be specified independently because the electron Zeeman interactions are much larger than the hyperfine interactions. Under these circumstances, the expression for the mixing interaction *M* becomes fairly simple:

$$M = \frac{1}{2}\left\{\Delta g \beta B_0 + \sum_i a_i m_i - \sum_i a_j m_j\right\}. \tag{2}$$

The radical containing the observed nucleus is denoted as radical 1, the other as radical 2. The first term of Eq. (2) describes the dephasing caused by the Zeeman interactions of the two electron spins (Δg, difference $g_1 - g_2$, with the g values of the two radicals; β, Bohr magneton; B_0, field of the magnet) and the last two terms the dephasing caused by the hyperfine interactions (a_k and m_k, hyperfine coupling constant and z component of nuclear spin for nucleus k; the positive sign and index i refer to the nuclei in radical 1, the negative sign and index j to those in radical 2).

Calculations of absolute CIDNP intensities are feasible by computing reaction yields for each nuclear spin state and obtaining NMR line intensities from the resulting nuclear spin populations in the products. As key steps of the computation, the coherent oscillation of the density matrix elements that is determined by Eq. (2) is weighted with the conditional probability of a first reencounter and averaged over time; multiple diffusive excursions are then taken into account by summing up a geometric series of matrices. Details can be found elsewhere [2, 3].

Uncertainties in these calculations arise from the fact that properties related to the diffusive excursions (e.g., starting and reencounter distances) and to the reactivities (e.g., probabilities of product formation upon reencounter of a pair with given multiplicity) are often not precisely known; this is aggravated by the considerable number of input parameters. As another obstacle, the number of radical pairs is often difficult to determine. It is, therefore, much safer to base mechanistic conclusions on relative CIDNP intensities, which removes or at least greatly reduces the described sources of errors.

One important quantitative result of these calculations is that the net polarization P_k of nucleus k in the products is determined by the square root $\sqrt{|M|}$ of the mixing matrix element of Eq. (2); by a first-order expansion one finds that for large Δg the net polarization is approximately proportional to the pertaining hyperfine coupling constant a_k divided by the square root of the Zeeman-frequency difference:

$$P_k \propto \frac{a_k}{(\Delta g \beta B_0)^{1/2}}. \tag{3}$$

This fact has long been known and exploited [21]; a series of recent studies retesting it has been performed [22–24].

Equation (3) has two implications. On one hand, the polarization pattern, meaning the relative polarization intensities of the different nuclei in a product, reflects the hyperfine coupling constants in the radical pair, that is, constitutes a sort of frozen EPR spectrum of that intermediate. The value of this for mechanistic investigations is obvious. On the other hand, CIDNP effects are seen to decrease not only with increasing Δg but also with increasing field strength B_0. Because the equilibrium populations increase linearly with B_0, CIDNP in today's strong magnets leads to much smaller signal enhancements as in the pioneer days.

In addition to the line intensities, the signs (absorption or emission), also called phases, of the polarizations carry important mechanistic information. Obviously one cannot distinguish whether the oscillation described by Eq. (2) is clockwise or counterclockwise, so only the magnitude of M is important for intersystem crossing. Moreover, CIDNP detects only differences of the intersystem crossing rates for pairs of nuclear spin states m_k and $m_{k'}$ that are connected by an NMR transition, so what counts is whether $|M|$ is larger or smaller for m_k and $m_{k'}$.

A radical pair with a positive Δg (i.e., $g_1 > g_2$), one proton in radical 1, and a positive hyperfine coupling constant serves as the simplest example for a CIDNP net effect. As is immediately seen from Eq. (2), the intersystem crossing rate is larger for $m = +1/2$ than for $m = -1/2$ because in the former case Zeeman and hyperfine interactions reinforce each other while in the latter case they partly compensate. With the mechanism of Chart 1 that leads to a surplus of molecules mit $m = +1/2$ for a diagonal pathway, and thus an enhanced absorption signal in a product formed via that pathway; as already mentioned, a vertical pathway yields the opposite result, i.e., an emission signal. Generalization of that example is easy; it emerges that only the relative signs of Δg and a_k are of importance for the CIDNP phases, in addition to the already defined (Chart 1) symbolic parameters μ and ε. Kaptein's rule for a CIDNP net effect [14] of nucleus k sums up the result:

$$\Gamma_k = \mu \times \varepsilon \times \mathrm{sign}(\Delta g) \times \mathrm{sign}\, a_k, \tag{4}$$

where $\Gamma_k = +1$ denotes enhanced absorption and $\Gamma_k = -1$ emission.

The above example also shows that a CIDNP net effect relies on both Δg and a_k being nonzero because either interaction on its own results in the same value of $|M|$ for either spin state.

An expression similar to Eq. (4) has been given for CIDNP multiplet effects [14]. However, in the high fields of modern NMR spectrometers, multiplet effects – which are largest if there is no differential Zeeman precession – are comparatively rare, so the equation will not be reproduced here.

Finally, interesting CIDNP effects arise if the reaction proceeds through two or more consecutive radical pairs, with nuclear spin sorting possible for each of them [25–28]. The result of such a "pair substitution" is not simply a superposition of CIDNP from the individual pairs weighted with their lifetimes because each pair inherits the complete state of the density matrix of its precursor pair, which includes a phase correlation that is not directly observable but can be converted into polarizations in the later pair.

3 Instrumentation and Pulse Sequences

3.1 *Hardware*

No modification to an ordinary modern NMR spectrometer is necessary but a pulsed field-gradient unit is almost essential, not least because the typical methods to get the light into the sample strongly disturb the field homogeneity, so shimming without gradients would be extremely difficult. For the frequent case of illumination by pulsed lasers, a spare line of the spectrometer (i.e., a fast digital output that can be addressed as part of the pulse sequence) is very helpful for synchronization; however, in the absence of a spare line an unused transmitter channel can also serve that purpose.

A continuous light source, such as a high pressure arc lamp, possibly combined with filters for wavelength-dependent experiments and with a shutter to allow gated illumination, is sufficient for almost all mechanistic experiments. Some additional flexibility is offered by a pulsed laser, which can accommodate not only that mode of measurement (quasi-continuous monochromatic illumination by a train of laser flashes in quick succession followed by an NMR pulse sequence) but also the less frequently employed time-resolved variant [29] of CIDNP (single laser flash – variable delay – NMR pulse).

To transport the light into the active region of the sample, a single quartz fiber can be used at lower power densities whereas a fiber bundle or quartz rod is almost indispensable when a pulsed laser is employed. The light guide is pushed down until to the top of the active region, so most of the light is absorbed in that part of the sample where the polarizations generated are detected with the maximum sensitivity.

A much more detailed review of these hardware aspects has recently been published [3].

3.2 *Pulse Sequences*

During the last decade, considerable progress has been made in the field of pulse sequences that address the two very different problems that the two modes of CIDNP experiments, time-resolved and stationary, are confronted with.

With time-resolved CIDNP, sensitivity is the main problem. The number of radical pairs produced by one laser flash, and thus the amount of polarization, is quite small, whereas the noise of the acquisition does not in any way decrease with the number of observed molecules but is fixed. Signal averaging will improve the signal-to-noise ratio in the usual way, by a factor of $\sqrt{n}$ when n signals are coadded, but because CIDNP means observation on a chemically reacting system, that approach is hampered by sample decomposition, which increases linearly with n. As a novel approach, pulse sequences have been devised [30, 31] that store and add up time-resolved CIDNP signals in the spin system itself, which is essentially noise-free, and use only a single acquisition per n flashes to read out the accumulated signal. In that way, the signal-to-noise ratio could ideally increase linearly with n. Because time-resolved experiments are not a necessity for mechanistic work, and because they are experimentally more demanding than stationary CIDNP experiments, a more detailed explanation of these pulse sequences is not warranted in this review, and the interested reader is referred to the literature cited.

In contrast, sensitivity is much less of an issue with stationary CIDNP experiments, but they face the problem of background signals. The situation of Fig. 1, where there is no background because the observed product is not there initially and conveniently decomposes fairly quickly, is comparatively rare. In those cases where CIDNP of the regenerated starting material, which can provide important mechanistic information, needs to be recorded, those signals are evidently superimposed on the equilibrium signals of the remaining reactants. As an even worse complication, polarized resonances of a product can fall into the same spectral range as the equilibrium signals of another compound that needs to be employed in high concentration, e.g., a quencher, and can thus be totally obscured by the latter.

Acquiring and subtracting spectra with and without illumination [32] is an obvious, and basically feasible, approach to solve this problem but incurs a noise penalty and suffers from subtraction artifacts. Unfortunately, those artifacts can be much more pronounced than, e.g., in NOE difference spectroscopy, because the temperature in the "light" spectrum is higher than in the "dark" spectrum and long-lived free radicals might be present in a higher amount during its acquisition, so small line shape changes and line shifts can result and strongly distort the difference spectrum.

Another long-known method is presaturation, i.e., destroying all the background magnetization by a pulse train, possibly combined with field gradients, prior to CIDNP generation. This works perfectly well in time-resolved CIDNP experiments where less than milliseconds elapse between the end of the presaturation sequence and the acquisition pulse. In stationary CIDNP experiments, however, one often illuminates for several hundred milliseconds; during that time the background recovers appreciably.

A much better remedy can be found [33] by recognizing that the presence of a background magnetization is no problem whatsoever at almost all points of time during a pulse sequence because it can only become observable at those points of time when a pulse converts magnetization into coherence. Based on that idea, one can use a string of π pulses such that the background is repeatedly allowed to relax partially and then inverted. It will thus be exactly zero at a certain time after each π pulse, and a sampling pulse (i.e., any pulse of flip angle different from π or a multiple thereof) applied at that moment will not sample the background. In Fig. 3 a double π pulse serves to illustrate that principle. As is clearly seen, the background is zero at the beginning and at the end of that building block (as well as at one point of time between the pulses). The delays before and after the pulses, Δ_1 and Δ_2, have to be chosen on the basis of T_1 of the background, with Δ_2 always being slightly smaller than Δ_1. It might be thought that such an element could only accommodate a background with a unique T_1 but that is a misconception. As displayed in the figure, a nucleus with a longer T_1 regains a smaller fraction of its equilibrium magnetization during Δ_1, so starts with a less negative initial magnetization at the beginning of Δ_2 such that it crosses zero at almost the same point of time as a faster relaxing nucleus. The building block is thus self-compensating, and one can

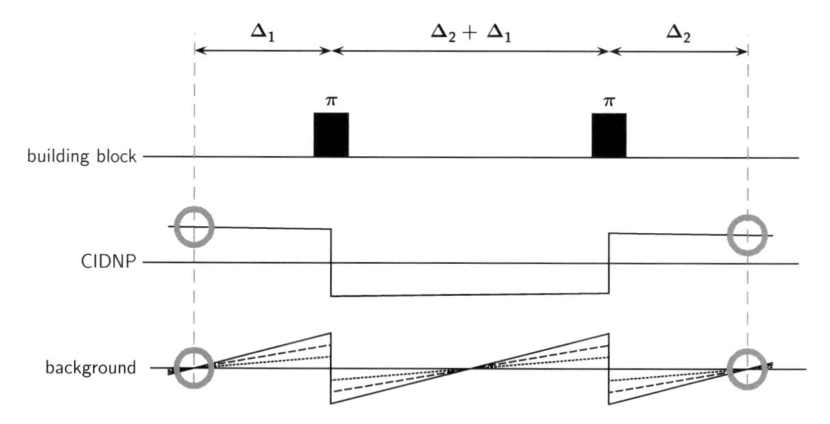

Fig. 3 Building block for background suppression (*top*) comprising two π pulses and delays Δ_1 and Δ_2, and effect of that block on CIDNP (*center*) and on a background signal that is zero at the beginning of the block (*bottom*). The *gray circles* at the start and end the block show that background-free CIDNP can be sampled at those times. The self-compensating effect of the block is illustrated by the lower trace, where the magnetization curves for three nuclei with different T_1 have been plotted in different line styles. In the center trace the effect of a relaxation of the polarization during the sequence has been greatly exaggerated for visibility

calculate the optimum values of Δ_1 and Δ_2 simply from the known or estimated shortest T_1 of all the background.

In contrast to a background magnetization starting out from zero at the beginning of the sequence, the behavior of the displayed building block towards a (nonzero) CIDNP signal present at that time is totally different. The double inversion restores its original state, so at the end of the sequence the CIDNP signal is basically completely preserved apart from a small loss by relaxation.

This strategy would already work with a single π pulse, but pairs of such pulses have the advantage that they are compatible with spin echoes. A spin echo demands that the delays before and after a π pulse be identical, while the background suppression constrains these delays to be unequal. However, as the delay between the two pulses of Fig. 3 is $\Delta_1 + \Delta_2$, both requirements can be met. Because the building block of Fig. 3 is thus basically a double spin echo, it can be used in all pulse sequences that implicitly contain such echoes, and inserted into the many others that tolerate spin echoes without deteriorating the performance.

The described building block opens up a further, completely independent method of background suppression by a phase cycle. In that case, the pulse sequence proper is simply prefaced by one or more such blocks, with illumination added during specific parts of each block. If illumination is restricted to the intervals before the first and after the second π pulse, polarizations generated during that time will emerge from the block(s) unchanged. In contrast, illumination restricted to the time between the two pulses will yield CIDNP that is inverted. Adding the two transients with opposite receiver phase will add CIDNP constructively but strongly suppress any residual background because that is unchanged by the illumination scheme and so is cancelled by the subtraction. Gating the illumination in the way described can be done with a shutter or by controlling the triggers of a pulsed laser. These experiments are essentially difference spectra as described above but have the advantages that the conditions (above all, temperature, but also concentration of paramagnetic transients) during acquisition of the two spectra to be subtracted are identical, and that each acquisition affords CIDNP signals (in contrast to simple subtraction of "light" and "dark" spectra, where there is only CIDNP in the former), so the sensitivity is higher by $\sqrt{2}$.

4 Using CIDNP for Mechanistic Investigations

4.1 What Makes CIDNP Unique for the Purpose?

As we teach our students in every introductory lecture on chemistry, a reaction mechanism is a black box with the reactants as its input, the products as its output, and external physical parameters (e.g., temperature) and chemical parameters (e.g., reaction medium) as global variables. Hence, elucidating a reaction mechanism literally means getting light into the interior of this black box in order to see what is going on there.

Before the advent of fast spectroscopic methods, this task could only be performed in a very indirect manner by changing the input and/or the global variables, monitoring the effects on the output, and constructing the simplest conceivable network of elementary reactions capable of accommodating all the experimental facts. Evidently, the very restricted number of observables necessitated frequent reinterpretations; perhaps more easily than elsewhere in science, the inexplicable new result of the "one experiment too many" was capable of ruining the carefully formulated hypothesis that explained a thousand previous experiments. As an asset, this procedure prompted the development of the kind of oblique thinking that chemists are famous for. One impressive example is the Hammond postulate that derives properties of an unobservable intermediate, such as stabilizing or destabilizing factors and – in the original formulation [34], which is often overlooked – the geometry, from the same properties of observable species directly preceding it or following it on the reaction coordinate.

On this elementary level, CIDNP is already well equipped for the task. As an NMR method, it is ideal for determining the structure of a product; the absence of signals from unpolarized nuclei in the products is not a real problem because coherence and magnetization transfer can be used to make these signals visible (see Sect. 4.2.3). Quantitative studies are also possible because the CIDNP intensities are proportional to the number of molecules that have reacted. The relatively low sensitivity of NMR is alleviated by the signal enhancements but, of course, CIDNP only responds to reactions through radical pairs and biradicals. This is probably the place to dispel a long-standing myth. Many organic chemists seem to think of CIDNP as so ultrasensitive that it will pick up strong signals from contributions of such a paramagnetic pathway to the products that in reality is quite minor and thus fool the investigator into regarding that pathway as the main one. However, while it is true that nonradical pathways are absolutely invisible to CIDNP, the enhancements, relative to the Boltzmann populations, in state-of-the-art NMR magnets are no longer huge [compare the discussion of Eq. (3) in Sect. 2.3], so with any sense of what a signal with a reasonable signal-to-noise ratio looks like, one can hardly fall into that trap today. In addition, by advanced CIDNP experiments, one can actually test whether or not the pathway giving rise to CIDNP is the main pathway of the reaction, as will be shown in Sect. 4.2.2 (Fig. 5).

The next level of precision in mechanistic work was reached with the ability to observe shorter-lived transients. In the above picture of a reaction mechanism this provides the investigator with a tool to cut holes into the black box. Obviously, the identification of a transient can rule out certain pathways beyond doubt and lend strong support to certain others. Again, CIDNP serves that purpose very well for reactions involving radical pairs or biradicals. As already mentioned, the polarization patterns carry information about the spin-density distribution in the paramagnetic intermediates, so one can basically employ CIDNP spectroscopy as one would employ EPR spectroscopy. Being magnetic resonance techniques, both are clearly much more specific than is optical spectroscopy, but CIDNP has one immediately evident advantage over EPR: the CIDNP effect arises only during the life of the spin-correlated intermediates (nanoseconds to subnanoseconds) but then persists in the products for a time on the order of the nuclear T_1 in diamagnetic molecules (seconds to tens of seconds). Hence, CIDNP is able to capture shorter-lived paramagnetic intermediates than is EPR, and to investigate them at leisure without any lifetime-related line broadening or similar effects.

Even with today's ultrafast spectroscopic methods, the orchestra of instruments for mechanistic investigations is still incomplete and one step away from perfection. To get an unambiguous picture of a reaction one would need the ability to obtain a map of the matrix of paths connecting reactants, intermediates, and products, in other words, the ability to lift the cover of the black box. As of now, one mostly has to resort to indirect methods to achieve that end, such as kinetic evidence relating the decay of one species and the buildup of another or preparing a presumed intermediate by another route and determining whether it affords the same products in the same yield as in the direct reaction. Labeling experiments can provide more direct evidence but always incur the risk of an unwanted modification of thermodynamic and/or kinetic parameters.

In this respect, CIDNP ranks highly. The polarizations can be regarded as almost ideal labels because their energies are tiny, so their influence on the course of the reaction – apart from the intersystem crossing step – may be safely neglected. When CIDNP that can be ascribed to a particular paramagnetic intermediate appears in a particular product, it can be concluded with certainly that that intermediate is a precursor to that product. What is more, each nucleus that has a reasonably strong hyperfine coupling constant is labeled with the value and sign of that constant, through the polarization it acquires, so multiple labels are attached to the molecules.

An example will be discussed in Sect. 4.4.2 where this feature is put to use to trace a complicated skeleton rearrangement of a biradical. Another consequence is a more reliable identification of the intermediates as compared to EPR. An EPR spectrum only yields a collection of unsigned coupling constants while a CIDNP spectrum directly relates one nucleus to one hyperfine coupling constant, including the relative sign of the latter. As a complement to other labeling methods, be they of a chemical nature or involve creating spin coherence in the reactants as in SCOTCH [36] and related experiments, the CIDNP labels are affixed at the paramagnetic stage, which is not accessible to the other labeling methods. Added to this comes the information about the pathway from the reactants to the intermediates, through the parameter μ, and from the intermediates to the products, through ε.

In summary, while CIDNP is only applicable to one important class of reactions, the list of its features closely resembles a list of requirements for an ideal method to elucidate the mechanisms of chemical reactions, especially complex ones.

4.2 Level 1: Focusing on the Reactants and Products

There is a plethora of general experimental techniques for thoroughly characterizing the starting materials and products of a chemical reaction, so at first glance yet another method would appear superfluous. However, bearing in mind that in the context of the radical pair mechanism "reactants" and "products" actually mean "direct precursors to the radical pairs" and "initial products after the radical pair stage," CIDNP can specifically address the following key issues.

4.2.1 Precursor Multiplicity

This question is central to our understanding of a photochemical reaction. CIDNP is able to furnish a fairly direct answer, through the dependence of the overall polarization phase on the multiplicity μ; see Eq. (4). In contrast, this information is much less easily obtainable by other means, e.g., sensitization and quenching studies.

The obvious difficulty in applying Eq. (4) to extract one parameter is that all other parameters entering this equation are needed. When the structure of the radical pair is known or has been identified (see Sect. 4.3.1), the exit channel ε is easiest in that respect (simple thermochemical calculations normally suffice to determine whether in-cage formation of a triplet product is feasible or not); specifying the signs of the hyperfine coupling constants a_i is also fairly straightforward; only the sign of Δg might be less trivial to predict. To minimize the danger of a misinterpretation, analyzing relative intensities and signs of the polarizations is thus preferable over analyzing absolute ones.

An instructive example is provided by the β-cleavage of (*N*-methylaniline)-acetone [37], which can occur both from the excited singlet state and from the excited triplet state. Its singlet cleavage is an activated process (E_a = 15–20 kJ/mol) whereas intersystem crossing to the much longer-lived triplet state is temperature independent. Thus at low temperatures singlet cleavage is slow, and formation of triplet pairs prevails. However, with increasing temperature the participation of singlet pairs becomes more and more important and ultimately dominates, so all CIDNP signals experience an inversion when the temperature is raised ($\mu = +1 \rightarrow \mu = -1$). The reaction mechanism and resulting temperature dependence of the CIDNP signals of two products formed through different exit channels are displayed in Fig. 4. Observing both a cage product, which must result from the singlet exit channel in that system, and an escape product (triplet exit channel), as well as observing the CIDNP inversion with temperature rather than observing CIDNP at a single temperature only, eliminates all ambiguities that might be caused by Eq. (4). Two other papers, where CIDNP was instrumental for detecting an unexpected precursor multiplicity, are [38, 39].

In an analogous way, one could use the dependence of the CIDNP intensities on the quencher concentration in cases where both singlet and triplet precursors can be quenched.

4.2.2 Do the Reactants Afford Radical Pairs at All?

Obviously the observation of CIDNP effects implies that radical pairs were formed. Unless one relies on tiny signals only, which would clearly be ill-advised with any experimental technique, this conclusion can be drawn with a substantial degree of certainty because the enhancement factors are no longer huge in today's NMR magnets, as already explained above. However, one moves on much more treacherous ground with the tempting reversal of the argument, namely, using the absence of CIDNP as evidence that no radical pairs are involved in a reaction. The simple reason is that there are so many factors that can reduce or suppress CIDNP. For instance, in the preceding example of Fig. 4 there are hardly any CIDNP effects at room temperature because CIDNP from singlet and triplet precursors accidentally compensates under these conditions; nevertheless, the reaction does involve radical pairs, as the large signals at lower or higher temperatures demonstrate.

Similar cancellation effects occur for product formation through both exit channels, e.g., in such electron transfer reactions where both singlet and triplet radical pairs can undergo reverse electron transfer to regenerate the (singlet or triplet) starting material. An example [40] can be found in Fig. 5. Electron-transfer quenching of the excited singlet state of 9-cyanoanthracene S by *trans*-anethole D produces radical ion pairs comprising the olefin radical cation and the sensitizer radical anion, which then undergo nuclear-spin selective intersystem crossing. As the bottom trace of Fig. 5 shows, no spin polarizations are observable in that system because the triplet energy of the sensitizer is so low that reverse electron transfer of the triplet pairs to give ^{3}S and D is efficient, in addition to the obviously feasible formation of both molecules in their ground state.

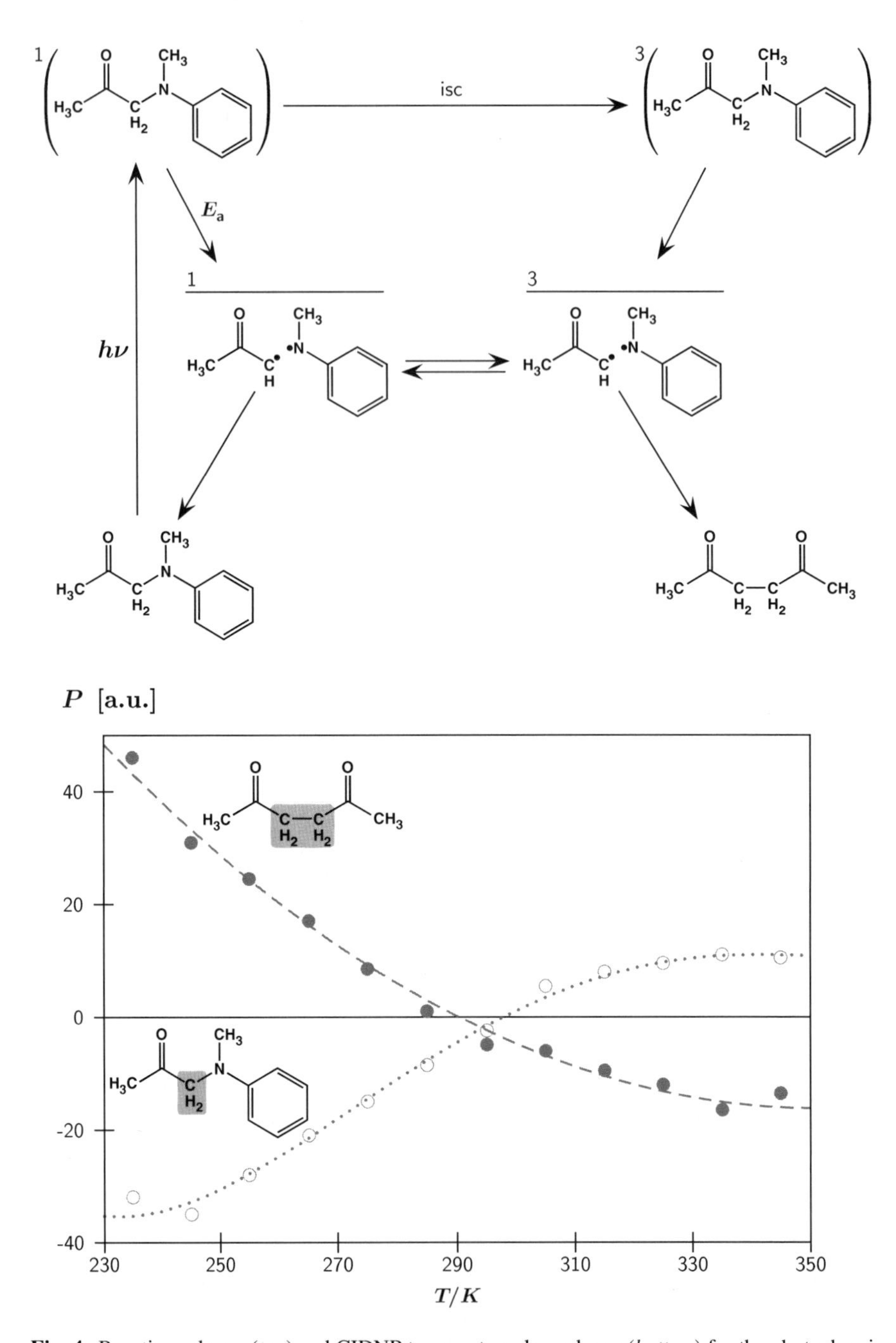

Fig. 4 Reaction scheme (*top*) and CIDNP temperature dependence (*bottom*) for the photochemical cleavage of (*N*-methylaniline)-acetone [37]. Radical pairs are formed from both multiplicities of the excited substrate. The starting material is regenerated from the singlet pairs, the triplet pairs afford an acetone dimer. The polarizations *P* of the methylene groups in both molecules

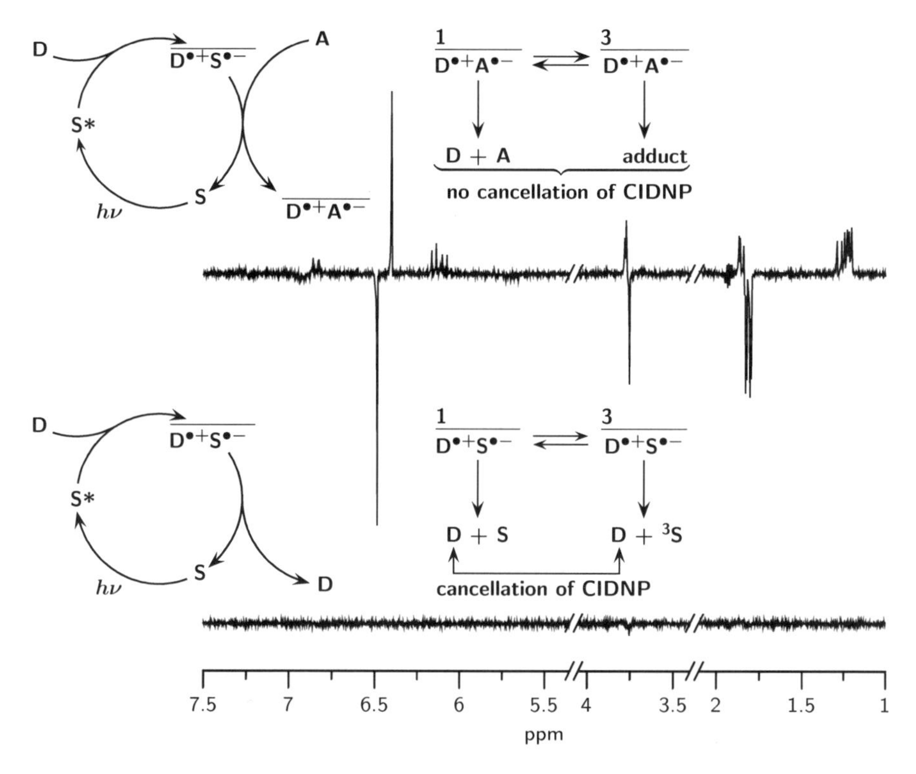

Fig. 5 CIDNP effects and pertaining reaction mechanisms [35] in the system 9-cyano-anthracene (S)/*trans*-anethole (D)/fumaronitrile (A). *Lower trace*, only D and S present; complete cancellation of CIDNP occurs because both exit channels of the radical pair $\overline{D^{\bullet+}S^{\bullet-}}$ regenerate the starting materials. *Upper trace*, in the three-component system PET sensitization transforms the primarily formed radical pair $\overline{D^{\bullet+}S^{\bullet-}}$ into the pair $\overline{D^{\bullet+}A^{\bullet-}}$ and leads to strong CIDNP effects because the secondary radical pair affords different products through the two exit channels. A direct formation of the second pair is impossible at the excitation wavelength used. For further explanation, see text

However, upon addition of a sufficient amount of a third substance, the electron acceptor fumaronitrile A, strong CIDNP signals appear, as can be perceived in the top trace of Fig. 5. The origin of that phenomenon is a replacement of the sensitizer radical anion in the pairs by the radical anion of the acceptor. This exchange occurs by a (nonphotochemical) exergonic electron transfer between $S^{\bullet-}$ and A during diffusive excursions of the primary pairs. The secondary pair is capable of producing CIDNP, in contrast to the primary one, because different products are formed through its two exit channels. In the example, it is easy to choose a wavelength such that only S can be

←

Fig. 4 (continued) (*open circles* and *dotted line*, reactant formed through the singlet exit channel; *filled circles* and *dashed line*, product through the triplet exit channel), as calculated from the CIDNP spectra in [37], invert near room temperature. The fit curves have been drawn to guide the eye and have no physical significance. For further explanation, see text

Chart 3 Cancellation of CIDNP in reverse electron transfer of radical ion pairs. For further explanation, see text

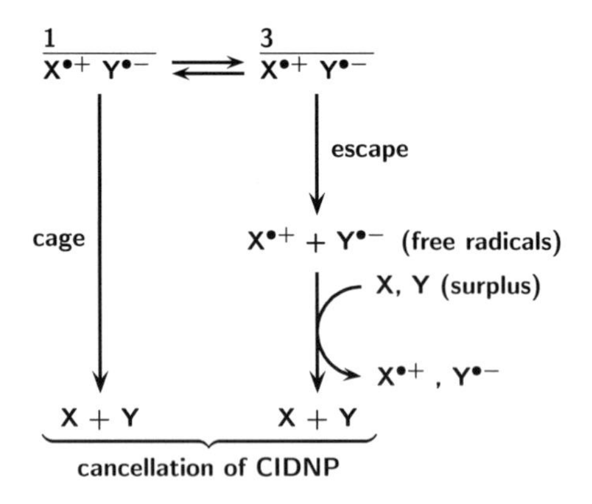

excited, so a direct formation of the secondary pair can be excluded with certainty; the name "photoinduced electron-transfer sensitization" (PET-sensitization) thus reflects the generation of the pair $\overline{D^{\bullet+}A^{\bullet-}}$ in an indirect way.

Not only does this again emphasize the advantages of devising CIDNP experiments such that their outcome under different conditions is compared as opposed to single measurements, but this mechanism can also be put to good use to rule out an exciplex as the precursor to the products, and so address the question whether or not the pathway giving rise to CIDNP is the main reaction route [35, 41]. The reasoning is as follows. In the PET-sensitized experiment, an exciplex between D and A is impossible because neither is ever excited. If CIDNP intensities in the direct photoreaction between D and A (which can be performed with a different excitation wavelength) and the PET-sensitized reaction are identical, then the reaction via radical pairs is the main reaction; a substantial involvement of exciplexes would remove these molecules from the ensemble observable by CIDNP and thus substantially change the CIDNP intensities.

With the mechanism at the bottom of Fig. 5, the polarizations from the triplet exit channel are initially hidden in the triplet molecules, which are unobservable by NMR, and only become detectable after their deactivation to give singlets. Hence, time-resolved experiments and studies with added triplet quenchers (which interact with the mechanism after the radical-pairs stage, in contrast to the PET-sensitization experiments) can provide information about the life and reactivity of the triplets [42, 43].

A further, less obvious compensation mechanism [21] of CIDNP (Chart 3) is realized in all "normal" electron transfer reactions where reverse electron transfer is only possible for one multiplicity (usually singlet). In a cage reaction, the singlet pairs regenerate the starting materials, which bear some spin polarizations. By want of a chemical deactivation channel, the triplet pairs necessarily terminate by escape to give free radicals; owing to the spin sorting, the polarizations of the latter are exactly opposite to those of the cage reaction. Subsequently, degenerate electron transfer reactions with surplus substrate molecules can occur. These processes are chemically nonproductive but lead to an appearance of the free radicals polarizations in the same molecules as formed by the cage reaction. This brings about a perfect compensation unless some fraction of the polarizations is lost by nuclear spin relaxation in the free radicals. Although such reactions exhibit little CIDNP, or none at all, in a stationary experiment, they can be investigated by time-resolved measurements because the cage reaction occurs on a timescale of a few nanoseconds while the reactions of free radicals are much slower. In a similar way to the above, the interaction with a quencher of the free radicals; can be utilized to obtain kinetic information; in [44] this was utilized to measure the deprotonation rate of guanosyl cation radicals.

Apart from the described three mechanisms that can suppress CIDNP despite the intermediacy of radical pairs, unfavorably small hyperfine coupling constants or too large g-value differences can also cause unobservably small CIDNP effects. Hence, the absence of CIDNP from a chemical reaction is much weaker evidence for the absence of radical intermediates than the presence of CIDNP is for an involvement of the latter.

Although protein chemistry falls outside the scope of this review, the routine application of CIDNP to that field [11] needs to be discussed at this point. Its essence is that certain amino acids develop polarizations after quenching a photo-excited sensitizer – for tryptophan, the most readily polarizable amino acid, the mechanism is that of Chart 3, with the residual polarizations due to relaxation in the free radicals being detected – but can only do so when they are solvent exposed such that the excited sensitizer can diffuse to within the quenching distance. This method intrinsically is a comparison experiment, which again illustrates the power of that approach in CIDNP work. The absence of CIDNP from a protein containing only one quenching site might not be very meaningful, but comparing the relative CIDNP intensities of the different quenching sites in the same protein provides an inner standard, and is thus strong enough even to allow a quantitative grading of the surface accessibilities.

4.2.3 Identification of Diamagnetic Products

Of particular interest in that respect are metastable transients. The enol of acetophenone was the first such example reported [45]; another example, a vinylamine [46], has already been shown in Fig. 1. Detecting such transients is straightforward when they are sufficiently long-lived for their free induction decay to be recorded. Gated or pulsed illumination allows a synchronization of their generation and observation, so their initial amount can be captured rather than the usually much lower steady-state concentrations in experiments with continuous generation.

No CIDNP arises for nuclei possessing negligible hyperfine coupling constants in the radicals. At first glance these missing signals might appear to make an identification problematic or impossible. However, this difficulty is easily overcome by NMR multipulse experiments [33, 47, 48]: Coherence of a polarized nucleus is generated, left to evolve into two-spin coherence under the influence of a scalar coupling to an unpolarized nucleus, and then converted into coherence of the latter nucleus. Similarly, magnetization can be transferred [49]. The resulting cross peaks exhibit the same polarization as the polarized first nucleus. Because all NMR multipulse methods, including NOE experiments, can be seamlessly combined with CIDNP generation, one may boldly state that unambiguous structural determination of a partly polarized molecule is always feasible.

Of particular interest in this respect are heteronuclear coherence transfers. The hyperfine coupling constants of heteronuclei are often much larger than those of protons, so substantially stronger polarizations can result. Against this, the lower sensitivity of heteronuclear NMR needs to be balanced. However, the subsequent transfer of these polarizations to protons allows a detection with the sensitivity of proton NMR, i.e., yields a further intensity gain by the factor $(\gamma_H/\gamma_X)^{3/2}$, which amounts to, e.g., almost an order of magnitude when ^{13}C is the heteronucleus. These CIDNP experiments are finding more and more favor recently [33, 50–52].

Two recent studies, where CIDNP featured prominently in identifying the products, including their exit channels, have addressed hydrogen abstractions from cyclic dienes by excited benzophenone [53] and β-phenylogous cleavage reactions of photoinitiators [54].

4.3 *Level 2: Information About the Intermediates*

4.3.1 Identifying the Species Giving Rise to CIDNP

As already stated, the polarization pattern in a product corresponds fairly directly to the EPR spectrum of the intermediate in which the polarizations arose. There is an approximate proportionality between the size of the polarization and the magnitude of the pertaining hyperfine coupling constant, so the EPR spectrum constructed from the polarization pattern has an uncalibrated horizontal axis. This small loss of information is more than overcompensated by the following two benefits. First, CIDNP yields the (relative) signs of the coupling constants. This information cannot be extracted from a normal EPR spectrum, but is of great diagnostic value for determining the intermediate structures. Second, and of even higher importance, an EPR spectrum merely yields an assortment of hyperfine coupling constants, which then have to be assigned to the nuclei in a process that can be very tedious and necessarily relies on further information and/or experiments (e.g., isotope substitution), whereas CIDNP unequivocally relates a particular hyperfine coupling constant to a particular nucleus [55].

The earliest, by now classical, example is the distinction between intermediates in sensitized hydrogen abstractions from triethylamine [46]. The expected result of a one-step process is the α-amino alkyl radical $R_2N-\dot{C}H-CH_3$ but the facile oxidation of amines also opens up the possibility of a two-step abstraction via the sequence electron transfer – proton transfer, with the amine radical cation

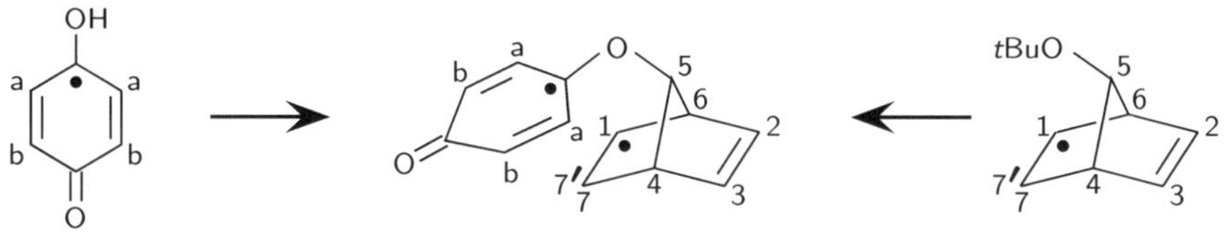

Chart 4 Estimating the hyperfine couplings in a biradical (*center*) from the hyperfine couplings in two radical moieties (*left* and *right*). For further explanation, see text

$R_2\overset{\bullet+}{N} - CH_2 - CH_3$ as a further intermediate. In the radical cation, the hyperfine coupling constants of both the α and β protons are positive but only those in the α position are strong. This translates into CIDNP of the same sign for α and β protons, but much stronger for the former. In contrast, the coupling constants in the α-amino alkyl radical are of comparable magnitude but have opposite signs. This results in the quite different up-down polarization pattern that can be seen in Fig. 1.

A more complex case is presented by the photocycloaddition of benzoquinone with quadricyclane [56], where the polarization pattern in the reaction products cannot be reconciled with any radical pair formed by a chemically conceivable reaction between the starting materials. However, a superposition of the hyperfine coupling constants of the two model radicals displayed at the left and right sides of Chart 4 is perfectly consistent with the polarization pattern, so the 1,5-biradical shown in the center of the figure – for which chemically reasonable pathways of formation from the reactants and decay to the observed products can be formulated; see below, Chart 7 – must be the source of the polarizations. This is one of the extremely rare reports of $S\text{–}T_0$-type CIDNP (i.e., spin sorting by the radical pair mechanism, as described in Sect. 2.1) from a biradical [57–60], an intermediate that would normally give rise to polarizations of one phase only, through $S\text{–}T_{\pm}$ intersystem crossing involving electron-nuclear spin flips. The unexpected behavior of that biradical is attributed to two factors. First, at the equilibrium geometry the orbitals bearing the two radical electrons are almost orthogonal to each other, so the exchange interaction (Fig. 2) vanishes even though the distance between the radical centers is fairly small. Second, and probably decisive, a spin-independent radical rearrangement (see Chart 7) leads to another biradical structure that undergoes very rapid intersystem crossing followed by a combination of the radical centers, so acts as the analog of an escape reaction.

The radical cations of cysteine derivatives are too short-lived for EPR observation, but the polarization pattern unambiguously shows them to be sulfur-centered, so electron transfer from the amino acid to the excited sensitizer occurs from sulfur, not nitrogen [61]. With methionine, the primarily formed open-chain radical cation can stabilize by the formation of a cyclic structure with a two-center three-electron bond between sulfur and nitrogen; both can be clearly distinguished by their CIDNP spectra [62]. Unusual structures of radical cations with cyclopropane moieties were found by CIDNP [63–65]. In the photoreduction of a ruthenium complex with different ligands, the unpaired electron was found to be localized on the expected ligand, i.e., that with the lowest energy in the metal-to-ligand charge transfer state [66].

4.3.2 Competitive Radical Pair Formation

If the reactants afford more than one radical pair through competing parallel reactions, and those pairs do not interconvert, the resulting polarizations are the sum of CIDNP characteristic for each type of radical pair (which is a constant parameter for each nucleus in the intermediate) weighted with the respective pair concentration (which is variable and can be modified, e.g., by the quencher concentration); when the analysis concentrates on a single product, the relative amount of polarization transfer to that product enters as a constant third factor. Hence, relative rate constants of radical pair formation can be obtained. As an attractive feature, relative CIDNP intensities depend only on the radical pairs whereas relative product yields – the analysis of which constitutes the usual experimental method – are also modified by processes after the radical-pair stage. For instance, deprotonation of an amine radical cation can occur for both the correlated radical pair and the free radicals, and the relative importance of these two routes depends on the substituents [67]. The above-mentioned probing of surface accessibilities of amino acids in a protein falls into that category of experiments. The same amino acid in different positions of a protein involves chemically identical radical pairs, so relative CIDNP intensities directly yield relative efficiencies of radical pair formation; when different amino acids are compared, one has to take into account the different intrinsic polarizabilities of the chemically different pairs, which enter as weights of the Stern–Volmer factors. Examples from the field of organic chemistry include relative reactivities of sensitized hydrogen abstractions from unsymmetrically substituted tertiary aliphatic amines [67], the competition of photocycloaddition with photoionization in sensitized reactions of electron-rich olefins [68], the competition between hydrogen abstraction and dissociative electron transfer in the photochemistry of the surfactant AOT with 2,6-anthraquinone disulfonate [69], the competition of hydrogen abstractions from different sites of the cavity in carbonyl photochemistry within β-cyclodextrin hosts [70], and the relative abstraction probabilities from the different rings of a polyphenol by a thioxanthone sensitizer [71].

4.3.3 Successive Radical Pairs

Compared to the preceding section, new phenomena arise when the reaction proceeds through a sequence of radical pairs. In that case, there is no longer a simple superposition of polarizations from each pair but rather a cooperative effect. The density matrix of a radical pair can be decomposed into the components population difference, phase correlation, and electron spin polarization, of which only the first and the last are directly observable. At the moment of transformation of one pair into another, the full density matrix – including the unobservable phase correlation – resulting from evolution under the first pair's Hamiltonian provides the initial condition for further evolution under the second pair's Hamiltonian.

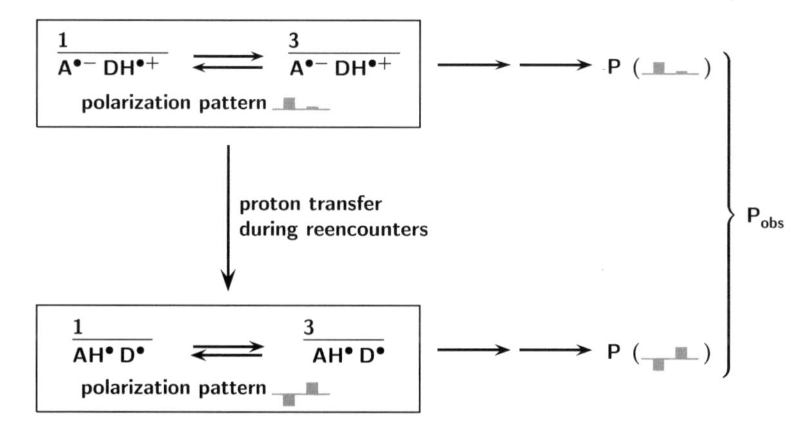

Chart 5 Different polarization patterns arising from two successive radical pairs in sensitized photoreactions of tertiary aliphatic amines **DH**. For further explanation, see text

This explains the seeming paradox that polarizations can be observed in reactions through a sequence of two radical pairs, where neither pair on its own is capable of generating a CIDNP net effect on its own, one because of vanishing Δg (but nonzero hyperfine coupling constants), the other because of vanishing hyperfine coupling constants (but nonzero Δg) [25].

The sensitized photoreactions of tertiary aliphatic amines (sensitizer, S; amines, DH) are one example of such a scenario. The reaction scheme is shown in Chart 5. The first radical pair comprises the sensitizer radical anion $A^{\bullet-}$ and the amine radical cation $DH^{\bullet+}$, the second radical pair the neutral radicals $AH^{\bullet}$ and $D^{\bullet}$ of these two molecules. Each pair ultimately affords the same product P, which is observed. The polarization pattern from each pair on its own is quite distinctive and characteristic (see Sect. 4.3.1), as symbolized in the chart. At each reencounter of the radicals of the first pair, a proton transfer from $DH^{\bullet+}$ to $A^{\bullet-}$ can occur regardless of the electron spin state, and transform the first pair into the second. From the resulting polarization pattern in the observed product, one can then obtain the rate constant of the proton transfer between the radicals. Again, comparison experiments are much more reliable than absolute CIDNP measurements, so the analysis focuses on the changes in the product polarizations when the rate constant of the proton transfer is varied by varying the energy difference between the two radical pairs (by using different sensitizers [72] or, allowing much more fine-grained steps, changing the dielectric constant of the reaction medium [73]). The bimolecular rate constants accessible by this analysis fall into a range that is chemically very important, from about diffusion control to two orders of magnitude slower [28].

The PET-sensitization experiments described in Sect. 4.2.2 are pair substitutions during diffusive excursions, and the rates of radical pair exchange have been measured by the dependence of the CIDNP intensities on the quencher concentration [35]. Other examples include the interconversion of the open-chain and cyclic forms of the radical cations of methionine [74, 75], and the loss of phenyl iodide from primary radical pairs in the sensitized photolysis of iodonium salts [76].

4.4 Level 3: Unraveling the Connecting Pathways

4.4.1 Polarizations as Labels to Molecules

The polarizations generated in a radical pair are simply passed on through all subsequent species until they arrive in the final diamagnetic products. Apart from the compensation when a secondary species receives opposite polarizations from both exit channels of the pair, the sole loss mechanism is nuclear spin relaxation, which is only significant in longer-lived free radicals. Hence, the distribution of the polarizations between the different products yields important qualitative and quantitative information about the pathways from the radical pairs to the products.

The reaction shown in Chart 6 serves as an example. In the photoreaction of excited naphthoquinone with *cis*- or *trans*-anethole [41], CIDNP arises in radical pairs composed of the quinone radical anion and the olefin radical cation; the latter is configurationally stable and does not undergo *cis–trans* isomerization. The only products of the singlet exit channel are the regenerated starting materials. However, the triplet pairs can react to give triplet biradicals by forming a bond between the atoms bearing the largest positive and negative charge densities in the radical ions. The configurations of the initial biradicals can be seen in the chart, with the relative orientations of the methyl and anisyl substituents necessarily being the same as in the starting olefin, and secondary orbital interactions being responsible for the relative orientation of the quinone-derived moiety and the anisyl substituent. Intersystem crossing of these biradicals followed by ring closure affords the two products P_1 and P_3. There is, however, a more stable configuration of the biradical, which ultimately leads to the product P_2. The ratio of CIDNP in the pair P_1/P_2 and P_3/P_2 (which is identical to the ratio of the product yields) is temperature dependent, with P_2 being favored at higher temperatures. The obvious reason is an activation barrier for conversion of the initial biradical (the precursor to P_1 or P_3) into the thermodynamically more stable one. By analyzing the polarization ratio as function of the temperature, the activation energies can be obtained. Once more, these CIDNP experiments are seen to be comparison experiments.

In the sensitized photoreactions of cysteine derivatives, the competition between the deactivation pathways of a secondary radical, scission or oxidation by surplus sensitizer, can be studied in a similar manner, and the relative rate constants can be determined [61]. The sensitized photolysis of triphenylsulfonium salts affords spin-polarized benzene in a cage reaction and oppositely polarized monodeuteriobenzene in an escape reaction; a line shape analysis for different temperatures and sensitizer concentrations yields the kinetic parameters of competitive scavenging of the free phenyl radicals by the solvent and by surplus sensitizer [77].

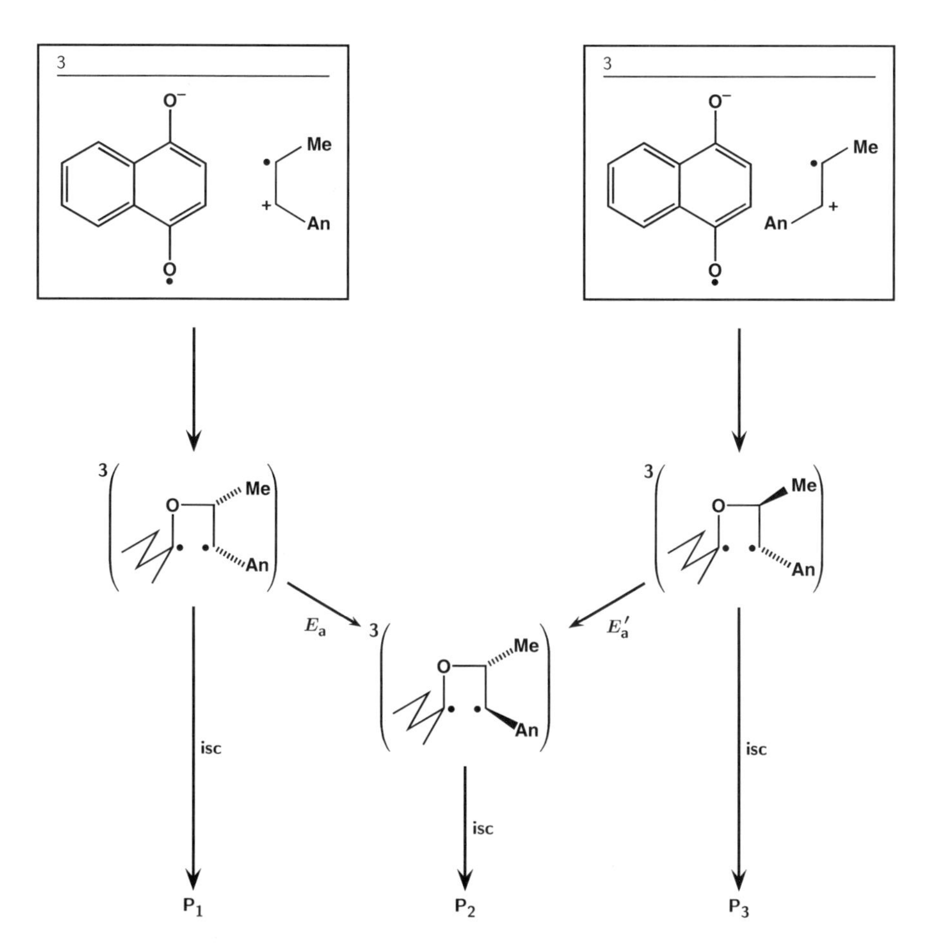

Chart 6 Photocycloadditions via radical ion pairs of naphthoquinone and anetholes (*left* and *right* formulas in the *boxes*) and subsequent interconverting biradicals. Starting olefin, *cis* anethole (*left box*) or *trans* anethole (*right box*). For further explanation, see text

4.4.2 Polarizations as Labels to Individual Nuclei

For the same reasons as above, the polarizations can also be utilized to trace out the relative arrangement of nuclei during chemical transformations following CIDNP generation. The complicated rearrangement [56] displayed in Chart 7 illustrates that aspect of CIDNP.

The main source of polarizations is the biradical that has already been identified in Chart 4 and is found on the left side of Chart 7. The protons bearing the strongest, and well distinguishable, polarizations from that biradical have been symbolized by gray and black rectangles and gray circles. Obviously the product formed through the singlet exit channel has these labels in the same place as in the biradical but the product of the triplet exit channel exhibits quite a different positioning of them, so a skeleton rearrangement must have taken place. With the reasonable assumption that each proton remains connected to its carbon during that rearrangement, these labels

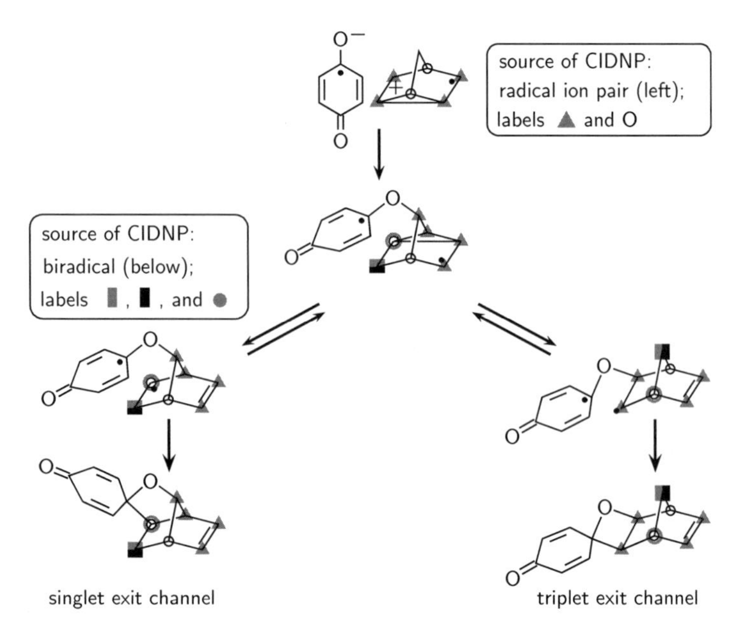

Chart 7 Polarizations as labels to individual nuclei in a sequence of biradical rearrangements. For further explanation, see text

establish a one-to-one correspondence between the atoms in the biradical and in the rearranged product, and allow the skeleton transformation to be identified as a cyclopropylmethyl-homoallyl rearrangement. The biradical bearing the cyclopropylmethyl moiety (see, center of the chart) is an intermediate with a higher energy and its ring opening yields the third biradical shown on the right side of Chart 7, which is the direct precursor to the product from the triplet exit channel of the first biradical. From the temperature dependence, the activation energy of that rearrangement was obtained [78].

Analyzing the differences between CIDNP in a polar and a nonpolar solvent, one finds that there is also a small contribution of polarizations from the radical ion pairs displayed at the top of Chart 7. Using the same reasoning, one can also trace out the pathway from the radical ion pair to the biradical world. As one can conclude from the distribution of the radical pair polarizations, the initially formed biradical is the high-energy one, and the route (not shown in detail in the chart) is a combination of the positions of highest charge density in the radical ions followed by another cyclopropylmethyl-homoallyl rearrangement.

The advantages of CIDNP labeling over chemical labeling are particularly clear in this case. Even neglecting the preparative difficulties of affixing a chemical or isotope label to the starting materials, and ignoring the possible changes in thermodynamics and/or kinetics caused by it, a two-stage labeling would never be possible for chemical labels. In contrast, CIDNP labels are affixed "automatically," within the lifetime of the spin-correlated intermediates, and at each of the two paramagnetic stages in that system, which gives a complete picture of this complicated rearrangement.

5 Conclusions

The preceding examples bear out that CIDNP yields a wealth of insight into diverse aspects of a photochemical reaction, and in many cases allows a complete elucidation of the mechanism without the need to resort to other experimental techniques. Especially intriguing is that it encodes information stemming from fast events that occur early on the reaction coordinate in the molecules such that it can be read out at leisure at some later point of the reaction. As the most useful side effect of this, connections between different successive species are established. Because of the many facets that can be probed, and because of the facile distinction between stages before the CIDNP-active intermediates, of the intermediates themselves, and after these intermediates, the method appears especially well suited for the application to complex reaction schemes.

References

1. Steiner UE, Ulrich T (1989) Magnetic field effects in chemical kinetics and related phenomena. Chem Rev 89:51–147
2. Goez M (1997) Photochemically induced dynamic nuclear polarization. Adv Photochem 23:63–163
3. Goez M (2009) Photo-CIDNP spectroscopy. Annu Rep NMR Spectrosc 66:77–147
4. Goez M (2010) Chemical transformations within the paramagnetic world investigated by photo-CIDNP. In: Forbes MDE (ed) Wiley series on reactive intermediates in chemistry and biology, vol 3. Wiley, Hoboken, pp 185–204
5. Berliner LJ, Bagryanskaya E (2011) Chemically induced electron and nuclear polarization. Multifrequency electron paramagnetic resonance 947–992
6. Hany R, Vollenweider JK, Fischer H (1988) Separation and analysis of CIDNP spin orders for a coupled multiproton system. Chem Phys 120:169–175
7. Bargon J, Fischer H, Johnsen U (1967) Nuclear magnetic resonance emission lines during fast radical reactions. I. Recording methods and examples. Z Naturforsch A 22:1551–1555
8. Ward HR, Lawler RG (1967) Nuclear magnetic resonance emission and enhanced absorption in rapid organometallic reactions. J Am Chem Soc 89:5518–5519
9. Closs GL (1969) Mechanism explaining nuclear spin polarizations in radical combination reactions. J Am Chem Soc 91:4552–4554
10. Kaptein R, Oosterhoff LJ (1969) Chemically induced dynamic nuclear polarization. II. Relation with anomalous ESR spectra. Chem Phys Lett 4:195–197
11. Hore PJ, Broadhurst RW (1993) Photo-CIDNP of biopolymers. Prog NMR Spectrosc 25:345–402
12. Bargon J (1977) CIDNP from geminate recombination of radical-ion pairs in polar solvents. J Am Chem Soc 99:8350–8351
13. Closs GL, Czeropski MS (1977) Amendment of the CIDNP phase rules. Radical pairs leading to triplet states. J Am Chem Soc 99:6127–6128
14. Kaptein R (1971) Simple rules for chemically induced dynamic nuclear polarization. Chem Commun 732–733
15. Parson WW (2007) Modern optical spectroscopy. Springer, Berlin
16. Adrian FJ (1972) Singlet-triplet splitting in diffusing radical pairs and the magnitude of chemically induced electron spin polarization. J Chem Phys 57:5107–5113

17. Closs GL, Forbes MDE, Piotrowiak P (1992) Spin and reaction dynamics in flexible polymethylene biradicals as studied by EPR, NMR, optical spectroscopy, and magnetic field effects. Measurements and mechanisms of scalar electron spin–spin coupling. J Am Chem Soc 114:3285–3294
18. Tsentalovich YP, Morozova OB, Avdievich NI, Ananchenko GS, Yurkovskaya AV, Ball JD, Forbes MDE (1997) Influence of molecular structure on the rate of intersystem crossing in flexible biradicals. J Phys Chem A 101:8809–8816
19. Wegner M, Fischer H, Grosse S, Vieth HM, Oliver AM, Paddon-Row MN (2001) Field dependent CIDNP from photochemically generated radical ion pairs in rigid bichromophoric systems. Chem Phys 264:341–353
20. Ward HR (1972) Chemically induced dynamic nuclear polarization (CIDNP). I. Phenomenon, examples, and applications. Acc Chem Res 5:18–24
21. Closs GL, Sitzmann EV (1981) Measurements of degenerate radical ion-neutral molecule electron exchange by microsecond time-resolved CIDNP. Determination of relative hyperfine coupling constants of radical cations of chlorophylls and derivatives. J Am Chem Soc 103:3217–3219
22. Kiryutin AS, Morozova OB, Kuhn LT, Yurkovskaya AV, Hore PJ (2007) 1H and 13C hyperfine coupling constants of the tryptophanyl cation radical in aqueous solution from microsecond time-resolved CIDNP. J Phys Chem B 111:11221–11227
23. Kiryutin AS, Ivanov KL, Morozova OB, Yurkovskaya AV, Vieth HM, Pirogov YA, Sagdeev ARZ (2009) TR-CIDNP as tool for quantitative analysis of hyperfine couplings in elusive radicals. Dokl Phys Chem 428:183–188
24. Morozova OB, Ivanov KL, Kiryutin AS, Sagdeev RZ, Koechling T, Vieth HM, Yurkovskaya AV (2011) Time-resolved CIDNP: an NMR way to determine the EPR parameters of elusive radicals. Phys Chem Chem Phys 13:6619–6627
25. DenHollander JA (1975) Radical pair substitution in chemically induced dynamic nuclear polarization. Cooperative effects. Chem Phys 10:167–184
26. DenHollander JA (1975) Radical pair substitution in CIDNP (chemically induced dynamic nuclear polarization). Chem Commun 352–353
27. DenHollander JA, Kaptein R (1976) Radical pair substitution in CIDNP. Spin-uncorrelated geminate radical pairs. Chem Phys Lett 41:257–263
28. Goez M, Sartorius I (2003) CIDNP determination of the rate of in-cage deprotonation of the triethylamine radical cation. J Phys Chem A 107:8539–8546
29. Closs GL, Miller RJ, Redwine OD (1985) Time-resolved CIDNP: applications to radical and biradical chemistry. Acc Chem Res 18:196–202
30. Goez M, Kuprov I, Hore PJ (2005) Increasing the sensitivity of time-resolved photo-CIDNP experiments by multiple laser flashes and temporary storage in the rotating frame. J Magn Reson 177:139–145
31. Goez M, Kuprov I, Mok KH, Hore PJ (2006) Novel pulse sequences for time-resolved photo-CIDNP. Mol Phys 104:1675–1686
32. Mok KH, Hore PJ (2004) Photo-CIDNP NMR methods for studying protein folding. Methods 34:75–87
33. Goez M, Mok KH, Hore PJ (2005) Photo-CIDNP experiments with an optimized presaturation pulse train, gated continuous illumination, and a background-nulling pulse grid. J Magn Reson 177:236–246
34. Hammond GS (1955) A correlation of reaction rates. J Am Chem Soc 77:334–338
35. Goez M, Eckert G (2006) Photoinduced electron transfer sensitization investigated by chemically induced dynamic nuclear polarization (CIDNP). Phys Chem Chem Phys 8:5294–5303
36. Pouwels PJW, Kaptein R (1993) Theory of the SCOTCH experiment for reactions involving radical pairs. J Phys Chem 97:13318–13325
37. Grimme S, Dreeskamp H (1992) Singlet- and triplet-state photodissociation of carbon-oxygen and carbon-nitrogen bonds in aromatic acetones studied by 1H-CIDNP spectroscopy. J Photochem Photobiol A 65:371–382

38. Pohlers G, Dreeskamp H, Grimme S (1996) The mechanism of photochemical C–O or C–S bond cleavage in aryl(thio)ethers. J Photochem Photobiol A 95:41–49
39. Gritsan NP, Tsentalovich YP, Yurkovskaya AV, Sagdeev RZ (1996) Laser flash photolysis and CIDNP studies of 1-naphthyl acetate photo-fries rearrangement. J Phys Chem 100:4448–4458
40. Goez M, Eckert G (2006) Olefin isomerization via radical-ion pairs in triplet states studied by chemically induced dynamic nuclear polarization (CIDNP). Helv Chim Acta 89:2183–2199
41. Eckert G, Goez M (1994) Photoinduced electron-transfer reactions of aryl olefins. 1. Investigation of the Paterno-Buechi reaction between quinones and anetholes in polar solvents. J Am Chem Soc 116:11999–12009
42. Schaffner E, Fischer H (1995) CIDNP from photogenerated geminate radical Ion pairs hidden in triplet-state products. J Phys Chem 99:102–104
43. Schaffner E, Fischer H (1996) Singlet and triplet state back electron transfer from photogenerated radical ion pairs studied by time-resolved CIDNP. J Phys Chem 100:1657–1665
44. Morozova OB, Saprygina NN, Fedorova OS, Yurkovskaya AV (2011) Deprotonation of transient guanosyl cation radical catalyzed by buffer in aqueous solution: TR-CIDNP study. Appl Magn Reson 41:239–250
45. Rosenfeld SM, Lawler RG, Ward HR (1973) Photo-CIDNP [chemically induced dynamic nuclear polarization] detection of transient intermediates. Enol of acetophenone. J Am Chem Soc 95:946–948
46. Roth HD, Manion ML (1975) Photoreactions of ketones with amines. CIDNP [chemically induced dynamic nuclear polarization] criteria for the intermediacy of aminoalkyl radicals and aminium radical ions. J Am Chem Soc 97:6886–6888
47. Scheek RM, Stob S, Boelens R, Dijkstra K, Kaptein R (1985) Applications of two-dimensional NMR methods in photochemically induced dynamic nuclear polarization spectroscopy. J Am Chem Soc 107:705–706
48. Goez M (1993) Coherence transfer by selective pulses in photo-CIDNP experiments. J Magn Reson A 102:144–150
49. Mok KH, Kuhn LT, Goez M, Day IJ, Lin JC, Andersen NH, Hore PJ (2007) A preexisting hydrophobic collapse in the unfolded state of an ultrafast folding protein. Nature 447:106–109
50. Sekhar A, Cavagnero S (2009) EPIC- and CHANCE-HSQC: two 15N-photo-CIDNP-enhanced pulse sequences for the sensitive detection of solvent-exposed tryptophan. J Magn Reson 200:207–213
51. Sekhar A, Cavagnero S (2009) 1H Photo-CIDNP enhancements in heteronuclear correlation NMR spectroscopy. J Phys Chem B 113:8310–8318
52. Lee JH, Sekhar A, Cavagnero S (2011) 1H-Detected 13C photo-CIDNP as a sensitivity enhancement tool in solution NMR. J Am Chem Soc 133:8062–8065
53. Andreu I, Neshchadin D, Rico E, Griesser M, Samadi A, Morera IM, Gescheidt G, Miranda MA (2011) Probing lipid peroxidation by using linoleic acid and benzophenone. Chemistry 17:10089–10096
54. Griesser M, Dworak C, Jauk S, Hoefer M, Rosspeintner A, Grabner G, Liska R, Gescheidt G (2012) Photoinitiators with β-phenylogous cleavage: an evaluation of reaction mechanisms and performance. Macromolecules 45:1737–1745
55. Roth HD (1993) Magnetic resonance studies and ab initio calculations as structure probes for radical cations - hexa-1,5diene systems. Z Phys Chem 180:135–158
56. Goez M, Frisch I (1995) Photocycloadditions of quinones with quadricyclane and norbornadiene. A mechanistic study. J Am Chem Soc 117:10486–10502
57. Doubleday CJ (1979) CIDNP and intersystem crossing in biradicals. Chem Phys Lett 64:67–70
58. Doubleday CJ (1981) A carbon-13 CIDNP study of 1,5-biradicals. Chem Phys Lett 79:375–380
59. Kaptein R, DeKanter FJJ, Rist GH (1981) CIDNP from a 1,4-biradical in the Norrish type II photoreaction of valerophenone. Chem Commun 499–501

60. DeKanter FJJ, Kaptein R (1982) CIDNP and triplet-state reactivity of biradicals. J Am Chem Soc 104:4759–4766
61. Goez M, Rozwadowski J, Marciniak B (1996) Photoinduced electron transfer, decarboxylation, and radical fragmentation of cysteine derivatives: a chemically induced dynamic nuclear polarization study. J Am Chem Soc 118:2882–2891
62. Goez M, Rozwadowski J, Marciniak B (1998) CIDNP spectroscopic observation of (S-N)+ radical cations with a two-center three-electron bond during the photooxidation of methionine. Ang Chem Int Ed Engl 37:628–630
63. Herbertz T, Roth HD (1997) Radical cations of ethano-bridged dicyclopropyl systems: remarkable stereoelectronic effects. J Am Chem Soc 119:9574–9575
64. Roth HD, Weng H, Herbertz T (1997) CIDNP study and ab-initio calculations of rigid vinylcyclopropane systems: evidence for delocalized "ringclosed" radical cations. Tetrahedron 53:10051–10070
65. Roth HD, Herbertz T, Lakkaraju PS, Sluggett G, Turro NJ (1999) Oxidation of arylcyclopropanes in solution and in a zeolite: structure and rearrangement of the phenylcyclopropane radical cation. J Phys Chem A 103:11350–11354
66. Perrier S, Mugeniwabagara E, Kirsch-DeMesmaeker A, Hore PJ, Luhmer M (2009) Exploring photoreactions between polyazaaromatic Ru(II) complexes and biomolecules by chemically induced dynamic nuclear polarization measurements. J Am Chem Soc 131:12458–12465
67. Goez M, Frisch I (1994) Deprotonation of aminium cations - a CIDNP study of relative group reactivities. J Photochem Photobiol A 84:1–12
68. Goez M, Eckert G (1996) Photoinduced electron transfer reactions of aryl olefins. 2. Cis–trans isomerization and cycloadduct formation in anethole-fumaronitrile systems in polar solvents. J Am Chem Soc 118:140–154
69. White RC, Gorelik V, Bagryanskaya EG, Forbes MDE (2007) Photoredox chemistry of AOT: electron transfer and hydrogen abstraction in microemulsions involving the surfactant. Langmuir 23:4183–4191
70. Krumkacheva OA, Gorelik VR, Bagryanskaya EG, Lebedeva NV, Forbes MDE (2010) Supramolecular photochemistry in β-cyclodextrin hosts: a TREPR, NMR, and CIDNP investigation. Langmuir 26:8971–8980
71. Neshchadin D, Levinn R, Gescheidt G, Batchelor SN (2010) Probing the antioxidant activity of polyphenols by CIDNP: from model compounds to green tea and red wine. Chemistry 16:7008–7016
72. Goez M, Sartorius I (1993) Photo-CIDNP investigation of the deprotonation of aminium cations. J Am Chem Soc 115:11123–11133
73. Goez M, Sartorius I (1994) Control of the deprotonation route of an aminium cation by the solvent polarity. Chem Ber 127:2273–2276
74. Goez M, Rozwadowski J (1998) Reversible pair substitution in CIDNP: the radical cation of methionine. J Phys Chem A 102:7945–7953
75. Morozova OB, Korchak SE, Sagdeev RZ, Yurkovskaya AV (2005) Time-resolved chemically induced dynamic nuclear polarization studies of structure and reactivity of methionine radical cations in aqueous solution as a function of pH. J Phys Chem A 109:10459–10466
76. Goez M, Eckert G, Mueller U (1999) Sensitized photolysis of iodonium salts studied by CIDNP - solvent dependence and influence of lipophilic substituents. J Phys Chem A 103:5714–5721
77. Eckert G, Goez M (1999) CIDNP investigation of radical decay pathways in the sensitized photolysis of triphenylsulfonium salts. J Am Chem Soc 121:2274–2280
78. Goez M, Frisch I (2002) Activation energy of a biradical rearrangement measured by photo-CIDNP. J Phys Chem A 106:8079–8084

Top Curr Chem (2013) 338: 33–74
DOI: 10.1007/128_2013_420

Published online: 28 March 2013

*para*hydrogen Induced Polarization by Homogeneous Catalysis: Theory and Applications

Lisandro Buljubasich, María Belén Franzoni, and Kerstin Münnemann

Abstract The alignment of the nuclear spins in *para*hydrogen can be transferred to other molecules by a homogeneously catalyzed hydrogenation reaction resulting in dramatically enhanced NMR signals. In this chapter we introduce the involved theoretical concepts by two different approaches: the well known, intuitive population approach and the more complex but more complete density operator formalism. Furthermore, we present two interesting applications of PHIP employing homogeneous catalysis. The first demonstrates the feasibility of using PHIP hyperpolarized molecules as contrast agents in ^{1}H MRI. The contrast arises from the J-coupling induced rephasing of the NMR signal of molecules hyperpolarized via PHIP. It allows for the discrimination of a small amount of hyperpolarized molecules from a large background signal and may open up unprecedented opportunities to use the standard MRI nucleus ^{1}H for, e.g., metabolic imaging in the future. The second application shows the possibility of continuously producing hyperpolarization via PHIP by employing hollow fiber membranes. The continuous generation of hyperpolarization can overcome the problem of fast relaxation times inherent in all hyperpolarization techniques employed in liquid-state NMR. It allows, for instance, the recording of a reliable 2D spectrum much faster than performing the same experiment with thermally polarized protons. The membrane technique can be straightforwardly extended to produce a continuous flow of a hyperpolarized liquid for MRI enabling important applications in natural sciences and medicine.

Keywords Homogeneous catalysis · Hyperpolarization · NMR signal enhancement · *para*hydrogen induced polarization · PHIP

In honor of Prof. Hans W. Spiess

L. Buljubasich, M.B. Franzoni, and K. Münnemann (✉)
Max Planck Institute for Polymer Research, Ackermannweg 10, 55128 Mainz, Germany
e-mail: muenne@mpip-mainz.mpg.de

Contents

1 Introduction

The hydrogenation of organic molecules with hydrogen enriched in the *para*-state creates a highly ordered spin state, resulting in the observation of largely enhanced NMR signals. In September 1987, Bowers and Weitekamp presented the first experimental verification of this effect. The title of the article, "*para*hydrogen and Synthesis Allow Dramatically Enhanced Nuclear Alignment" [1], inspired the popular name PASADENA for experiments where the hydrogenation reaction and the NMR spectrum acquisition are accomplished within the strong magnetic field of the spectrometer. Large *antiphase multiplets* in the ^{1}H NMR spectra were reported. These signals appeared to be approximately 100 times larger than the intensity of the signals acquired with hydrogen in thermal equilibrium at room temperature.

A few months later in the same year an independent work by Eisenschmid et al., reported similarly enhanced spectra in two different hydrogenation reactions [2]. The title "Para Hydrogen Induced Polarization in Hydrogenation Reactions" represented the inception of the widespread acronym for all kind of experiments involving enriched *para*hydrogen i.e., PHIP.

In the next year (1988) another significant contribution was published by Pravica and Weitekamp [3], reporting the hydrogenation of styrene to ethylbenzene with enriched *para*hydrogen. The reaction was carried out at low magnetic field and subsequently the sample tube was adiabatically transported into the high field of the spectrometer to perform the NMR experiment. The resulting spectra markedly differed from the PASADENA counterpart. It displayed *two separate multiplets of opposite phases* for the two proton sites of the product molecule occupied by *para*hydrogen. They named this effect ALTADENA (Adiabatic Longitudinal Transport After Dissociation Engenders Net Alignment).

Since these initial publications the hydrogenation with *para*hydrogen has become a promising technique to boost the low sensitivity of NMR. Hence, during the last two and a half decades the technique was used in a wide range of applications encompassing: the investigation of the kinetics of inorganic reactions [4–6]; to explore heterogeneous reactions [7–9]; the observation of the spatial distribution of hyperpolarized gases by MRI [10, 11]; the use as contrast agent in MRI [12–15]; the transfer of the accomplished hyperpolarization using either suitable pulse sequences [16, 17], adiabatic field cycling [18] or transport through level avoiding crossing [19–21]; the study of long lived states [22–24] originating from p-H_2 [25–28] and, far from chemical applications, the particular p-H_2 spin state has been used in the context of quantum information processes [29–32].

This chapter does not represent a thorough review of the applications of p-H_2 and therefore the list of references presented here is necessarily incomplete. For a more complete discussion of the manifold applications of PHIP we refer the reader to the excellent review recently published by Green et al. [33] and to the contribution written by Duckett and Mewis within this book.

The emphasis of this chapter is on the theoretical aspects of PHIP, based mainly on a numerical approach, along with two experimental examples. In the theoretical part, first a short summary of the physics of *para*hydrogen is given, followed by a treatment of NMR with PHIP in a population model approach. Next, the density operator formalism is introduced and the major features of PASADENA and ALTADENA are explained in this context. Finally, hyperpolarization transfer to a third spin is treated. In the experimental part we include two practical applications of PHIP. The first example shows the potential use of PHIP to create a novel contrast in ^{1}H MRI. The second example is related to the achievement of continuous hyperpolarization with PHIP by means of hollow fiber membranes. Two model compounds are presented: one is soluble in water while the other one is soluble in organic solvents.

2 Theory

2.1 Spin Isomers of the Hydrogen Molecule

Among the applications of quantum mechanics in chemistry, one of the first triumphs was the successful calculation of the structure of very simple molecules. The simplest of all molecules is the hydrogen molecule-ion, H_2^+, composed of two hydrogen nuclei and one electron. Within a year after the development of quantum mechanics, a description of the normal state of the hydrogen molecule-ion was obtained, in complete agreement with experiments. The next molecule, in terms of simplicity, is the hydrogen molecule. The contribution by Heitler and London in 1927 [34] is considered as the inception of the application of quantum mechanics to problems of molecular structure. However, the full treatment is rather complicated.

We include here only a short summary of the quantum mechanical properties of the hydrogen molecule (a detailed treatment can be found, for instance, in [35]).

The total wave function of the hydrogen molecule can be represented as the product of five functions:

$$\psi = \psi_{\mathrm{e}}^{\mathrm{orb}} \psi_{\mathrm{e}}^{\mathrm{s}} \psi_{\mathrm{n}}^{\mathrm{vib}} \psi_{\mathrm{n}}^{\mathrm{rot}} \psi_{\mathrm{n}}^{\mathrm{s}}, \tag{1}$$

describing the orbital motion of electrons, the electron spin state, the vibrational state of the nuclei, the rotational motion of the nuclei, and the nuclear spin state, respectively [35]. The first wave function, corresponding to the electronic ground state, ${}^{1}\Sigma_{\mathrm{g}}^{+}$, is symmetric with respect to the electrons [36], the second is antisymmetric, and the rest are independent of the electrons' variables and symmetric. This makes the entire wave function antisymmetric in the two electrons, as required by Pauli's principle [37]. On the other hand, $\psi_{\mathrm{e}}^{\mathrm{orb}}$ is symmetric with respect to the nuclei [35]; $\psi_{\mathrm{e}}^{\mathrm{s}}$ is also symmetric because it is independent of nuclear coordinates. As the positions of both nuclei can be interchanged without affecting the vibrational state, the third function is symmetric. The overall symmetry of the total wave function depends, thus, on the symmetry of the product $(\psi_{\mathrm{n}}^{\mathrm{rot}} \psi_{\mathrm{n}}^{\mathrm{s}})$.

Interchanging the two nuclei transforms $\psi_{\mathrm{n}}^{\mathrm{rot}}$ to [35, 37]

$$P_{12}(\psi_{\mathrm{n}}^{\mathrm{rot}}) = (-1)^{J} \psi_{\mathrm{n}}^{\mathrm{rot}}, \tag{2}$$

where P_{12} represents the permutation operator that interchanges the nuclei's positions and J is the rotational quantum number. Hence, the rotational wave function is symmetric for even rotational states ($J = 0, 2, 4, \ldots$) and antisymmetric for odd rotational states ($J = 1, 3, 5, \ldots$). The nuclear spin function can be either symmetric or antisymmetric. Following the rules for adding angular momenta in quantum mechanics, it can be shown that the combination of the two spins corresponding to each nucleus gives four possible functions. They can be written as linear combinations of the direct product of the two possible spin orientations for each single spin, i.e., as

$$\begin{aligned} \psi_{+1}^{\mathrm{T}} &= |\alpha\alpha\rangle, \\ \psi_{0}^{\mathrm{T}} &= \tfrac{1}{\sqrt{2}}(|\alpha\beta\rangle + |\beta\alpha\rangle), \\ \psi_{-1}^{\mathrm{T}} &= |\beta\beta\rangle, \\ \psi_{0}^{\mathrm{S}} &= \tfrac{1}{\sqrt{2}}(|\alpha\beta\rangle - |\beta\alpha\rangle). \end{aligned} \tag{3}$$

The functions are grouped according to their total angular momentum number. The first three functions have total spin $S = 1$ (with the z-projection indicated in the subindices) while the fourth function possesses total spin $S = 0$. They are commonly termed as triplet and singlet, respectively; the triplet is symmetric on the nuclei and the singlet is antisymmetric.

According to Pauli's principle, the symmetric rotational functions must be combined with the singlet nuclear function, whereas each antisymmetric rotational function has to be associated with the three symmetric spin functions, to yield a total wave function being antisymmetric on the nuclei. Hence, there are two hydrogen isomers, one called *para*hydrogen (p-H_2), having an antisymmetric nuclear spin function and existing only in even rotational states, and the other called *ortho*hydrogen (o-H_2), having a symmetric nuclear spin function and existing only in the odd rotational states. Moreover, as the transition between even and odd rotational states implies a transition between singlet and triplet nuclear spin states, which is symmetry forbidden, the proportion of *ortho*- and *para*hydrogen is quasistable at any given temperature. It was Dennison, back in 1927, who showed that the interconversion of o-H_2 to p-H_2 is extremely slow [38]. Two years later it was discovered by Bonhoeffer and Harteck [39] that catalysts such as charcoal accelerate the achievement of thermodynamic equilibrium, thus permitting the fast modification of the ratio o-H_2/p-H_2 by cooling or heating the gas.

2.2 Hydrogen Described from Statistical Mechanics

In order to quantitatively describe the state of an ensemble of hydrogen molecules it is necessary to resort to statistical mechanics. The partition function of the system [78] reads

$$Z = \sum_i d_i \, \mathrm{e}^{-E_i/kT}, \tag{4}$$

where d_i represents the degeneracy of the ith energy level E_i, k is the Boltzmann constant, and T the absolute temperature. Vibrational and electronic modes are not activated in hydrogen gas at room temperature or below [40] and therefore only the vibrational and electronic ground states are populated and contribute to the partition function. Thus, we need to focus only on the rotational and nuclear spin states. Using the Born–Oppenheimer rigid rotor approximation, the rotational energy levels can be expressed as

$$E_J = J(J+1)\frac{\hbar^2}{2I}, \tag{5}$$

where I is the moment of inertia of the hydrogen molecule [35]. It can be clearly seen from Eq. (5) that the energy gap increases with the quantum number J, being the first (i.e., $E_{J=0} \rightarrow E_{J=1}$) [40]:

$$\Delta E_{J=0\rightarrow J=1} = h \times 3.7 \times 10^{12}\,\mathrm{Hz}. \tag{6}$$

The energy differences between nuclear states are much smaller (in the order of Hz) and therefore the partition function can further be simplified by considering only the rotational energy levels with the extra degeneracy produced by the nuclear states.

The rotational energy levels are populated according to the Boltzmann equilibrium distribution, with the population of each state given by

$$P_J = \frac{1}{Z} d_J d_S \, \mathrm{e}^{-E_J/kT}. \tag{7}$$

The factor d_J corresponds to the degeneracy of the rotational states, which is $(2J + 1)$ in the absence of electric and magnetic fields, and d_S is the degeneracy due to nuclear spin states. The population of each level can be recast as

$$P_J = \frac{1}{Z}(2J + 1) d_S \, \mathrm{e}^{-J(J+1)\theta_r/T}, \tag{8}$$

where the *rotational temperature* has been defined as $\theta_r = \hbar^2/2IK$ [30, 41].

In Fig. 1a, b the populations of the first six rotational levels are plotted against the absolute temperature for *para*hydrogen ($J = 0, 2, 4$) and *ortho*hydrogen ($J = 1, 3, 5$), respectively. At very low temperatures ($T < 20\,\mathrm{K}$) only the level corresponding to $J = 0$ is observably populated and, consequently, almost 100% of the hydrogen molecules in the gas will be in the *para*-state. As the temperature increases, more molecules initially at the lowest level populate the next levels. At room temperature the first four states are substantially more populated and contribute to the *para* and *ortho*hydrogen fractions.

The amount of *para*hydrogen at any temperature in thermal equilibrium is obtained by accounting the populations of all rotational levels with even quantum number:

$$N_{para} = \frac{1}{Z} \sum_{J=\text{even}} (2J + 1) \, \mathrm{e}^{-J(J+1)\theta_r/T}. \tag{9}$$

This curve is displayed in Fig. 1c. The dashed vertical line points out the conversion temperature of $T = 77\,\mathrm{K}$, the boiling temperature of nitrogen, where the gas is comprised of equal amounts of p-H_2 and o-H_2. At room temperature the fraction of p-H_2 is ~1/4.

2.3 Preparation of Enriched parahydrogen

The plot in Fig. 1c suggests the method of enriching hydrogen in the *para*-state. As stated above, the thermal equilibrium in an ensemble of hydrogen molecules involves transitions between nuclear states with different symmetry. Such

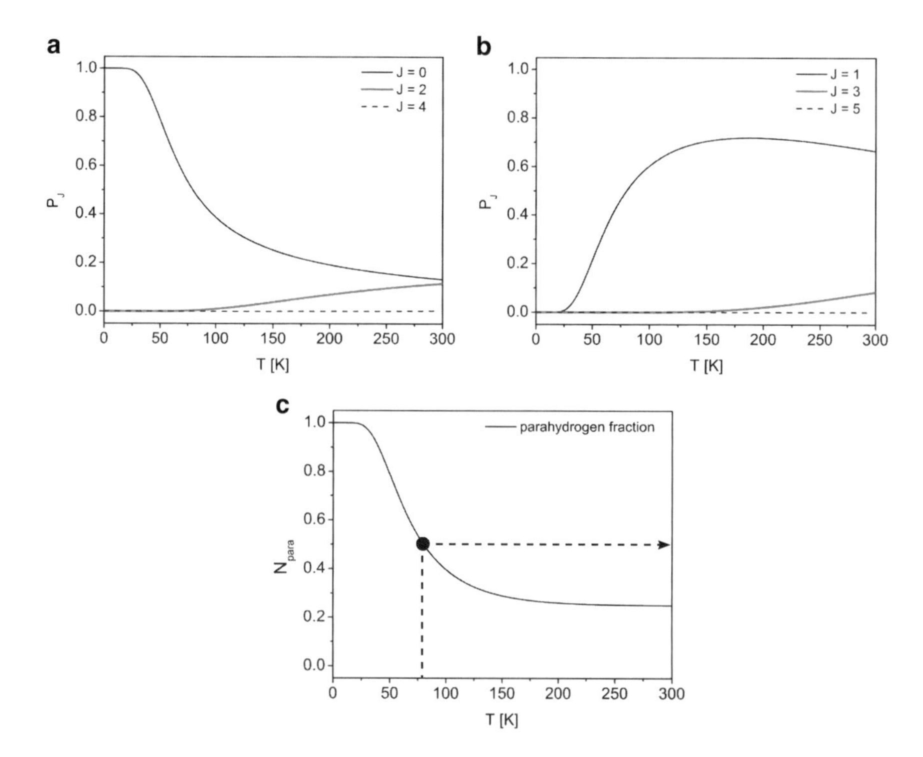

Fig. 1 (**a**) Population of the rotational levels $J = 0, 2, 4$, corresponding to *para*hydrogen; (**b**) same as in (**a**) for *ortho*hydrogen ($J = 1, 3, 5$); (**c**) *para*hydrogen fraction present in an ensemble of hydrogen molecule in the gas phase at thermal equilibrium vs temperature. The *dashed line* depicts the nitrogen boiling temperature, i.e., 77 K

transitions are possible in the presence of perturbations involving the nuclear spins, but these perturbations are extremely small in pure hydrogen. In other words, it is possible to have hydrogen gas for a reasonably long time out of thermal equilibrium (ranging from days to weeks, depending on the container). In contact with a catalyst like charcoal, however, the paramagnetic centers make the transitions more likely, causing thermal equilibrium to occur very rapidly.

If a volume of hydrogen is cooled with, for instance, liquid nitrogen in the presence of a charcoal catalyst, it will quickly reach the situation marked with a dot in Fig. 1c. If the catalyst is removed afterwards and the gas is warmed up to room temperature the *para*hydrogen amount is ~50%, in contrast to the ~25% expected in thermal equilibrium at the same temperature.

Note that if the cooling process is performed at temperatures lower than 20 K the gas will contain almost 100% *para*hydrogen.

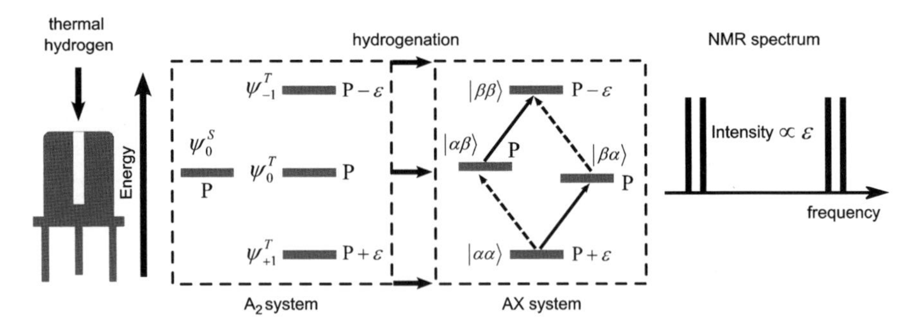

Fig. 2 NMR spectrum after a hydrogenation with natural abundance p-H_2. The four eigenstates in the initial A_2 spin system are equally populated. The transferred protons form an AX system in the target molecule

2.4 *NMR Signal Enhancement with Enriched* para*hydrogen: A Population-Oriented Approach*

In order to understand intuitively the mechanism leading to the NMR signal enhancement when performing a hydrogenation reaction with hydrogen enriched in the *para*-state, it is useful to analyze the different NMR spectra focusing on the levels' population differences. This is the most utilized approach described in several publications [36, 42, 43], and we will shortly summarize it here for the sake of completeness. We will focus our attention on a homogeneous hydrogenation reaction where the hydrogen nuclei are transferred *pairwise* to a target molecule. The catalyst used for the hydrogenation reaction is crucial in these processes, because the enhanced signal arises only from those reactions capable to preserve the singlet symmetry of p-H_2. We assume that the p-H_2 protons in the target molecule form, at high fields, an AX system isolated from the rest of the molecule. An example of such a reaction is the hydrogenation of propiolic acid-d_2 catalyzed by $[Rh(COD)dppb]^+BF_4^-$ [42].

When performing the reaction with thermal hydrogen at room temperature (i.e., hydrogen at thermal equilibrium) the NMR spectrum is independent of whether the reaction occurs at low or high magnetic field. We assume that the gas is introduced in the bore and the reaction is carried out inside the magnet, as depicted in Fig. 2. The hydrogen forms an A_2 spin system described by the eigenfunctions of Eq. (3), and the four eigenstates are equally populated ($P = 0.25$) out of the magnet. Inside the magnet the energy levels are accordingly modified and the population distribution follows the Boltzmann distribution, changing by an amount known as the Boltzmann factor, $\varepsilon = \hbar\gamma B_0/kT_{\text{exp}}$, of the order of 10^{-5} for protons at room temperature and ordinary magnetic fields. After the reaction with thermal hydrogen, the transferred protons become an AX system described by the eigenfunctions of the Zeeman basis [44], $B = \{|\alpha\alpha\rangle, |\alpha\beta\rangle, |\beta\alpha\rangle, |\beta\beta\rangle\}$. The corresponding NMR spectrum consists of four lines of identical intensities, proportional to the Boltzmann factor, as pictorially shown in Fig. 2.

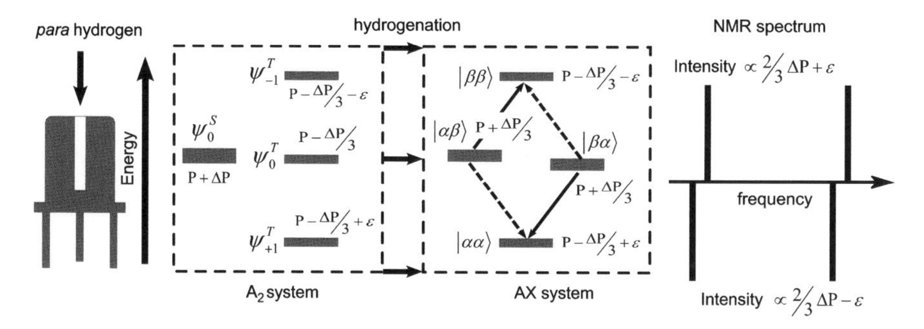

Fig. 3 PASADENA experiment for an AX spin system. The chemical reaction with hydrogen gas enriched in the *para*-state occurs inside the magnet. The *p*-H2 enrichment is evidenced in the overpopulation of the singlet state, eigenstate of the A_2 system

In contrast, if the experiment is performed with hydrogen enriched in the *para*-state, the final result strongly depends on the magnetic field at which the reaction is accomplished. There are two main procedures distinguished in the literature: PASADENA, where the gas is transported to the magnet and the reaction is carried out at the same magnetic field strength at which the NMR experiment is performed (typically high magnetic fields [1, 45]), and ALTADENA, where the reaction is conducted at low field and the product is subsequently moved into the magnet to start the NMR experiment.

Let us consider hydrogen gas with an excess of population in the *para*-state denoted as ΔP, ranging from 0 (thermal hydrogen) to $3P$ (pure *para*hydrogen). Then the total population of the singlet and triplet states are $P + \Delta P$ and $P - \Delta P$, respectively. Inside the magnet the populations of the levels corresponding to ψ^{T}_{+1} and ψ^{T}_{-1} are corrected by ε due to the Boltzmann equilibrium distribution (see Fig. 3).

Right after the reaction the excess population of ψ^{S}_{0} equally populates the levels labeled as $|\alpha\beta\rangle$ and $|\beta\alpha\rangle$ in the product molecule. As a result, the NMR spectrum presents two antiphase doublets with intensities of the order of $\frac{2}{3}\Delta P \pm \varepsilon$, as shown in Fig. 3. Note that ΔP is close to unity, whereas $\varepsilon \sim 10^{-5}$, demonstrating the large signal enhancement that can be accomplished by hyperpolarization. The intensity of the PASADENA spectrum is, for the case treated here and neglecting relaxation, between four and five orders of magnitude larger than the intensity obtained in the reaction with thermal hydrogen.

If, alternatively, the reaction is performed at low field (typically at the earth field) and the sample is adiabatically transported to the magnet in the sense that the initial eigenstate ψ^{S}_{0} follows the corresponding eigenstates for each magnetic field, then the total excess of population initially in ψ^{S}_{0} ends up in $|\beta\alpha\rangle$, the eigenstate of the AX system. Consequently, the NMR spectrum looks different compared to the PASADENA spectrum, displaying only two peaks with opposite phases which have twice the intensities of the PASADENA peaks (Fig. 4).

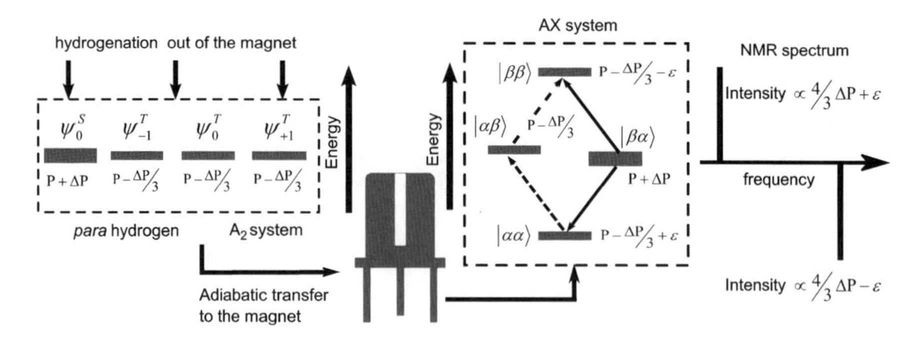

Fig. 4 ALTADENA experiment for an AX spin system. The chemical reaction is performed outside the magnet, followed by an adiabatic transport of the sample into the NMR observation field

The population oriented explanation helps in understanding the major features of PHIP experiments, such as the shape of the spectra and the signal enhancement. However, the analysis becomes extremely cumbersome when trying to extend it to strongly coupled (e.g., AB or A_2) and larger spin systems. A more general treatment of PHIP, applicable to arbitrary spin systems, can be performed by using the enormous power of the density matrix formalism to calculate NMR signals.

2.5 Density Operator Formalism Applied to PHIP

The density operator formalism is commonly used in the NMR community to calculate signals after the application of a given r.f. pulse sequence [44, 46]. We briefly describe here the formalism applied to the calculation of experiments involving hydrogenations with *para*hydrogen.

The formal definition of the density operator is

$$\rho = \sum_i w_i |\psi_i\rangle\langle\psi_i|, \tag{10}$$

where w_i represents the fraction of the ith state in the ensemble. In the absence of an external magnetic field the natural basis to express the matrix associated to the density operator of a generic ensemble of hydrogen molecules is:

$$B' = \{\psi_{+1}^{\mathrm{T}}, \psi_0^{\mathrm{T}}, \psi_{-1}^{\mathrm{T}}, \psi_0^{\mathrm{S}}\}, \tag{11}$$

since the vectors are the eigenvectors [see Eq. (3)]. However, it is usually preferred to express the density operator in the Zeeman basis (alternatively called computational basis), $B = \{|\alpha\alpha\rangle, |\alpha\beta\rangle, |\beta\alpha\rangle, |\beta\beta\rangle\}$. The components of the density operator are in this basis:

$$|\psi_{+1}^{\mathrm{T}}\rangle\langle\psi_{+1}^{\mathrm{T}}| = \begin{pmatrix} 1 & 0 & 0 & 0 \\ 0 & 0 & 0 & 0 \\ 0 & 0 & 0 & 0 \\ 0 & 0 & 0 & 0 \end{pmatrix}, \quad |\psi_{0}^{\mathrm{T}}\rangle\langle\psi_{0}^{\mathrm{T}}| = \frac{1}{2}\begin{pmatrix} 0 & 0 & 0 & 0 \\ 0 & 1 & 1 & 0 \\ 0 & 1 & 1 & 0 \\ 0 & 0 & 0 & 0 \end{pmatrix},$$
$$|\psi_{-1}^{\mathrm{T}}\rangle\langle\psi_{-1}^{\mathrm{T}}| = \begin{pmatrix} 0 & 0 & 0 & 0 \\ 0 & 0 & 0 & 0 \\ 0 & 0 & 0 & 0 \\ 0 & 0 & 0 & 1 \end{pmatrix}, \quad |\psi_{0}^{\mathrm{S}}\rangle\langle\psi_{0}^{\mathrm{S}}| = \frac{1}{2}\begin{pmatrix} 0 & 0 & 0 & 0 \\ 0 & 1 & -1 & 0 \\ 0 & -1 & 1 & 0 \\ 0 & 0 & 0 & 0 \end{pmatrix}. \tag{12}$$

The use of this basis permits the straightforward decomposition in product operators [44, 47], as

$$\begin{aligned} |\psi_{+1}^{\mathrm{T}}\rangle\langle\psi_{+1}^{\mathrm{T}}| &= \tfrac{1}{4}\mathbb{I} + \tfrac{1}{2}(I_1^z + I_2^z + 2I_1^zI_2^z), \\ |\psi_{0}^{\mathrm{T}}\rangle\langle\psi_{0}^{\mathrm{T}}| &= \tfrac{1}{4}\mathbb{I} + \tfrac{1}{2}(2I_1^xI_2^x + 2I_1^yI_2^y - 2I_1^zI_2^z), \\ |\psi_{-1}^{\mathrm{T}}\rangle\langle\psi_{-1}^{\mathrm{T}}| &= \tfrac{1}{4}\mathbb{I} + \tfrac{1}{2}(-I_1^z - I_2^z + 2I_1^zI_2^z), \\ |\psi_{0}^{\mathrm{S}}\rangle\langle\psi_{0}^{\mathrm{S}}| &= \tfrac{1}{4}\mathbb{I} + \tfrac{1}{2}(-2I_1^xI_2^x - 2I_1^yI_2^y - 2I_1^zI_2^z). \end{aligned} \tag{13}$$

Slightly modifying notation, we define one density operator for pure *ortho*hydrogen and one density operator for pure *para*hydrogen, setting in Eq. (10) $w = 1/3$ for every component of o-H_2 and $w = 1$ for p-H_2, to obtain an accurate normalization. Thus,

$$\begin{aligned} \rho^{ortho} &= \tfrac{1}{3}\{|\psi_{+1}^{\mathrm{T}}\rangle\langle\psi_{+1}^{\mathrm{T}}| + |\psi_{0}^{\mathrm{T}}\rangle\langle\psi_{0}^{\mathrm{T}}| + |\psi_{-1}^{\mathrm{T}}\rangle\langle\psi_{-1}^{\mathrm{T}}|\} = \tfrac{1}{4}\mathbb{I} + \tfrac{1}{3}(I_1^xI_2^x + I_1^yI_2^y + I_1^zI_2^z), \\ \rho^{para} &= |\psi_{0}^{\mathrm{S}}\rangle\langle\psi_{0}^{\mathrm{S}}| = \tfrac{1}{4}\mathbb{I} - (I_1^xI_2^x + I_1^yI_2^y + I_1^zI_2^z). \end{aligned} \tag{14}$$

The above expressions can be more compactly written as

$$\begin{aligned} \rho^{ortho} &= \tfrac{1}{4}\mathbb{I} + \tfrac{1}{3}\mathbf{I}_1 \cdot \mathbf{I}_2, \\ \rho^{para} &= \tfrac{1}{4}\mathbb{I} - \mathbf{I}_1 \cdot \mathbf{I}_2. \end{aligned} \tag{15}$$

Denoting by ρ^{hydr} the density matrix of an ensemble of hydrogen molecules with an arbitrary fraction of *para*hydrogen represented by N_{para}, we obtain

$$\rho^{\mathrm{hydr}} = (1 - N_{para})\rho^{ortho} + N_{para}\rho^{para} = \frac{1}{4}\mathbb{I} - \xi\mathbf{I}_1 \cdot \mathbf{I}_2 - \frac{\varepsilon}{4}(I_1^z + I_2^z), \tag{16}$$

where $\xi \equiv (4N_{para} - 1)/3$. The third factor above, associated with thermal equilibrium in the presence of a strong magnetic field, was added for completeness. The fraction

N_{para} depends on the temperature at which the hydrogen is enriched, as shown in Fig. 1c. Note that $N_{para} = \frac{1}{4}$ at room temperature and therefore the second term of the right-hand side of Eq. (16) vanishes (i.e., $\xi = 0$). Consequently, the term associated with the Boltzmann factor will dominate, recovering the density matrix for a two spin system in thermal equilibrium at that temperature. On the other hand, at low temperatures, typically $T < 20$ K, $N_{para} \approx 1$ (i.e., $\xi = 1$), the density operator is dominated by the second term on the right side of Eq. (16), and the thermal factor can be safely neglected.

In PHIP experiments the hydrogen gas is always cooled down and the *para*hydrogen state is sufficiently enriched such that $1/2 \leq N_{para} \leq 1$. Therefore the thermal equilibrium is usually neglected in Eq. (16). For the rest of the chapter, and without loss of generality, we will set $\xi \equiv 1$ in the calculations. The overall behavior of the NMR signals will still be reproduced except for a scaling factor ξ influencing the intensity of the NMR spectra.

The initial density operator ρ^{para} is proportional to the term $\mathbf{I}_1 \cdot \mathbf{I}_2$ [see Eq. (15)] which is invariant under rotations and consequently no NMR observable magnetization is achieved in principle. However, if the two protons of the same p-H_2 molecule are transferred pairwise to a molecule during a hydrogenation reaction, the thermal equilibrium density operator of the product molecule will be perturbed. Assuming that the precursor of the target molecule (educt) consists of N spins-1/2, the thermal equilibrium density operator in the high temperature approximation can be expressed as [44, 46, 48]

$$\rho^{\text{ed}}(0) = \frac{1}{2^N}\mathbb{I} - \frac{1}{2^N}\sum_{k=1}^{N}\varepsilon_k I_k^z, \tag{17}$$

where $\mathbb{I}$ is the identity operator with the proper dimensions, and ε_k the Boltzmann factor of the kth nucleus (to include the possibility of heteronuclei). This is the density operator for the educt molecule just before the beginning of the reaction. The density operator for the product molecule at $t = 0$ is constructed as usual:

$$\rho^{\text{pr}}(0) = \rho^{\text{hydr}} \otimes \rho^{\text{ed}}(0). \tag{18}$$

The symbol $\otimes$ denotes the direct product of the matrices associated with the density operators. A very useful approximation, commonly introduced to simplify the calculations, consists in neglecting the Boltzmann terms in Eq. (17), invoking the argument given above, in Eq. (16) [49, 50]. Therefore the product's density operator just after the hydrogenation results in

$$\rho^{\text{pr}}(0) = \rho^{\text{hydr}}(0) \otimes \frac{1}{2^N}\mathbb{I}. \tag{19}$$

Immediately after the reaction, there evolves according to the Liouville–von Neumann equation,

$$\frac{\mathrm{d}\rho^{\text{pr}}(t)}{\mathrm{d}t} = -i[H^{\text{pr}}, \rho^{\text{pr}}(t)], \tag{20}$$

where H^{pr} is the spin Hamiltonian of the product molecule. The Hamiltonian can, in general, be written as [44, 46]

$$H^{\mathrm{pr}} = 2\pi \sum_{k=1}^{N} \nu_k I_k^z + 2\pi \sum_{k,l;k}^{N} J_{k,l} \mathbf{I}_k \cdot \mathbf{I}_l. \tag{21}$$

The terms on the right-hand side of the equation represent the *chemical shift* and the *J-coupling* interactions, respectively. For this time-independent Hamiltonian the formal solution of Eq. (20) is [46, 51]

$$\rho^{\mathrm{pr}}(t) = \mathrm{e}^{(-\mathrm{i}H^{\mathrm{pr}}t)} \rho^{\mathrm{pr}}(0) \mathrm{e}^{(+\mathrm{i}H^{\mathrm{pr}}t)}. \tag{22}$$

This description is appropriate for the time evolution of a single reacting molecule. However, in order to treat the whole hydrogenation process, an ensemble of molecules must be considered. Let us assume that the chemical reaction is carried out in a total reaction time t_{r}. Several molecules in the ensemble will react at different times τ_i, distributed in the time interval $0 < \tau_i < t_{\mathrm{r}}$, prior to the beginning of the NMR experiment. Thus, at the time of the NMR measurement, there will be an ensemble of density operators ρ_i^{pr} which have individually evolved for different time periods t_{r}-τ_i. If the reaction takes longer than the characteristic times of any internal evolution, i.e., $t_{\mathrm{r}} \gg 1/\nu_k, 1/J_{k,l}$ for all ν_k and $J_{k,l}$ present in the Hamiltonian, the average density operator can be obtained by time averaging [4, 43]:

$$\bar{\rho}^{\mathrm{pr}}(t_{\mathrm{r}}) = \frac{1}{t_{\mathrm{r}}} \int_{t=0}^{t_{\mathrm{r}}} \rho^{\mathrm{pr}}(t)\,\mathrm{d}t. \tag{23}$$

As a consequence of the reaction and subsequent evolution, hyperpolarization can spread from the *para*hydrogen to the rest of the molecule, depending only on the molecular Hamiltonian [4, 49]. Under favorable conditions even the signals of heteronuclei can be significantly enhanced by hyperpolarization transfer [14, 50].

2.6 Two Spin Systems

The time evolution of the density operator along with the different features of the PHIP NMR spectra is revised here for the case where the *para*hydrogen protons are deposited into two magnetically inequivalent sites. It is further assumed that there exists *no* coupling to the rest of the molecule. Thus the transferred protons will form either a weakly coupled AX spin system or a strongly coupled AB spin system depending on the comparison between their coupling strength and their chemical shift difference. If both spins are chemically equivalent, i.e., forming an A_2 spin system, the former *para*hydrogen protons maintain the singlet nature which is NMR silent, and therefore it is excluded from this treatment [21]. It is important to note that the case AA′ is nonexistent in an isolated two spin system.

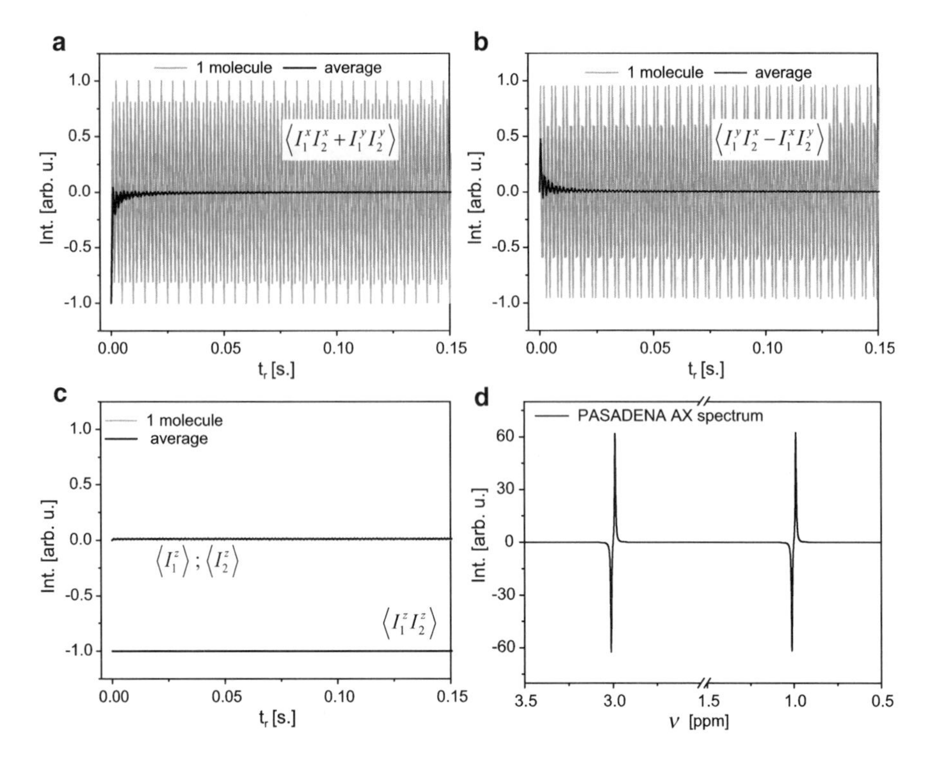

Fig. 5 PASADENA AX system. Evolution of the expectation values: (**a**) ZQ_x; (**b**) ZQ_y; (**c**) I_1^z, I_2^z, and $I_1^z I_2^z$. The results for one single molecule (*gray*) and for the ensemble average (*black*) are shown. Note that only the longitudinal order expectation value remains different from zero in the ensemble. (**d**) Simulated spectrum

2.6.1 PASADENA in an AX Spin System

We consider here a reaction with enriched *para*hydrogen performed at high magnetic field, such that the protons form an isolated AX spin system in the product molecule. Before the reaction, the density operator is $\rho^{\mathrm{pr}}(0) = \mathbb{I}/4 - \mathbf{I}_1 \cdot \mathbf{I}_2$. The plots in Fig. 5 display the evolution of the density operator's components under the high field Hamiltonian during the reaction, for a single molecule $\rho^{\mathrm{pr}}(t_{\mathrm{r}})$ (gray line), superimposed with the ensemble average evolution $\bar{\rho}^{\mathrm{pr}}(t_{\mathrm{r}})$ (black line). The contribution of the identity operator is neglected.

The expectation value of the zero-quantum operator, $\mathrm{ZQ}_x = -(I_1^x I_2^x + I_1^y I_2^y)$ [4], initially present in the density operator $\rho^{\mathrm{pr}}(0)$, oscillates and produces the zero quantum term $\mathrm{ZQ}_y = (I_1^y I_2^x - I_1^x I_2^y)$ that oscillates as well with zero mean expectation value, as shown in Fig. 5a, b. However, the time average of both expectation values vanishes after sufficiently long reaction times. The expectation value of the longitudinal order term $I_1^z I_2^z$, on the other hand, does not evolve and the time average yields exactly the initial value. The terms $\langle I_1^z \rangle$ and $\langle I_2^z \rangle$ maintain their zero-value during the process (Fig. 5c).

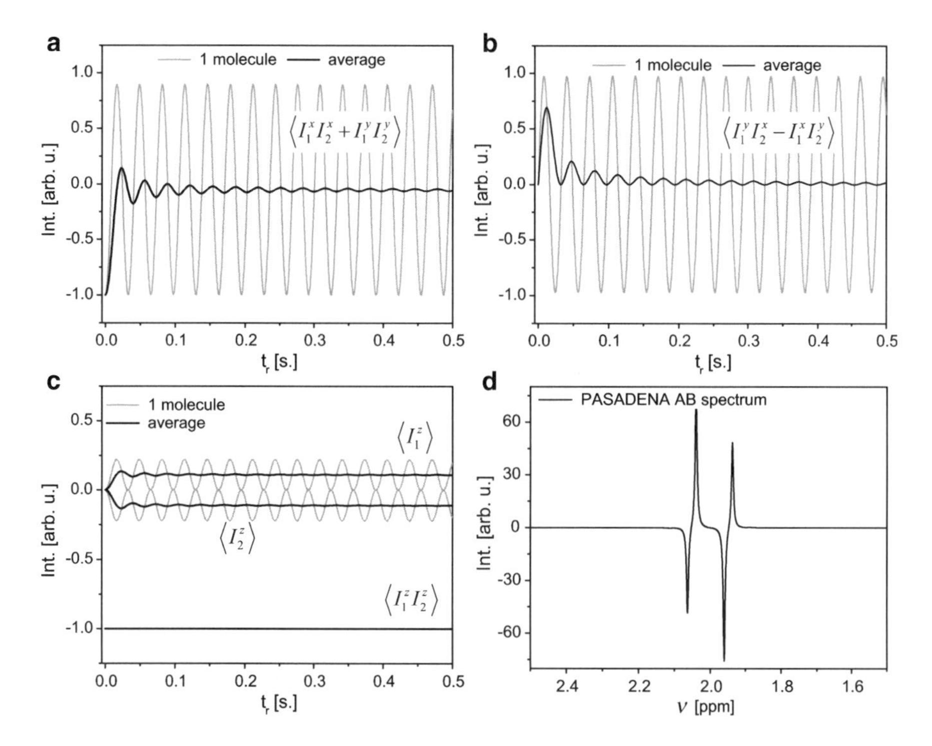

Fig. 6 PASADENA AB spin system. Evolution of the expectation values for the operators: (**a**) ZQ_x; (**b**) ZQ_y; (**c**) I_1^z, I_2^z, and $I_1^z I_2^z$. The results for one single molecule (*gray*) and for the ensemble average (*black*) are shown. The polarization expectation values also remain different from zero in the ensemble. (**d**) Simulated spectrum

In order to clarify the notation, it is convenient to introduce the variable t_f to indicate the reaction time which is long enough to fulfill the condition: $t_\mathrm{f} > 1/\nu_k$, $1/J_{k,l}$ for all ν_k and $J_{k,l}$. Then we can write $\bar{\rho}^{\mathrm{pr}}(t_\mathrm{f}) = \mathbb{I}/4 - I_1^z I_2^z$, which results, after a 45° radiofrequency pulse, in the spectrum shown in Fig. 5d. The following numerical data were introduced in the calculations: $B_0 = 7\,\mathrm{T}$, $\Delta\nu = 2\,\mathrm{ppm}$ and $J_{1,2} = 7\,\mathrm{Hz}$. Note that the spectrum's shape is similar to that predicted by the population approach (see Fig. 3).

2.6.2 PASADENA in an AB Spin System

If the chemical shift difference between the *para*hydrogen protons in the product molecule is not much larger than $J_{1,2}$, they form an isolated AB spin system. The smaller chemical shift difference reduces the oscillation frequency of the zero quantum expectation values $\langle ZQ_x \rangle$ and $\langle ZQ_y \rangle$ and consequently it takes longer to reach the steady state, as can be seen in Fig. 6a, b (i.e., $t_\mathrm{f} > \frac{1}{\Delta\nu}$). The longitudinal order is conserved in the reaction as before, but the evolution of the zero-quantum terms generates non-zero average polarization in both sites, $\langle I_1^z \rangle$ and $\langle I_2^z \rangle$ (Fig. 6c) [52]. The average density operator after reaching the steady state differs considerably from its counterpart in the AX system: $\bar{\rho}^{\text{pr}}(t_{\text{f}}) = \mathbb{I}/4 - I_1^z I_2^z + 0.1(I_1^z - I_2^z)$. The corresponding spectrum also presents the characteristic anti-phase doublets, with slightly different intensities. The spectrum was calculated setting: $B_0 = 7\,\text{T}$, $\Delta\nu = 0.1$ ppm, and $J_{1,2} = 7\,\text{Hz}$.

Although the PASADENA-AX and PASADENA-AB NMR spectra from Figs. 5, 6 do not differ substantially, it is important to note that in both cases a 45° pulse was employed before signal calculation. In general, a θ pulse acts differently on longitudinal order operators and polarization operators. Typically, a 90° pulse applied to a polarization operator (I_1^z or I_2^z) produces maximum signal while being applied to a longitudinal order term ($I_1^z I_2^z$), giving zero signal. On the other hand, a 45° pulse produces maximum signal on longitudinal order operators and only a fraction of the maximum ($\sqrt{2}/2$) on polarization operators. Therefore, the PASADENA-AB spectrum shows a strong shape dependence on the radiofrequency pulse while the PASADENA-AX spectrum is just rescaled [26, 43].

2.6.3 General PASADENA

We can make the situations treated above more general by considering the former *para*hydrogen protons being transferred to a molecule where they form an isolated two spin system described by $\zeta = \Delta\nu/J$. In the lower limit case, $\zeta = 0$, the spins form an A_2 system. The initial density operator commutes with the Hamiltonian and, consequently, it does not evolve, preserving the magnetic equivalence during the reaction $\bar{\rho}^{\text{pr}}_{(\zeta=0)}(t_{\text{f}}) = \mathbb{I}/4 - \mathbf{I}_1 \cdot \mathbf{I}_2$ and remaining NMR silent. From the calculations it can be observed that, for any $\zeta \neq 0$, the steady state averaged density operator can be expressed as $\bar{\rho}^{\text{pr}}_{(\zeta\neq0)}(t_{\text{f}}) = \mathbb{I}/4 - \eta_{\text{ZQ}_x}(I_1^x I_2^x + I_1^y I_2^y) - I_1^z I_2^z + \eta_{\text{pol}}(I_1^z - I_2^z)$. Analytical expressions for η_{ZQ_x} and η_{pol} can be found, for instance, in [4, 43]. We maintain here our numerical treatment. The latter expression can be rearranged (noting that $\mathbf{I}_1 \cdot \mathbf{I}_2 = I_1^x I_2^x + I_1^y I_2^y + I_1^z I_2^z$), to yield

$$\rho^{\text{pr}}_{(\zeta\neq0)}(t_{\text{f}}) = \mathbf{I}/4 - \eta_{\text{ZQ}_x}(\mathbf{I}_1 \cdot \mathbf{I}_2) - (1 - \eta_{\text{ZQ}_x})I_1^z I_2^z + \eta_{\text{pol}}(I_1^z - I_2^z). \tag{24}$$

In Fig. 7 the longitudinal order and the expectation values of the polarizations after the reaction are plotted vs ζ, ranging from 0 to 35. The dotted lines represent the coefficient accompanying the term $\mathbf{I}_1 \cdot \mathbf{I}_2$.In the upper limit case, when $\zeta \to \infty$, the protons form an AX system. The terms for the coherences and polarizations are averaged out during the reaction as shown above ($\eta_{\text{ZQ}_x} = \eta_{\text{pol}} = 0$), and only the longitudinal order survives.

In both limiting cases, $\zeta = 0$ and $\zeta \to \infty$, $\langle I_1^z - I_2^z \rangle$ has a zero mean value. On the other hand, we have seen that for an AB system the latter mean value is nonzero. This suggests, assuming continuous behavior, the existence of a particular value ζ^m

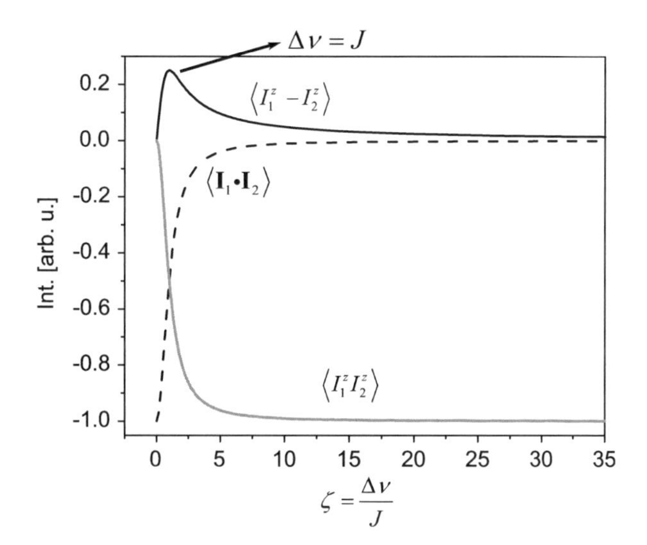

Fig. 7 PASADENA. Expectation values of the longitudinal order, polarization difference and isotropic term after the reaction calculated as a function of the parameter $\zeta = \Delta\nu/J$

for which the term $\langle I_1^z - I_2^z \rangle$ reaches a maximum mean value. From Fig. 7 it can be seen that in fact the maximum occurs if $\zeta^m = 1$, i.e., if the condition $\Delta\nu = J$ is satisfied.

2.6.4 Hydrogenation at Low Field and Transfer to High Magnetic Field

The plot in Fig. 7 can be interpreted not only in the sense that ζ varies by changing the product molecule, i.e., the internal parameters, but also in a very different way. While the chemical shift difference between the two proton sites linearly depends on the external magnetic field (B_0), the J-coupling constant is related to the chemical bond and is independent of B_0 [44]. This means that the variation of ζ can also be achieved by changing the magnetic field at which the reaction is performed. In that case, the sample must subsequently be transported from the reaction to the acquisition field. This transportation step plays an important role in the final shape of the NMR signal.

To satisfy the ALTADENA condition, for instance, the transportation must be adiabatic with respect to the internal molecular dynamics [3, 20].

2.6.5 Sudden vs Adiabatic Transport

In order to illustrate the effect of the transportation step on the NMR signal, we treat here two limiting cases: adiabatic and sudden magnetic field changes. These two cases were described in the frame of hyperpolarization transfer in [19, 53], and specifically related to *para*hydrogen in [20].

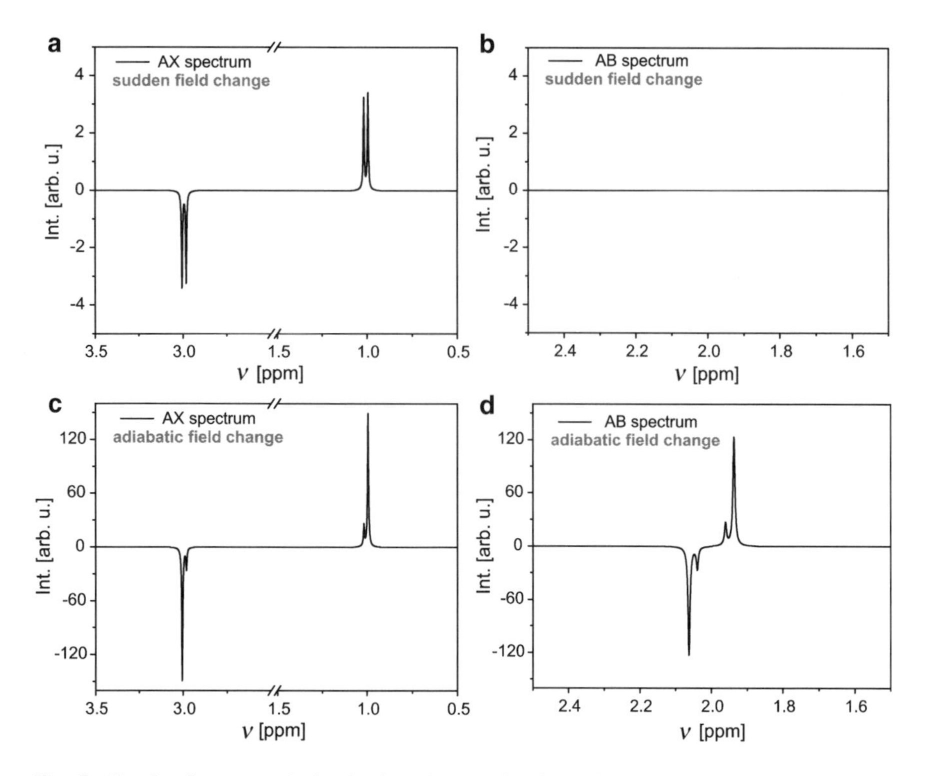

Fig. 8 Simulated spectra obtained when the reaction is performed at low field (3 mT) and suddenly (**a**, **b**) or adiabatically (**c**, **d**) moved to the observation field (7 T). In (**a**, **c**) an AX spin system ($\Delta\nu = 0.2$ ppm, $J = 7$ Hz) and in (**b**, **d**) an AB spin system ($\Delta\nu = 0.1$ ppm, $J = 7$ Hz) were considered

For the sudden field change the transport is assumed to be *instantaneous*, in the sense that no spin evolution is allowed between the reaction and the NMR measurement. Let us assume that the reaction is performed at $B_0 = 3\,\text{mT}$ and the NMR experiment at $B_0 = 7\,\text{T}$. If the protons form an AX system at high field ($\Delta\nu = 2\,\text{ppm},\ J_{1,2} = 7\,\text{Hz}$), the averaged density operator after the reaction is $\bar{\rho}^{\text{pr}}_{\text{AX}}(t_{\text{f}}) = \mathbf{I}/4 - 0.9987(I_1^xI_2^x + I_1^yI_2^y) - I_1^zI_2^z + 1.8\times10^{-2}(I_1^z - I_2^z)$, or equivalently $\bar{\rho}^{\text{pr}}_{\text{AX}}(t_{\text{f}}) = \mathbf{I}/4 - 0.9987(\mathbf{I}_1\cdot\mathbf{I}_2) - 1.3\times10^{-3}I_1^zI_2^z + 1.8\times10^{-2}(I_1^z - I_2^z)$.

At the end of the reaction almost 100% of the initial p-H_2 density operator is preserved. After a *sudden* transport to the observation field only the tiny terms corresponding to longitudinal order and polarization difference will contribute to the NMR spectrum (see Fig. 8a).

The spectrum's shape is dominated by $\langle I_1^z - I_2^z\rangle$. On the other hand, if the protons form an AB system in the product molecule ($\Delta\nu = 0.1\,\text{ppm},\ J_{1,2} = 7\,\text{Hz}$), after the hydrogenation the initial *para*hydrogen density operator remains unchanged, $\bar{\rho}^{\text{pr}}_{\text{AB}}(t_{\text{f}}) = \mathbb{I}/4 - \mathbf{I}_1\cdot\mathbf{I}_2$. The subsequent *sudden* transport does not affect it, and the system is NMR silent, as can be seen in Fig. 8b.

Similar calculations for both spin systems (with the hydrogenation performed again at $B_0 = 3\,\mathrm{mT}$), but the sample *adiabatically* transported to the observation magnetic field (7 T), reveal remarkable differences in the corresponding NMR spectra.

While the averaged density operators after the reaction are exactly the same as for the sudden field change case, their evolutions with the full Hamiltonian at different magnetic field strengths transform them into $\bar{\rho}^{\mathrm{pr}}_{\mathrm{AX}}(t_\mathrm{f}) = \mathbb{I}/4 - 1.1 \times 10^{-2}(\mathbf{I}_1 \cdot \mathbf{I}_2) - 0.9885 I_1^z I_2^z + 0.485(I_1^z - I_2^z)$ for the AX system and $\bar{\rho}^{\mathrm{pr}}_{\mathrm{AB}}(t_\mathrm{f}) = \mathbb{I}/4 - 0.23(\mathbf{I}_1 \cdot \mathbf{I}_2)$ $-0.77 I_1^z I_2^z + 0.495(I_1^z - I_2^z)$ for the AB system, respectively.

The values displayed above correspond to simulations in which the sample evolved for long times at every magnetic field value from the reaction to the observation field (in 1,000 steps). The resulting spectra can be seen in Fig. 8c, d, where the differences of the *sudden* field transport are clearly manifested. The shapes of the spectra observed after adiabatic transport agree with the description given in the population approach for the ALTADENA experiment, as expected. The fact that in ALTADENA not only the reaction at low field but also the adiabaticity of the transport step is crucial to allow the coherences present in the term $\mathbf{I}_1 \cdot \mathbf{I}_2$ to be transformed into polarization difference now becomes evident.

2.6.6 Different Field Variations During the Transport Step

The two limiting cases treated above serve to grasp the importance of the transport step. In practice, however, during the experiments the magnetic field change will be neither completely adiabatic nor perfectly sudden. In general, we will have a spatial field variation between the place where the reaction is performed and the center of the NMR apparatus. Furthermore, the sample will be transported through the magnetic field profile with a certain velocity. In order to include those variables in the calculations, we note that

$$\frac{\mathrm{d}B_0^z}{\mathrm{d}t} = \frac{\mathrm{d}B_0^z}{\mathrm{d}z}\frac{\mathrm{d}z}{\mathrm{d}t}. \tag{25}$$

The first factor on the right-hand side of the equation is the magnetic field profile, which can be easily measured. In Fig. 9a we have included such a profile obtained in our laboratory (the data shown were interpolated). The z-component of the magnetic field falls down from 7 T to about 3 mT in ~1.4 m. The second factor on the right-hand side of Eq. (25) represents the z-component of the sample's velocity. By controlling both terms one can obtain the time dependency of the external magnetic field change and, therefore, the time dependence of the Hamiltonian, through the chemical shift part [see Eq. (21)]. The Liouville–von Neumann equation should be solved in this case step by step, because the Hamiltonians at different times do not commute (i.e., $[H^{\mathrm{pr}}(t_i), H^{\mathrm{pr}}(t_i + \Delta t_i)] \neq 0$).

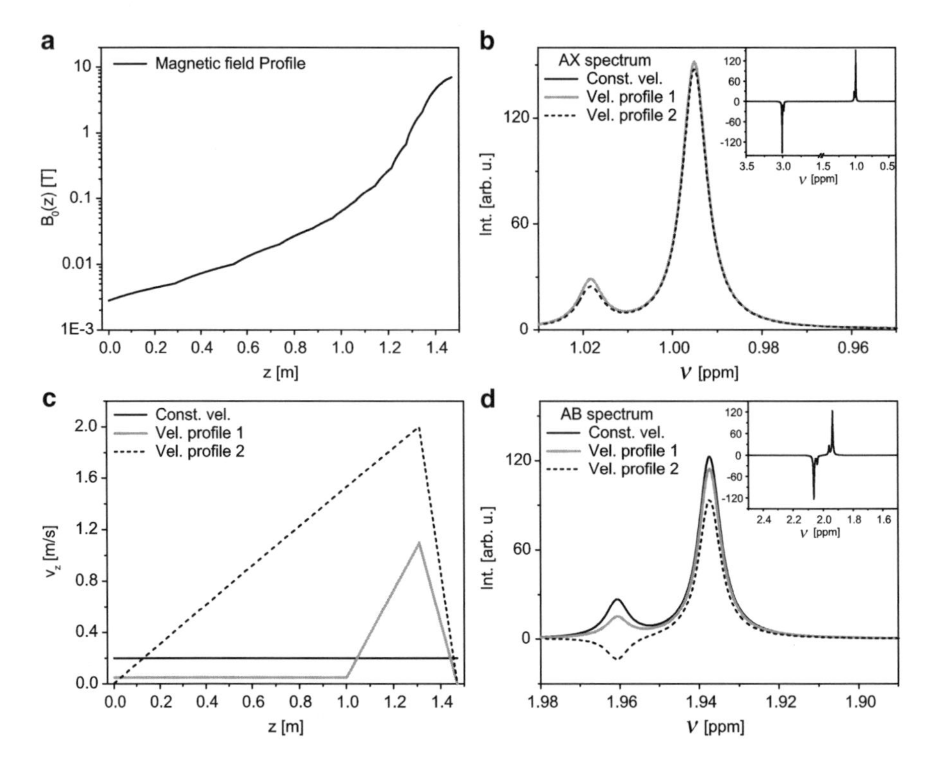

Fig. 9 (**a**) Interpolated magnetic field profile in *z*-direction, obtained in our laboratory; (**c**) three different velocity profiles considered in the simulations for the sample transportation; (**b**) AX spectrum and (**d**) AB spectrum obtained for each of the velocity profiles

As an illustration, we have chosen three different velocity profiles. First, we assumed that the sample travels 1.4 m at constant velocity $v_z = 0.2\,\mathrm{m/s}$. Second, the sample is transported at lower constant velocity, $v_z = 0.05\,\mathrm{m/s}$ for the first meter and then accelerates until being ~0.15 m apart from the coil, followed by a deceleration until it settles at 7 T. Finally, we have considered a velocity profile which includes acceleration from the reaction place to ~0.15 m from the magnet's center followed by a deceleration until the measurement spot. Considering these three cases we treat constant velocity, acceleration only near the top of the magnet, and acceleration in a wide field range, respectively. The three profiles can be seen in Fig. 9c.

Again, an AX and an AB spin system were introduced in the calculations and the corresponding spectra are shown in Fig. 9b, d. The spectra corresponding to the weakly coupled system seem not to be affected by variations in the velocity profile, despite some little intensity changes. On the other hand, the strongly coupled system shows a more pronounced variation in the shape of the spectra with respect to the velocity profile. Accelerating at low magnetic field values, for instance, inverts a part of the spectrum.

It can be concluded that, although visible changes in the resulting NMR spectra are observed in some cases when switching between velocity profiles, the transport step seems not to be critical for a two spin system, as the overall shape of the spectra is not grossly perturbed. However, we can anticipate more pronounced changes in larger spin systems.

2.6.7 Enhancement Factor

Before finishing the treatment of two spin systems we will calculate the theoretical enhancement in a PASADENA experiment for an AX spin system, using the results obtained with the density operator formalism. The enhancement is derived by considering, on one hand, the maximum signal resulting from an experiment where the hydrogenation is carried out with hydrogen in thermal equilibrium at room temperature, and, on the other hand, the maximum signal obtained from a PHIP-PASADENA experiment with enrichment factor ξ.

Assuming that the NMR experiment is carried out at temperature T_{exp}, the thermal equilibrium density operator is expressed as $\rho^{\mathrm{th}} = \mathbb{I}/4 - \varepsilon/4(I_1^z + I_2^z)$ [44, 46, 48]. After a 90° r.f. pulse, with phase $-y$ for instance, we get $\rho^{\mathrm{th}} = \mathbb{I}/4 - \varepsilon/4$ $(I_1^x + I_2^x)$, yielding $|S_{90^\circ}^{\mathrm{th}}| = \varepsilon/4$.

Regarding the experiment with enriched *para*hydrogen, after the hydrogenation and before the application of the r.f. pulse, the density operator is $\bar{\rho}^{\mathrm{pr}}(t_{\mathrm{f}}) = \mathbb{I}/4 - \xi I_1^z I_2^z$. In contrast to the thermal equilibrium case, the maximum signal results after a 45° r.f. pulse [43]. The corresponding density operator just after the r.f. pulse is $\bar{\rho}^{\mathrm{pr}}(t_{\mathrm{f}}) = \mathbb{I}/4 - (\xi/2)\{I_1^z I_2^z + I_1^x I_2^x + (I_1^x I_2^z + I_1^z I_2^x)\}$. The NMR signal intensity is initially zero because none of the terms involved in the density operator are directly observable. However, the term $(I_1^x I_2^z + I_1^z I_2^x)$ becomes observable after evolving with the J-coupling Hamiltonian, giving $|S_{45^\circ}^{\mathrm{PHIP}}| = \xi/4$. Here, we have supposed that the spectrum's line width is much smaller than the J-coupling constant.

Combining both expressions, the signal enhancement as a function of the *para*hydrogen fraction is

$$S_{\mathrm{enh}}(N_{para}) = \frac{\xi}{\varepsilon} = \frac{(4N_{para} - 1)kT_{\mathrm{exp}}}{3\gamma\hbar B_0}. \tag{26}$$

Invoking Eq. (9), and after some algebraic manipulation, we can recast the signal enhancement as function of the temperature at which the *para*hydrogen is enriched, as

$$S_{\mathrm{enh}}(T) = \frac{3\sum_{J=0}(2J+1)\,\mathrm{e}^{J(J+1)\theta_{\mathrm{r}}/T}}{Z(3\gamma\hbar B_0/kT_{\mathrm{exp}})}. \tag{27}$$

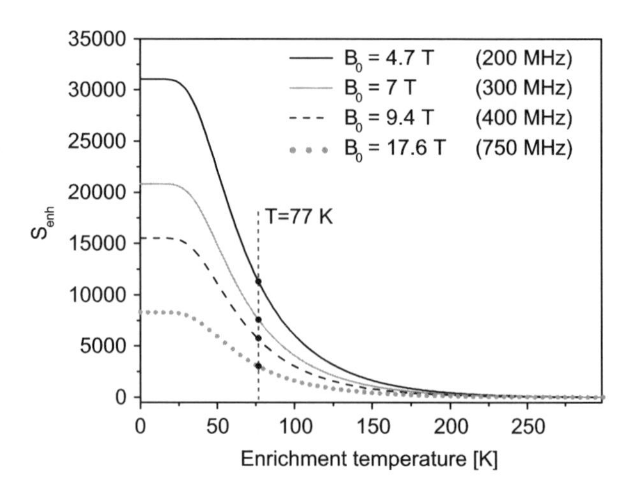

Fig. 10 Enhancement factor for different observation fields B_0 at room temperature, as a function of the p-H_2 enrichment temperature. Note that the predicted enhancement is larger for lower B_0 fields

In Fig. 10 the signal enhancement is plotted for different magnetic field strengths B_0 at which the PASADENA experiment is performed, setting $T_{exp} = 300\,K$. Note that the enhancement is larger for smaller magnetic fields. Remarkably, the liquid nitrogen boiling temperature falls close to the point where $|dS_{enh}/dT|$ is maximum.

At this point it must be emphasized that the enhancement shown above is the maximum enhancement theoretically achievable. In practice, however, lower (and even much lower) values are observed associated with relaxation effects and partial peak cancellation due to the magnetic field inhomogeneity across the sample (NMR line width).

2.7 *Larger Spin Systems and Hyperpolarization Transfer*

If the former *para*hydrogen protons are deposited into sites exhibiting nonzero interaction with the rest of the product molecule (or parts of the molecule), a new issue might appear: the hyperpolarization can be partially transferred to other involved nuclei, depending on the Hamiltonian H^{pr} [14, 20, 49].

During a PASADENA experiment, the hyperpolarization might or might not be transferred from one of the former *para*hydrogen protons to a third nucleus, depending on the coupling network. If the third nucleus, labeled as k, is strongly coupled to at least one of the *para*hydrogen protons, the transfer will occur. In contrast, if the third nucleus is weakly coupled to both *para*hydrogen protons the hyperpolarization will remain confined to the p-H_2 sites [14].

By lowering the magnetic field at which the reaction is carried out, the weakly coupled spins can become strongly coupled and then allow the hyperpolarization to be redistributed. At *very low* magnetic field values, all spins are strongly coupled, even to heteronuclei. In this case, the hyperpolarization will spread over the whole molecule [43, 49].

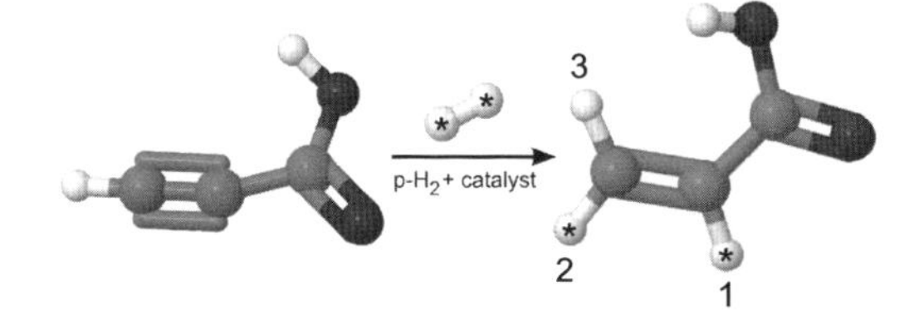

Fig. 11 Scheme of the hydrogenation of propiolic acid with p-H_2. The former p-H_2 protons are marked with asterisks in the product molecule

We will concentrate here on a particular three spin system, the hydrogenation of propiolic acid with *para*hydrogen (see Fig. 11), to illustrate the major features of PASADENA and ALTADENA experiments. The choice of the system fulfills two purposes: first, it represents an interesting three-spin system, with a particularly large coupling between the third proton and one of the *para*hydrogen protons; second, as ALTADENA experiments on this system were already reported [26] and a set of theoretical spectra were published recently [20], it appears as a suitable well known system to be used as an example. The molecular parameters are: $\nu_1 = 6.0$, $\nu_2 = 5.8$, and $\nu_3 = 6.25$ ppm; $J_{1,2} = 10.2$, $J_{1,3} = 17.2$, and $J_{2,3} = 1.8$ Hz, respectively (data extracted from [20]).

2.7.1 PASADENA in a Three Spin System

At 300 MHz Larmor frequency, the protons labeled as 1 and 2 (stemming from *para*hydrogen) are relatively strongly coupled, with $\zeta_{1,2} \sim 6$, as well as the protons labeled as 1 and 3, with $\zeta_{1,3} \sim 4.5$. In contrast, the protons labeled as 2 and 3 are weakly coupled, with $\zeta_{2,3} \sim 75$. Therefore, one would expect a considerable amount of hyperpolarization to be transferred to the third proton. However, the dynamics are dominated by the spins 1 and 2, and the hyperpolarization mostly remains on the former *para*hydrogen protons, at this Larmor frequency. This can be observed from the simulation shown in Fig. 12, where the peaks of the former p-H_2 protons are marked with asterisks. Although a tiny amount of polarization is observed at the site of spin 3, the calculation shows that more than 95% of the initial polarization is still at the sites 1 and 2 after the hydrogenation (of the form $I_1^z I_2^z$).

2.7.2 ALTADENA in a Three Spin System

A very different result is obtained with the same spin system in an ALTADENA experiment. Carrying out the hydrogenation at, for instance, $B_0 = 3\,\mathrm{mT}$, the chemical shift to J-coupling ratios are: $\zeta_{1,2} \sim 2.5 \times 10^{-3}$, $\zeta_{1,3} \sim 1.9 \times 10^{-3}$, and $\zeta_{2,3} \sim 3.2 \times 10^{-2}$, respectively. Starting with $\rho^{\mathrm{pr}}(0) = \mathbb{I}/4 - \mathbf{I}_1 \cdot \mathbf{I}_2$, the following averaged density operator results after the reaction:

$$\bar{\rho}^{\mathrm{pr}}(t_{\mathrm{f}}) = \mathbb{I}/4 - 0.335(\mathbf{I}_1 \cdot \mathbf{I}_2) - 0.363(\mathbf{I}_1 \cdot \mathbf{I}_3) - 0.302(\mathbf{I}_2 \cdot \mathbf{I}_3). \tag{28}$$

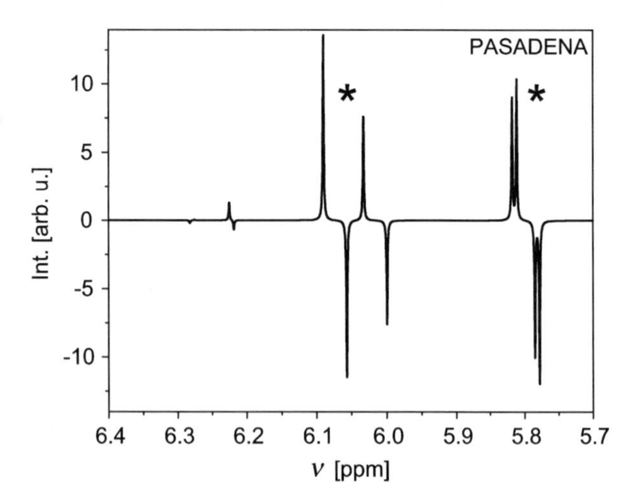

Fig. 12 Spectrum simulated for the three spin system under PASADENA conditions. The marked peaks refer to the former p-H_2 protons. The third proton is not influenced by the hyperpolarization

The initial hyperpolarization is almost equally distributed in the three spin system. The expectation values are directly related to the ratios of ζ. If the averaged density operator is expressed as $\bar{\rho}^{\mathrm{pr}}(t_{\mathrm{f}}) = \mathbb{I}/4 - \sum_i c_{i,j}\mathbf{I}_i \cdot \mathbf{I}_j$ (with $\sum_i c_{i,j} = 1$), it is observed that $c_{1,3} > c_{1,2} > c_{2,3}$ when $\zeta_{1,3} < \zeta_{1,2} < \zeta_{2,3}$. This means the stronger the coupling the larger the hyperpolarization fraction obtained.

The hyperpolarization distribution can be directly seen in the plot in Fig. 13a. The NMR spectrum (45º r.f. pulse) acquired in a 7-T magnet after an adiabatic sample transportation, i.e., the typical ALTADENA experiment, is shown in Fig. 13b.

Despite the hyperpolarization transfer, which is almost absent in PASADENA experiments, there is another remarkable difference between ALTADENA and PASADENA spectra: while the signals in PASADENA are in antiphase, the adiabatic sample transport to the acquisition magnetic field in ALTADENA produces net polarization in the spins 2 and 3, keeping the antiphase character only on spin 1. If the transport step is modified, however, the overall shape of the spectrum is strongly affected. The five spectra in Fig. 14 show the differences when varying the way in which the sample is transported to the magnet from the same reaction field. The labels correspond to the velocity profiles introduced in Fig. 9. Moving the sample constantly at 0.05 m/s seems not to affect the adiabaticity of the transport, while doing it at 0.2 m/s nearly destroys the signal, as expected in the case of a sudden transport (see Fig. 8). On the other hand, accelerating the sample either from the reaction field or only from relatively high field produces a great discrepancy with the adiabatic case.

The results described in this section can generally be extended to cases when the protons originating from *para*hydrogen are deposited into sites coupled to larger spin systems. The most striking difference will appear in ALTADENA experiments where, depending on the reaction field, the hyperpolarization might migrate even to heteronuclei. At the Earth's magnetic field or lower, for example, hyperpolarization transfer to ^{13}C is very well feasible [14, 50, 54].

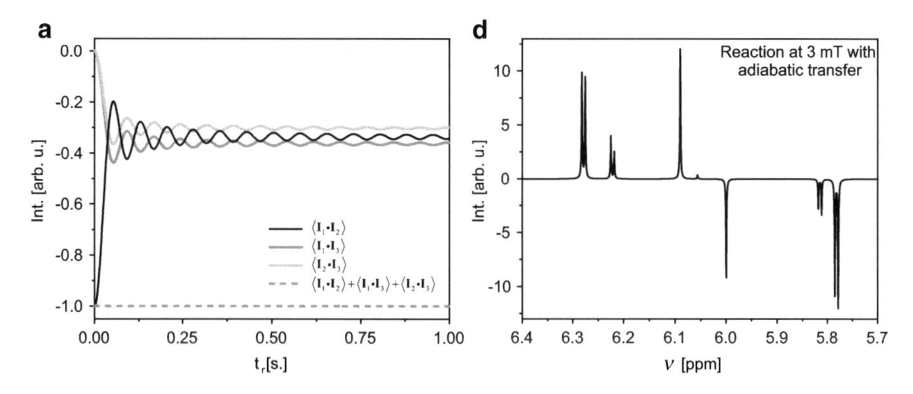

Fig. 13 Three spin system under ALTADENA conditions. (**a**) Expectation values of the isotropic operators during the chemical reaction at 3 mT magnetic field strength; (**b**) simulated NMR spectrum after an adiabatic transport to 7 T magnetic field strength

3 Experimental Results and Applications

In this section, two practical applications and experiments using *para*hydrogen Induced Polarization employing homogeneous catalysis to enhance NMR signals are presented. There are numerous fields where nuclear magnetic resonance (NMR) has nowadays proved to be essential. However, the poor NMR signal has always been a drawback which can be overcome by means of PHIP [55]. The first application presented in this chapter shows the capability of using PHIP in ^{1}H magnetic resonance imaging (MRI) where not only the enhanced signal can be exploited to generate MRI contrast but also the difference in the initial spin state of the PHIP polarized protons compared to the thermal polarization of the background. The second application chosen for this chapter shows a method to provide the enhanced hyperpolarized signal in a continuous fashion, making it possible to perform traditional 2D NMR experiments much faster.

In this chapter, two different substrates are used as model compounds for the hydrogenation with p-H_2. Both of them show an excellent acceptance of the p-H_2 molecule. The model compounds are chosen because of their different properties: one is soluble in water and the other one in acetone, thus providing examples for experiments in aqueous and organic systems. As explained in the preceding section, the experiments shown here correspond to a homogeneous reaction with enriched p-H_2, which enables a pairwise transfer of the p-H_2 protons to the substrate.

3.1 Catalytic Systems

The catalytic systems presented here for PHIP are homogeneous catalyst complexes with rhodium as metal center. They are cationic in order to prevent isomerization as a side reaction. Unfortunately, most commercially available homogeneous catalysts

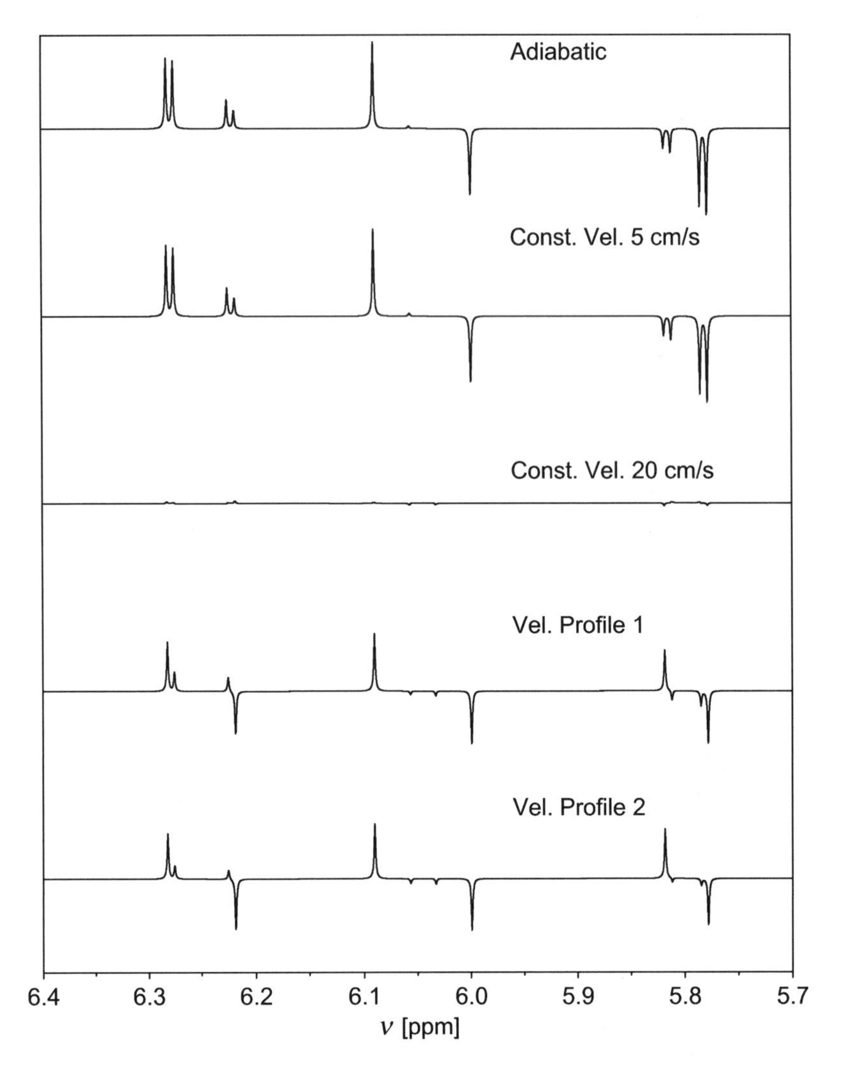

Fig. 14 Hydrogenation of propiolic acid: simulated spectra were obtained by using different velocity profiles during the transportation to high magnetic field

are soluble only in organic solvents like acetone or methanol, not in water. This is the case for the catalyst shown in Fig. 15a.

To date, water-soluble PHIP catalysts are not commercially available, but the most popular one can be easily synthesized [56]. In order to enhance the solubility in aqueous solvents, polar groups, e.g., sulfonate groups, are added to the ligand system of the catalyst. One drawback of this intervention into the electron density of the ligand system is the possible deactivation of the total catalyst. Fortunately, this is not always the case and, in Fig. 15b a water-soluble catalyst is shown whose capability to enable the pairwise transfer of the protons to the substrate had been proved [56].

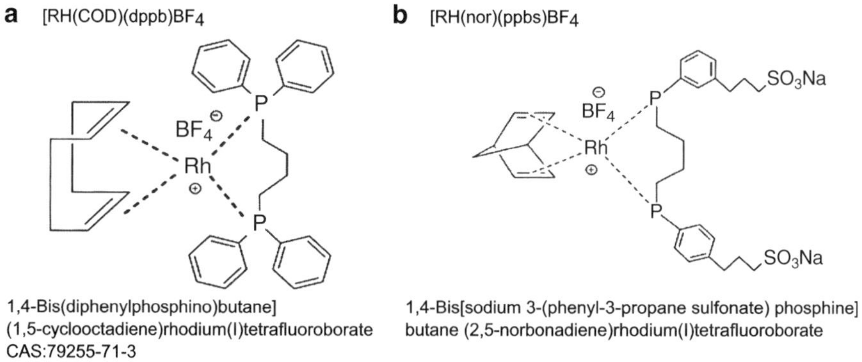

a [RH(COD)(dppb)BF4]

1,4-Bis(diphenylphosphino)butane]
(1,5-cyclooctadiene)rhodium(I)tetrafluoroborate
CAS:79255-71-3

b [RH(nor)(ppbs)BF4]

1,4-Bis[sodium 3-(phenyl-3-propane sulfonate) phosphine]
butane (2,5-norbonadiene)rhodium(I)tetrafluoroborate

Fig. 15 Homogeneous PHIP catalysts: (**a**) water insoluble catalyst; (**b**) water soluble catalyst

Both catalyst systems presented here have shown a high activity and low substrate specificity; moreover they ensure the pairwise transfer of H_2 to the substrate molecule. The embedding of stereo information is possible. Unfortunately, the most important disadvantage of them is that both are toxic and they must be removed if a medical application is considered [57].

3.2 *The Model Compounds*

As mentioned above, two model compounds are used in the presented experiments. Whenever possible, the experiments are shown in both compounds in order to allow for a comparison: 1-hexyne is used for reactions in organic solvents and 2-hydroxyethylacrylate for reactions under aqueous conditions.

3.2.1 1-Hexyne

1-Hexyne is a highly flammable liquid, with a boiling point between 71°C and 72°C, and it is commercially available. As can be seen in Fig. 16, hydrogenation of the terminal triple bond results in 1-hexene.

1-Hexyne is barely sterically hindered, which makes it a good model compound allowing excellent interaction to several catalysts. The hyperpolarized double bond shows a long spin-lattice relaxation time T_1 of approximately 30 s, whereas the relaxation times of all other protons in the molecule are around 15 s.

3.2.2 2-Hydroxyethyl Acrylate

2-Hydroxyethyl acrylate is the water soluble model compound used in the experiments. It is a toxic liquid having its boiling point between 210°C and

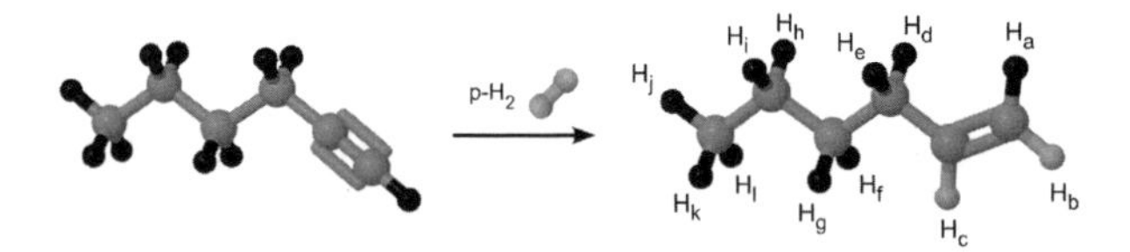

Fig. 16 Hydrogenation of 1-hexyne leads to 1-hexene

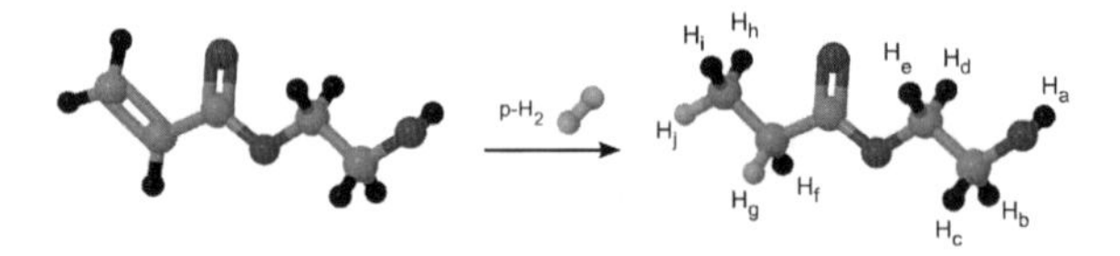

Fig. 17 Hydrogenation of 2-hydroxyethyl acrylate leads to 2-hydroxyethyl propionate

215°C, and it is also commercially available. In Fig. 17 the hydrogenation of 2-hydroxyethyl acrylate is shown, resulting in the reaction product 2-hydroxyethyl propionate. The T1 relaxation time of the hyperpolarized protons is approx. 5 s, which is substantially shorter than the one of 1-hexene.

3.3 PHIP Hyperpolarized Substances as Contrast Agents in ^{1}H MRI

Magnetic Resonance Imaging (MRI) is one of the most interesting targets for PHIP applications. So far, medical imaging with hyperpolarized compounds was realized with ^{13}C or ^{15}N hyperpolarized substances, which have the advantage of negligible background signal, very long T_1 times, and large chemical shift ranges [58].

Very exciting medical applications of substances with hyperpolarized heteronuclei have already been demonstrated like tumor diagnosis [59–61] or in vivo pH mapping [62]. However, MRI of heteronuclei is technically demanding and thus expensive, because it requires the use of additional hardware (e.g., broadband amplifiers and coils). Moreover, co-registration with a high-resolution proton image is necessary to obtain anatomical information. Thus, using the standard NMR and MRI nucleus—the proton—for molecular imaging would be beneficial. However, the huge amount of protons present in the body gives rise to an NMR signal far exceeding that of a small amount of hyperpolarized protons. As a consequence the hyperpolarized signal is difficult to distinguish from the thermal background. In addition, the antiphase character of the PHIP signal can produce signal cancellation, making the desired contrast even more difficult to achieve. However, there is a simple method recently proposed [15] which manages to take advantage of the special PHIP spin state, thus enabling the use of PHIP hyperpolarized compounds as contrast agents in ^{1}H MRI. Because of its practical interest and novelty this was chosen to be the first application presented in this chapter.

The idea exploits the differences between the initial spin state of the p-H_2 and the thermally polarized protons (see the theoretical section). The procedure was applied to both model compounds introduced above. In both cases good contrast was observed, manifesting the general validity of the proposed method. Below, the method is introduced and each experiment is preceded by a short theoretical description of the main concepts involved. As the theoretical descriptions are intended to be only pictorial, some helpful assumptions are made:

1. For the PHIP compound an AX spin system is assumed, even though the real systems are more complicated (see Figs. 16 and 17) and, as explained in the theoretical section, more than two spins should be used when performing exact calculations. However, the main features of the PHIP signal observed here can be explained using this simplification.
2. The thermally polarized molecules of the background, i.e., the system from which the PHIP signal should be distinguished, possess a single resonance line (for example water).
3. The pulses of the NMR sequence are set on-resonance with the thermal background.

All experiments treated here were performed under PASADENA conditions (although the contrast works also for ALTADENA) using a clinical 1.5-T MRI system (Magnetom Sonata, Siemens Medical, Erlangen, Germany) and a basic FLASH imaging pulse sequence with centric reordering.

There are several cases regarding the spatial confinement of the hyperpolarized substance we should distinguish when discussing these experiments. These different cases depend not only on the T_2^* decay times of each component but also on the relative signal amplitude differences. According to the relative T_2^* relationship the following two cases should be separately analyzed.

3.3.1 Case 1: Thermal Signal Decays Faster than Hyperpolarized Signal

Let us first consider the case where the decay of the thermal signal of the background molecules is faster than the decay of the hyperpolarized signal. Experimentally this condition can be achieved by confining the hyperpolarized molecules to a small volume placed in the middle of the magnet and by surrounding them with a large volume of thermally polarized molecules. As the magnetic field is more homogeneous close to the center of the magnet, T_2^* in this region should be longer than in the outer part. In Fig. 18 the phantom used for this experiment is shown together with a representation of the FIDs corresponding to the AX system (Fig. 18a) and the experimentally data obtained with 1-hexene (Fig. 18b).

In a PASADENA experiment, the initial state of an AX spin system is $\bar{\rho}^{\mathrm{pr}}(t_{\mathrm{f}}) = \mathbb{I}/4 - \xi I_1^z I_2^z$, as theoretically explained in the first section, and after application of a 45° pulse it is converted into an oscillating observable signal of the form

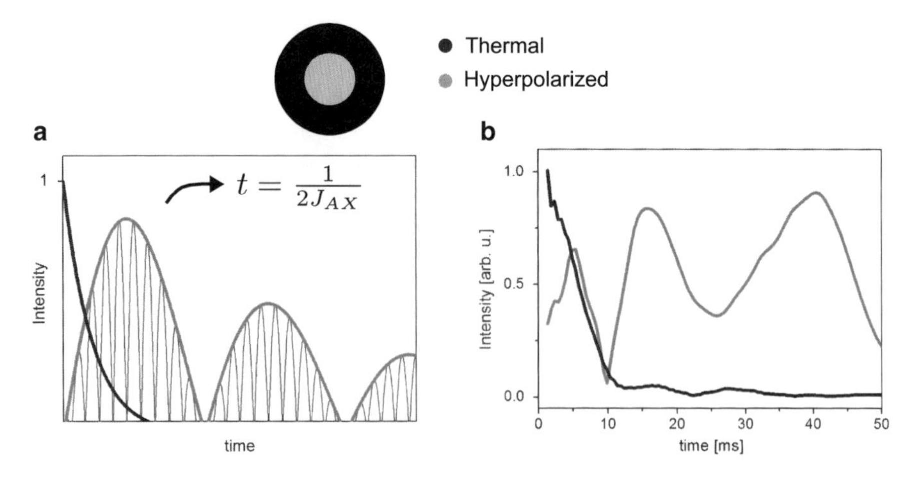

Fig. 18 Experimental configuration for case 1 where the hyperpolarized compound is confined to a small area. (**a**) Calculated fast decaying thermal signal (*black line*) compared to the oscillating hyperpolarized signal (*gray line*). (**b**) Experimental data for water and 1-hexene

$$\rho^{45^\circ_{-y}}(t) = \mathbb{I}/4 - \xi \, \sin(\pi J_{AX} t)(I_1^y + I_2^y). \tag{29}$$

From this equation it can be shown that the maximum signal of the hyperpolarized compound occurs at $t = (2J_{AX})^{-1}$, as indicated in Fig. 18a. The possibility of exploiting this oscillatory behavior to generate the desired contrast can be understood from the figure: at the time when the amplitude of the hyperpolarized signal is maximal, the thermal signal is almost negligible. In Fig. 18b the experimental data obtained for a phantom with hyperpolarized 1-hexene in the inner tube and water in the outer tube are shown. Even though 1-hexene is a large spin system as shown in Fig. 16, the oscillatory behavior of the two p-H_2 protons dominates the FID and the position of the maximum signal ($\approx 16\,\text{ms}$) can easily be determined.

Once the position of this maximum difference between both signals is known this time can be set as the echo time T_E in the imaging sequences to highlight the difference, i.e., maximize the contrast. In Fig. 19, images obtained for eight different echo times (T_E) in the imaging pulse sequence are shown for the two model compounds. The oscillation of the signal amplitude of the inner tube can be clearly recognized for both hyperpolarized substances, whereas the signal of the outer tube, containing water, decays with increasing echo time. As predicted, the best contrast is observed when the echo time corresponds to the maximum of the hyperpolarized signal (for 1-hexene $T_E \sim 16$ ms and for 2-hydroxyethyl propionate $T_E \sim 15$ ms). Of course, the spin systems of the two hyperpolarized components are different but both show the J-coupling induced refocusing of the signal as theoretically predicted when starting from a PASADENA experiment. This demonstrates that the method can easily be applied to different molecules which can be hyperpolarized via PHIP and highlights its general applicability.

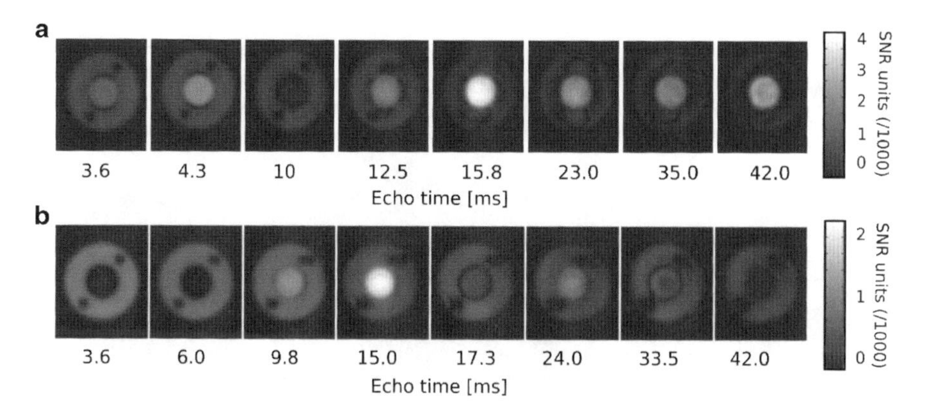

Fig. 19 Images acquired with different echo times (T_E) in the MRI pulse sequence, for 1-hexene (*top*) and 2-hydroxyethyl propionate (*bottom*)

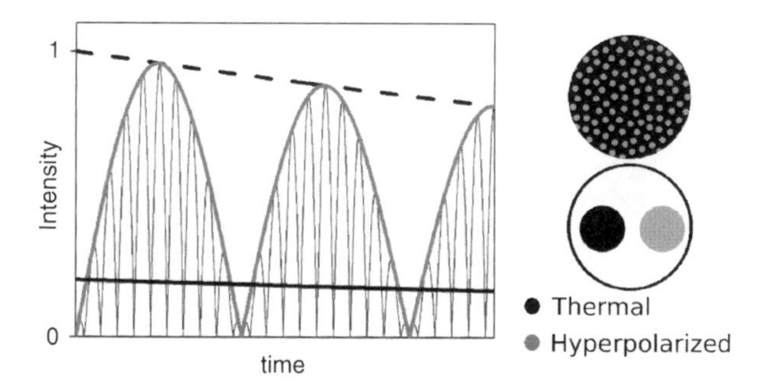

Fig. 20 Identical T_2^* decays for the hyperpolarized and the thermally polarized substance. Two different signal amplitudes for the thermal signal are plotted (*straight* and *dotted lines*)

3.3.2 Case 2: Similar or Identical Decay Times

Another interesting case occurs when the hyperpolarized and the thermal signal both decay with the same T_2^* time constant. In Fig. 20 the theoretical representations of the FID signals are schematized for this case. The experimental set up can be either two tubes at the same position relative to the center of the magnet or a mixture of both (thermally polarized and hyperpolarized) compounds contained in the same tube.

In Fig. 20 two possible scenarios for the amplitude of the thermal signal are plotted. The dotted line corresponds to the case where the amplitude of the thermal signal is comparable to the amplitude of the hyperpolarized signal. The solid line resembles the case where the hyperpolarized signal is noticeably larger than the thermal signal.

In the latter case we can once more obtain a good MRI contrast by choosing the optimal echo time for the hyperpolarized substance as can be concluded from the experimental data representing this case shown in Fig. 21.

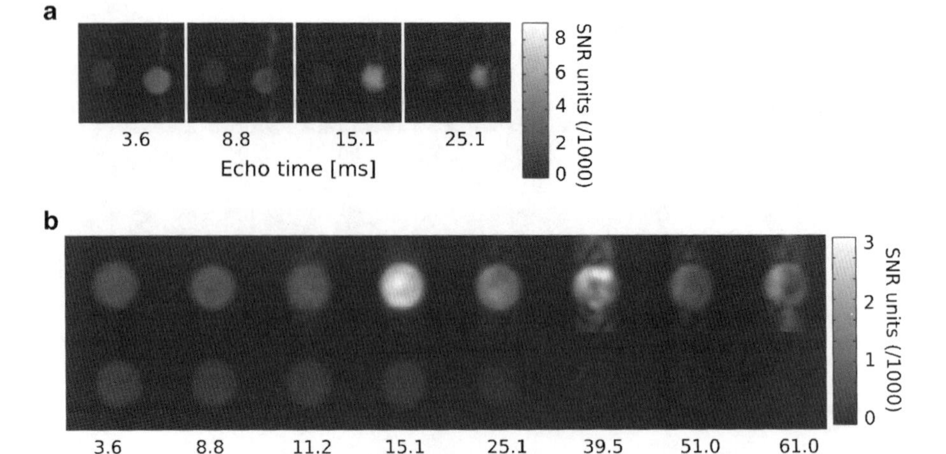

Fig. 21 (**a**) The left tube contains water and the right one hyperpolarized 1-hexene. (**b**) One single tube contains hyperpolarized 1-hexene dissolved in a non-deuterated solvent. For both phantom geometries the best contrast is observed for 15 ms echo time

The phantom consisting of two tubes contained water and hyperpolarized 1-hexene, whereas the experiments with only one tube (containing both the hyperpolarized fluid and the thermally polarized background molecules) were performed with 1-hexene dissolved in non-deuterated acetone. The spin density of the proton background is somewhat smaller in this case, as for the pure water phantom, but still much higher than the spin density of the hyperpolarized protons. Because 1-hexene can be hyperpolarized very efficiently by using PHIP it is nonetheless easy to fulfill the condition that the hyperpolarized signal is exceeding the thermal one. The experiments in Fig. 21 show once more the good contrast that can be obtained for the echo time agreeing with the position of the maximal hyperpolarized signal, $T_E = 15$ ms, for both phantom geometries.

In the less preferable case, where both signal amplitudes are identical, it is not possible to generate a contrast by using an echo time which corresponds to the maximum of the hyperpolarized signal. Certainly it is still possible to obtain a negative contrast if the chosen echo time corresponds to a minimum of the hyperpolarized signal. In this case the hyperpolarized signal cannot be observed and its location appears dark in the images.

However, a negative contrast is not optimal because dark regions in MRI images can have many causes (e.g., susceptibility differences, etc.). Therefore we developed a method to obtain a positive contrast even in this case, which is explained in the following.

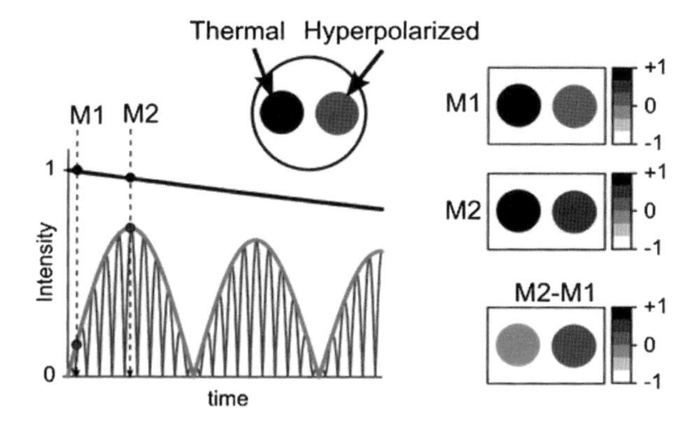

Fig. 22 Two images with different echo times are subtracted to generate the desired contrast

3.3.3 Contrast by Subtraction Method

The method requires the acquisition of two images with different echo times and is schematized in Fig. 22. The echo times should be chosen such that for the first echo time the signal of the thermally polarized background is maximal whereas the contribution of the hyperpolarized substance is almost zero and for the second echo time the hyperpolarized signal should be maximal while the thermal signal remains almost unaltered. These echo times are labeled as M1 and M2 in Fig. 22. Subtraction of these images results in a new image where the hyperpolarized signal is positive in contrast to the signal of the thermally polarized background which is always negative due to its decay with T_2^*. This allows for an unambiguous differentiation of the two areas by sign.

In Fig. 23 the subtraction method is applied for the two model compounds studied in this section. In both cases an excellent contrast is obtained and the differentiation of the two areas by signal sign could be verified. This method would allow for a substantial reduction of the concentration of the hyperpolarized component in the experiments. It reaches its limits when the signal of the hyperpolarized component becomes comparable to the noise level of the images.

In this section we introduced a novel MRI contrast which allows for the discrimination of a small amount of PHIP hyperpolarized protons from a huge amount of surrounding thermally polarized protons. The contrast arises from the different time evolution of the PHIP hyperpolarized proton signal compared to the evolution of the normal (thermally polarized) proton signal and can be simply implemented by using basic product pulse sequences (FLASH, TrueFISP, EPI), which are varied in a minor way by only choosing the optimal echo time for the hyperpolarized substance. The optimal echo times can simply be found by recording an FID of the hyperpolarized substance prior to image acquisition.

The new method might be applied for metabolic imaging, perfusion MRI, or catheter visualization during MRI guided interventions using only conventional proton pulse sequences and equipment (NMR coils), which reduces the technical

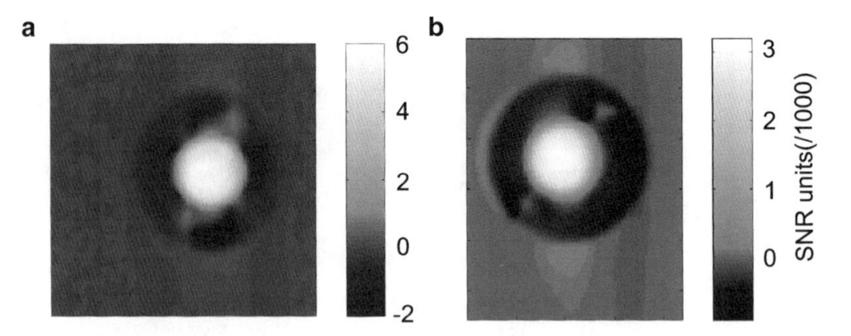

Fig. 23 Subtraction images of images acquired with different echo times for (**a**) 1-hexene and (**b**) 2-hydroxyethyl propionate

demands that arise from MRI of heteronuclei. Moreover, Most heteronuclei (especially ^{13}C and ^{15}N) have a low gyromagnetic ratio, which makes it difficult to provide images with very high spatial resolution because the gradient strengths of conventional MRI systems is limited and usually optimized for protons. Our method can be used not only for our model substances but for every molecule which can be hyperpolarized via *para*hydrogen Induced Polarization, which includes a large number of biological relevant molecules, e.g., succinate [63] (component of the citrate cycle) or barbiturates [64] (anesthetics). This variability concerning molecules and different clinical applications in combination with the simplicity of our method may result in widespread usage in MRI.

3.4 Continuous Generation of a Hyperpolarized Fluid Using PHIP and Hollow Fiber Membranes

Despite the many important applications hyperpolarization techniques have found in natural sciences and medicine, several general problems remain. The most severe limitation is the limited lifetime of the hyperpolarized state caused by T_1 relaxation. This problem is less pronounced for hyperpolarized noble gases (^{129}Xe, ^{3}He) which exhibit T_1 times of hours [65]. In liquids efficient relaxation processes restrict the hyperpolarization to last typically from seconds to, at best, a few minutes. Fortunately, this drawback can be at least partially overcome by storing the fast decaying hyperpolarization in slowly relaxing singlet states [22, 23, 28, 66, 67]. Another shortcoming is the partial destruction of the hyperpolarization by the application of r.f.-pulses, rendering the usage of complex pulse sequences for multi-dimensional NMR experiments difficult. This can be circumvented via stepwise use of the generated hyperpolarization by applying only small flip angles [68] or by using specially designed sampling strategies [69, 70]. Another very intriguing concept to avoid these severe limitations is to use the hyperpolarization methods in a continuous flow fashion providing a continuous supply of hyperpolarized molecules or atoms.

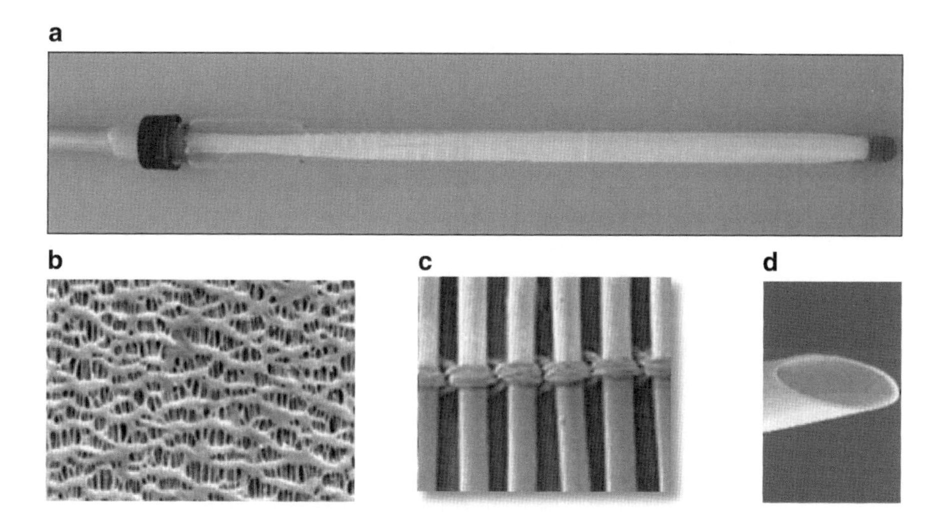

Fig. 24 (**a**) Hollow fiber membranes implemented in an NMR tube; (**b–d**) different magnifications of the membrane fibers

In the following section a very efficient method to generate continuously a hyperpolarized fluid via PHIP using hollow fiber membranes is presented. The technique employs commercially available hollow fiber membranes for the dissolution of the p-H_2 gas in the liquid sample [71]. The idea was originally developed for the dissolution of hyperpolarized ^{129}Xe into liquids and was introduced in 2006 under the name XENONIZER [72, 73]. The microfibers of the membranes are thin-walled, opaque, and made from polypropylene (see Fig. 24).

Polypropylene is a hydrophobic polymer and therefore a polar solvent is unable to pass through the membrane walls. By implementing hollow fiber membranes into an NMR tube containing the PHIP precursor and catalyst dissolved in a polar solvent (e.g., water) and connecting the membranes to a reservoir of p-H_2 gas, hyperpolarization of the sample can be achieved over a long time until all precursor molecules are consumed. This continuously generated hyperpolarization can easily be used either for averaging (to obtain a high SNR in very diluted samples) or to perform 2D experiments. The membranes exhibit very large gas–liquid interfaces, providing the opportunity to bring gas molecularly into solution, thereby preventing the formation of bubbles and foam. Therefore a high spectral resolution can be maintained with the membrane setup while the p-H_2 gas is continuously delivered to the sample. This is normally a problem when the gas is introduced into the sample by bubbling where strong susceptibility artifacts occur. Because of this absence of foaming and bubbles, it is not necessary to stop the gas flow in the membrane experiments and wait before the measurement, thus avoiding unnecessary T_1 relaxation of the sample.

It was proved that the dissolution of the gas into the solvent via the membranes happens on a time scale of milliseconds and also that the amount of dissolved gas

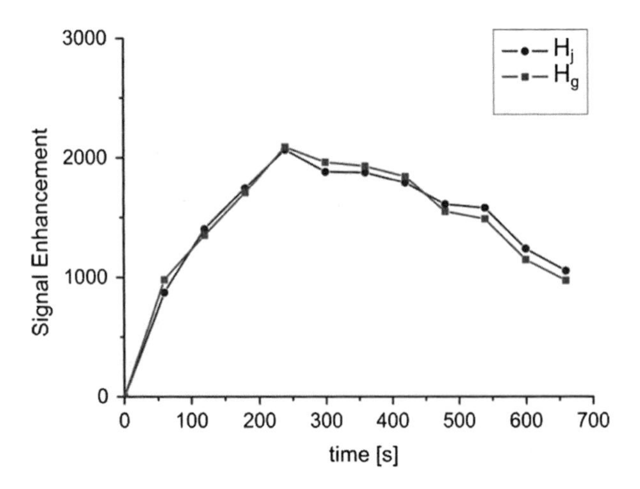

Fig. 25 Time evolution of the signal enhancement of the hyperpolarized protons

comes close to the theoretical Ostwald solubility [72]. It is, however, important to mention a drawback of the membrane setup: it can only be used in aqueous solutions. This fact limits the possibility of using the membranes in all PHIP applications since the majority of PHIP catalysts are not soluble in water.

We present below the hydrogenation of the model compound 2-hydroxythyl acrylate (Fig. 17) by continuously dissolving *p*-H_2 in an aqueous solution employing such hollow fiber membranes under PASADENA conditions. To enhance the conversion of the PHIP reaction the experiments are performed at 80°C. The generated hyperpolarized product shows an anti-phase spectrum free of artifacts. In Fig. 25 the signal amplitude of the hyperpolarized protons is plotted as a function of time.

At the beginning of the experiment, a build-up of the hyperpolarized signal can be observed, due to the increasing conversion of the hydrogenation reaction after the *p*-H_2 flow is switched on. In the course of the reaction the intensity of the hyperpolarized signal decreases owing to lower amounts of starting material. However, during a time of approximately 7 min (much longer than the spin-lattice relaxation of the hyperpolarized protons of ~5 s) the achieved hyperpolarized signal remains almost constant.

3.4.1 PHIP ^{1}H–^{1}H COSY

As an example of the possibilities arising when the hydrogenation is performed in a continuous fashion through the membranes, we present here an interesting application: the fast acquisition of a 2D NMR spectrum, namely a ^{1}H–^{1}H COSY [74, 75] spectrum. This kind of routine experiment, which is very useful and widely employed in spectroscopic NMR, also provides a rigorous test in order to check whether or not

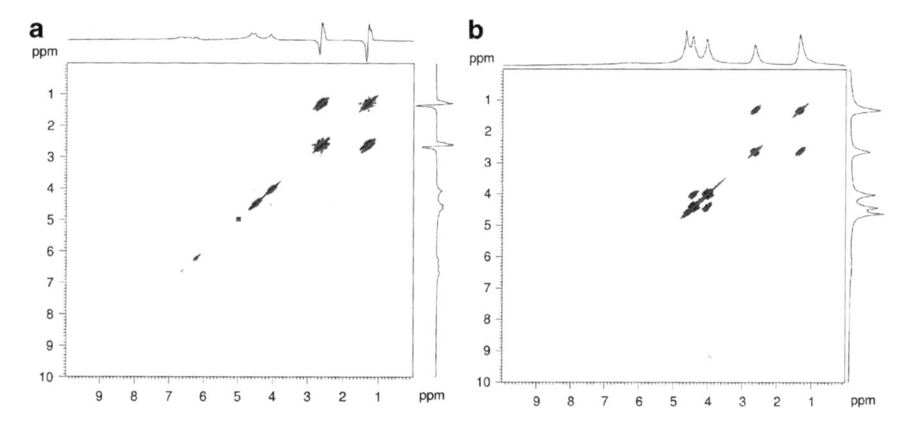

Fig. 26 (**a**) PHIP ^{1}H–^{1}H COSY spectrum, one scan, duration: 7 min 18 s. (**b**) Reference ^{1}H–^{1}H COSY spectrum, eight scans, duration: 57 min 45 s

constant hyperpolarization is obtained with the described setup. In the presented PHIP experiments the first 90° pulse of the COSY sequence was replaced by a 45° pulse [76] to ensure an optimal excitation of the PASADENA spin state.

In Fig. 26a, the PHIP ^{1}H–^{1}H COSY spectrum acquired with one scan is shown; the experimental time was only 7 min 18 s. After full conversion of the hydrogenation reaction a reference experiment was performed and the thermally polarized ^{1}H–^{1}H COSY spectrum obtained is shown in Fig. 26b. The COSY experiment of the thermally polarized sample required eight scans and lasted 57 min 45 s. Two main differences are observed between the two 2D spectra: in the PHIP case only the cross-peaks of the hyperpolarized protons at around 1.5 and 3.0 ppm are visible due to the enhanced signal of these protons and their particular initial spin state. Additionally, the acquisition during the ongoing reaction led to the observation of small signals between 6 and 7 ppm stemming from the double bond of the starting material. These peaks are not observed in the thermal case because the latter was acquired with the fully converted sample. A small artifact (at 5 ppm) can be identified in the middle of the PHIP 2D spectrum, which arises from the lack of phase cycling in the one scan PHIP COSY experiment.

The experiments presented above show the possibility of recording a reliable 2D spectrum with chemical selectivity in a much shorter time than by measuring a sample with thermal polarization only. From the comparison between the PHIP and reference COSY experiments it can also be concluded that the enhanced signal achieved with the membrane setup remained almost constant at least for the time required for the COSY experiment.

There are many possible applications for PHIP experiments employing hollow fiber membranes. It is worth emphasizing, for example, that by using appropriated pulse sequences (like the PH-INEPT^{+}sequence [16]) the accomplished proton polarization can be continuously transferred to heteronuclei, like ^{13}C, allowing for experiments with continuously hyperpolarized heteronuclei [71]. The presented technique opens up the possibility of investigating otherwise elusive reaction intermediates because of the possibility of accumulating several scans during the ongoing reaction. Thus, the full site selectivity provided by PHIP can be exploited. The membrane technique can easily be extended to produce a continuous flow of a hyperpolarized liquid by using a slightly different setup which we described in [72]. This method, especially when combined with an important technique recently developed by Adams et al. [77], which overcomes the restriction of the PHIP technique to unsaturated molecules, would allow for the continuous production of hyperpolarized molecules giving rise to new applications in natural sciences and medicine.

The two applications presented here are of course only examples of the many possibilities in the still open and constantly developing field of NMR and PHIP.

4 Conclusion

In this chapter we have presented an overview of the major features of *para*hydrogen Induced Polarization (PHIP) employing homogeneous catalysis to enhance NMR signals. An introductory theoretical approach was depicted and illustrated with two interesting applications.

First, we have explained in simple terms how hydrogen can be enriched in the *para*-state. In order to explain the NMR signal enhancement associated with PHIP experiments, two different approaches were used. The first was the well known population approach, commonly found in the PHIP literature [36]. In this context, the former *para*hydrogen protons are considered as a weakly coupled two spin system (AX) in the target molecule. The overall shape and intensity of the different PHIP spectra, i.e., PASADENA and ALTADENA, can be understood from the population differences of the two-spin energy levels. Even though this model is useful to understand the physics of PHIP in an intuitive fashion, it represents a simplification. If, for instance, the p-H_2 protons form a more strongly coupled spin system in the target molecule, such as an AB spin system, the model is no longer suitable. Moreover, if couplings to other nuclei are present, the model rapidly becomes extremely complicated. With this in mind, we have introduced a second theoretical approach in this chapter: the treatment of PHIP experiments with the density operator formalism. This method is more flexible and can be used to describe more complex spins systems, although it is mathematically more demanding. In particular, we have adopted a numerical approach here. Features such as the dependence of the shape of the spectra on the pulse duration, hyperpolarization transfer to a third nucleus, different ways to transport the sample from the hydrogenation place to the NMR apparatus, and the response to pulse sequences can be included here in a relatively simple way.

In the second part of the chapter, two examples of recent applications of PHIP are presented. The first example shows the feasibility of using PHIP hyperpolarized molecules as contrast agents in ^{1}H MRI. The method exploits the fact that thermal in-phase and hyperpolarized anti-phase signals originate from different initial spin states and, as a consequence, evolve differently during the period between excitation and acquisition in NMR experiments. Because of this difference, a simple variation of a waiting time in the MRI sequence is sufficient to generate outstanding contrast between a small amount of hyperpolarized molecules and a large excess of thermally polarized molecules. Optimal waiting times can simply be found by recording an FID of the hyperpolarized substance prior to image acquisition. Possible applications of this method include metabolic imaging, perfusion MRI, or catheter visualization during MRI guided interventions. Additionally, it could be of considerable help in chemistry, for instance for the investigation of different reactor designs, or for the optimization of lab-on-a-chip devices like micromixers, where the flow and mixing of two components must be studied on very small length scales.

The second example shows the possibility of producing hyperpolarization with PHIP in a continuous fashion by employing hollow fiber membranes. The continuous generation of hyperpolarized samples can overcome the problem of fast relaxation times commonly involved in liquid state NMR. The use of hollow fiber membranes to dissolve molecularly p-H_2 gas in the PHIP reaction mixture in a continuous fashion enables one to generate hyperpolarization in a sample over at least some minutes. This allows for the recording of a reliable 2D spectrum much faster than when performing the same experiment with thermally polarized protons. Furthermore, the described membrane setup opens up new possibilities in the field of reaction intermediates research because of the feasibility of accumulating several scans during the reaction. The membrane technique can be easily extended to produce a continuous flow of a hyperpolarized liquid for MRI, with direct applications in chemistry, biology, or medicine.

Acknowledgments The authors gratefully acknowledge financial support by the Max Planck Society, the German research foundation (DFG, FOR474), and the Federal Ministry of Education and Research (VIP0199). We thank Membrana GmbH for provision of the hollow fiber membranes.

References

1. Bowers CR, Weitekamp DP (1987) J Am Chem Soc 109(18):5541–5542
2. Eisenschmid TC, Kirss RU, Deutsch PP, Hommeltoft SI, Eisenberg R, Bargon J, Lawler RG, Balch AL (1987) J Am Chem Soc 109(26):8089–8091
3. Pravica MG, Weitekamp DP (1988) Chem Phys Lett 145(4):255–258
4. Natterer J, Schedletzky O, Barkemeyer J, Bargon J, Glaser SJ (1998) J Magn Reson 133(1):92–97
5. Duckett SB, Sleigh CJ (1999) Prog Nucl Magn Reson Spectrosc 34(1):71–92
6. Duckett SB, Blazina D (2003) Eur J Inorg Chem 2003(16):2901–2912
7. Koptyug IV, Kovtunov KV, Burt SR, Anwar MS, Hilty C, Han S-I, Pines A, Sagdeev RZ (2007) J Am Chem Soc 129(17):5580–5586

8. Kovtunov KV, Beck IE, Bukhtiyarov VI, Koptyug IV (2008) Angew Chem Int Ed 120(8):1514–1517
9. Kovtunov KV, Zhivonitko VV, Corma A, Koptyug IV (2010) J Phys Chem Lett 1(11):1705–1708
10. Bouchard LS, Burt SR, Anwar MS, Kovtunov KV, Koptyug IV, Pines A (2008) Science 319:442–445
11. Bouchard L-S, Kovtunov KV, Burt SR, Anwar MS, Koptyug IV, Sagdeev RZ, Pines A (2007) Angew Chem Int Ed 119(22):4142–4146
12. Goldman M, Jóhannesson H (2005) C R Phys 6(4–5):575–581
13. Goldman M, Jóhannesson H, Axelsson O, Karlsson M (2005) Magn Reson Imaging 23(2):153–157
14. Reineri F, Viale A, Giovenzana G, Santelia D, Dastrù W, Gobetto R, Aime S (2008) J Am Chem Soc 130:15047–15053
15. Dechent JF, Buljubasich L, Schreiber LM, Spiess HW, Münnemann K (2012) Phys Chem Chem Phys 14(7):2346
16. Haake M, Natterer J, Bargon J (1996) J Am Chem Soc 118(36):8688–8691
17. Sengstschmid H, Freeman R, Barkemeyer J, Bargon J (1996) J Magn Reson Ser A 120(2):249–257
18. Jóhannesson H, Axelsson O, Karlsson M (2004) C R Phys 5(3):315–324
19. Ivanov KL, Yurkovskaya AV, Vieth H-M (2008) J Chem Phys 128:154701
20. Korchak SE, Ivanov KL, Yurkovskaya AV, Vieth HM (2009) Phys Chem Chem Phys 11:11146
21. Buljubasich L, Franzoni MB, Spiess HW, Münnemann K (2012) J Magn Reson 219:33–40
22. Carravetta M, Johannessen O, Levitt M (2004) Phys Rev Lett 92(15):153003–153004
23. Carravetta M, Levitt MH (2004) J Am Chem Soc 126(20):6228–6229
24. Carravetta M, Levitt MH (2005) J Chem Phys 122(21):214505
25. Jonischkeit T, Bommerich U, Stadler Jr, Woelk K, Niessen HG, Bargon J (2006) J Chem Phys 124(20):201109
26. Canet D, Bouguet-Bonnet S, Aroulanda C, Reineri F (2007) J Am Chem Soc 129:1445–1449
27. Vinogradov E, Grant AK (2008) J Magn Reson 194(1):46–57
28. Franzoni MB, Buljubasich L, Spiess HW, Münnemann K (2012) J Am Chem Soc 134(25):10393–10396
29. Anwar M, Blazina D, Carteret H, Duckett S, Halstead T, Jones J, Kozak C, Taylor R (2004) Phys Rev Lett 93(4):040501–040504
30. Anwar M, Xiao L, Short A, Jones J, Blazina D, Duckett S, Carteret H (2005) Phys Rev A 71(3):032327–032306
31. Anwar MS (2005) NMR quantum information processing with parahydrogen. arXiv:quant-ph/0509046v1
32. Hübler P, Bargon J, Glaser SJ (2000) J Chem Phys 113(6):2056
33. Green RA, Adams RW, Duckett SB, Mewis RE, Williamson DC, Green GGR (2012) Prog Nucl Magn Reson Spectrosc 67:1–48
34. Heitler W, London F (1927) Zeitsch Phys 44:455–472
35. Pauling L, Wilson EB Jr (1935) Introduction to quantum mechanics, with applications to chemistry, New edition. Dover, Mineola, New York, United States
36. Bowers CR (2007) Sensitivity enhancement utilizing parahydrogen. In: Harris RK (ed) Encyclopedia of magnetic resonance. Wiley, Chichester
37. Sakurai JJ, Tuan SF (1994) Modern quantum mechanics. Addison Wesley, Boston, United States
38. Dennison DM (1927) Proc R Soc Lond A Math Phys 115(771):483–486
39. Bonhoeffer KF, Harteck P (1929) Die Naturwissenschaften 17(11):182
40. Jonischkeit T, Woelk K (2004) Adv Synth Catal 346(8):960–969
41. Silvera I (1980) Rev Mod Phys 52(2):393–452

42. Canet D, Aroulanda C, Mutzenhardt P, Aime S, Gobetto R, Reineri F (2006) Concept Magn Reson A 28A:321–330
43. Natterer J, Bargon J (1997) Prog Nucl Magn Reson Spectrosc 31(4):293–315
44. Levitt MH (2008) Spin dynamics: basics of nuclear magnetic resonance, 2nd edn. John Wiley and Sons Ltd, Chicester, United Kingdom
45. Bowers CR, Weitekamp DP (1986) Phys Rev Lett 57(21):2645–2648
46. Ernst RR, Bodenhausen G, Wokaun A (1987) Principles of nuclear magnetic resonance in one and two dimensions. Oxford University Press, Published in New York United States
47. Sørensen OW, Eich GW, Levitt MH, Bodenhausen G, Ernst RR (1984) Prog Nucl Magn Reson Spectrosc 16:163–192
48. Abragam A (1983) Principles of nuclear magnetism. Oxford University Press, Published in New York United States
49. Aime S, Gobetto R, Reineri F, Canet D (2003) J Chem Phys 119:8890
50. Aime S, Gobetto R, Reineri F, Canet D (2006) J Magn Reson 178:184–192
51. Goldman M (1991) Quantum description of high-resolution NMR in liquids. Oxford University Press, Published in New York United States
52. Bouguet-Bonnet S, Reineri F, Canet D (2009) J Chem Phys 130:234507
53. Miesel K, Ivanov KL, Yurkovskaya AV, Vieth HM (2006) Chem Phys Lett 425(1–3):71–76
54. Theis T, Ganssle P, Kervern G, Knappe S, Kitching J, Ledbetter MP, Budker D, Pines A (2011) Nat Phys 7(7):571–575
55. Münnemann K, Spiess HW (2011) Nat Phys 7(7):522–523
56. Hoevener J-B, Chekmenev EY, Harris KC, Perman WH, Robertson LW, Ross BD, Bhattacharya P (2009) Magn Reson Mater Phys 22(2):111–121
57. Roth M (2010) ArchiMeD – sensitivity enhancement in NMR by using parahydrogen induced polarization. Roth, Meike. http://www.ubm.opus.hbz-nrw.de/volltexte/2010/2280/
58. Mansson S, Johansson E, Magnusson P, Chai CM, Hansson G, Petersson JS, Stahlberg F, Golman K (2006) Eur Radiol 16(1):57–67
59. Golman K, In't Zandt R, Lerche M, Pehrson R, Ardenkjaer-Larsen JH (2006) Cancer Res 66(22):10855–10860
60. Wilson DM, Keshari KR, Larson PEZ, Chen AP, Hu S, Van Criekinge M, Bok R, Nelson SJ, Macdonald JM, Vigneron DB, Kurhanewicz J (2010) J Magn Reson 205(1):141–147
61. Golman K, In't Zandt R, Thaning M (2006) Proc Natl Acad Sci USA 103(30):11270–11275
62. Gallagher FA, Kettunen MI, Day SE, Hu D-E, Ardenkjaer-Larsen JH, In't Zandt R, Jensen PR, Karlsson M, Golman K, Lerche MH, Brindle KM (2008) Nature 453(7197):940–943
63. Chekmenev EY, Hoevener J, Norton VA, Harris K, Batchelder LS, Bhattacharya P, Ross BD, Weitekamp DP (2008) J Am Chem Soc 130(13):4212–4213
64. Roth M, Kindervater P, Raich H-P, Bargon J, Spiess HW, Münnemann K (2010) Angew Chem Int Ed 122(45):8536–8540
65. Acosta RH, Blümler P, Münnemann K, Spiess H-W (2012) Prog Nucl Magn Reson Spectrosc 66:40–69
66. Warren WS, Jenista E, Branca RT, Chen X (2009) Science 323(5922):1711–1714
67. Vasos PR, Comment A, Sarkar R, Ahuja P, Jannin S, Ansermet JP, Konter JA, Hautle P, van den Brandt B, Bodenhausen G (2009) Proc Natl Acad Sci U S A 106(44):18469–18473
68. Wild JM, Paley MNJ, Viallon M, Schreiber WG, van Beek EJR, Griffiths PD (2002) Magn Reson Med 47(4):687–695
69. Hu S, Lustig M, Chen AP, Crane J, Kerr A, Kelley DAC, Hurd R, Kurhanewicz J, Nelson SJ, Pauly JM, Vigneron DB (2008) J Magn Reson 192(2):258–264
70. Mishkovsky M, Frydman L (2008) Chemphyschem 9(16):2340–2348
71. Roth M, Koch A, Kindervater P, Bargon J, Spiess HW, Münnemann K (2010) J Magn Reson 204(1):50–55
72. Baumer D, Brunner E, Blümler P, Zänker PP, Spiess HW (2006) Angew Chem Int Ed 45 (43):7282–7284

73. Amor N, Zaenker PP, Bluemler P, Meise FM, Schreiber LM, Scholz A, Schmiedeskamp J, Spiess HW, Muennemann K (2009) J Magn Reson 201(1):93–99
74. Jeener J (1971) Lecture notes from Ampere summer school in Basko Polje, Yugoslavia. unpublished
75. Aue WP (1976) J Chem Phys 64(5):2229
76. Messerle BA, Sleigh CJ, Partridge MG, Duckett SB (1999) J Chem Soc Dalton Trans (9):1429
77. Adams RW, Aguilar JA, Atkinson KD, Cowley MJ, Elliott PIP, Duckett SB, Green GGR, Khazal IG, Lopez-Serrano J, Williamson DC (2009) Science 323(5922):1708–1711
78. Toda M, Kubo R (1992) Statistical physics. I. Equilibrium statistical mechanics, 2nd edn. Springer, Berlin

Top Curr Chem (2013) 338: 75–104
DOI: 10.1007/128_2012_388

Published online: 9 November 2012

Improving NMR and MRI Sensitivity with *Para*hydrogen

Simon B. Duckett and Ryan E. Mewis

Abstract *Para*hydrogen induced polarisation (PHIP) has wide utility in NMR and MRI as it can increase the sensitivity of both techniques. The transfer of spin order from *para*hydrogen to nuclei in the analyte leads to an increased magnetic response following interrogation by RF pulses. This spin transfer is catalysed by a homogeneous or heterogeneous catalyst. The increased magnetic response not only reduces the number of transients required to obtain the spectrum or image, but can also illuminate previously undetectable species present in solution. From its theoretical prediction to its experimental validation, PHIP has been applied in a range of different areas such as the structural analysis of complexes, understanding reaction mechanisms involving hydrogen and for the production of contrast agents for use in MRI. PHIP can also be readily combined with other techniques such as photochemistry which widens its field of applicability. In this review, we detail the properties of *para*hydrogen and the methods for its preparation and utilisation in homogeneous and heterogeneous based hydrogenation and non-hydrogenative reactions. Specific examples are explained for the application of PHIP in photochemical and hydroformylation reactions. Pulse sequences designed to be compatible with PHIP are described to exemplify how the increase in sensitivity can be increased even further by the interrogation of the magnetic states optimally. Finally, a section on the use of PHIP in the production of contrast agents suitable for MRI, and the monitoring of hydrogenation reactions using imaging techniques is discussed.

Keywords Heterogeneous catalysis · Homogeneous catalysis · Hyperpolarisation transfer · NMR · *Para*hydrogenation · PHIP

S.B. Duckett and R.E. Mewis (✉)
Department of Chemistry, University of York, Heslington, York YO10 5DD, UK
e-mail: ryan.mewis@york.ac.uk

Contents

Abbreviations

ALTADENA	Adiabatic longitudinal transport after dissociation engenders nuclear alignment
CIDNP	Chemically induced dynamic nuclear polarisation
COD	Cyclooctadiene
DEPT	Distortionless enhancement by polarisation transfer
DNP	Dynamic nuclear polarisation
dpae	1,2-Bis(diphenylarsino)ethane
dppe	1,2-Bis(diphenylphosphino)ethane
dppm	1,2-Bis(diphenylphosphino)methane
FISP	Fast imaging with steady state precession
HFM	Hollow fibre membrane
HMBC	Heteronuclear multiple bond correlation
HMQC	Heteronuclear multiple quantum correlation
HOHAHA	Homonuclear Hartman–Hahn spectroscopy
IMes	1,3-Bis(2,4,6-trimethylphenyl)imidazole-2-ylidene
INADEQUATE	Incredible natural abundance double quantum transfer experiment
INEPT	Insensitive nuclei enhanced by polarisation transfer
MAA	Methyl 2-acetamidoacrylate
MAP	Methyl 2-acetamidopropanote
MRI	Magnetic resonance imaging
NMR	Nuclear magnetic resonance

OPSY	Only *para*hydrogen spectroscopy
PASADENA	*Para*hydrogen and synthesis allow dramatically enhanced nuclear alignment
PET	Positron emission tomography
PHIP	*Para*hydrogen induced polarisation
RF	Radio frequency
S/N	Signal-to-noise
SABRE	Signal amplification by reversible exchange
SDS	Sodium dodecyl sulphate
SEPP	Selective excitation of polarisation using PASADENA
TOCSY	Total correlation spectroscopy

1 Introduction

1.1 Para*hydrogen Induced Polarisation*

Nuclear magnetic resonance (NMR) is an extremely well regarded and heavily utilised spectroscopic technique which is used on a routine basis to obtain structural information at the molecular level. Information from the corresponding NMR spectra allows for structural characterisation through chemical shift values, spin–spin coupling (*J*-coupling) and integral values. The utilisation of 1D and higher dimensional approaches can provide the information required to elucidate the structure of proteins, for example. Similarly, magnetic resonance imaging (MRI) is used to obtain clinical information to assist in the treatment and management of disease states. However, both of these techniques suffer inherently from low sensitivity because the detected signal strength depends on the population difference that exists between the probed nuclear spin states in a magnetic field. Numerous techniques have been developed to counter this sensitivity problem such as the brute force approach [1], dynamic nuclear polarisation (DNP) [2] and *para*hydrogen induced polarisation (PHIP) [3–7]. It is the latter which will be the subject of this chapter and herein its application to the improvement of sensitivity in NMR and MRI will be discussed.

The sensitivity issue is exemplified by considering an ensemble of nuclear spins at ambient temperatures in the applied magnetic field of an NMR spectrometer. The equilibrium magnetisation obtained in this way is less than 10^{-4} of the value that could be obtained if the spins were all parallel. Weitekamp and Bowers first explored the use of *para*hydrogen in order to obtain large non-equilibrium magnetisations in 1986 [8]. They suggested that the phenomena that they predicted "should have wide utility in the study of the chemical reactions of molecules created by hydrogen addition or subsequent reactions." A year later, Bowers and Weitekamp published a paper entitled "*Para*hydrogen and Synthesis Allow

Dramatically Enhanced Nuclear Alignment" [9]. This would later become abbreviated to PASADENA and be used to describe the formation of non-equilibrium magnetisations using *para*hydrogen at high field. A similar observation was made at the same time by the group of Eisenberg who was studying the hydrogenation reactions of alkynes using $Rh_2H_2(CO)_2(dppm)_2$ (dppm = 1,2-bis (diphenylphosphino)methane) as a catalyst [3]. Initially they had thought that CIDNP (chemically induced dynamic nuclear polarisation) was responsible for increased NMR absorptions, but it was the storing of the sample in liquid nitrogen overnight in the presence of hydrogen gas which led to the formation of ~50% *para*hydrogen. By shaking the sample and inserting into the NMR spectrometer, the incorporation of *para*hydrogen into the olefin led to the observation of the increased NMR absorptions. In 1988, Pravica and Weitekamp demonstrated that net NMR alignment results by adiabatic transport of *para*hydrogen addition products to high magnetic field; hydrogenation of the substrate occurs at low magnetic field [10]. This gave rise to the phenomenon that would later be called adiabatic longitudinal transport after dissociation engenders nuclear alignment (ALTADENA).

PASADENA and ALTADENA are typically used to refer to reactions in which the *para*hydrogen molecule is incorporated into an unsaturated olefin. However, once this has occurred and the magnetic states associated with the hydrogenated product are read out, the fully saturated molecule can no longer be polarised. An alternative methodology termed signal amplification by reversible exchange (SABRE) [11] overcomes this. This methodology does not require the incorporation of *para*-hydrogen into the molecule of interest; instead it is the scalar coupling network which facilitates transfer of polarisation from *para*hydrogen to the analyte which is established at a metal centre. This will be covered in greater detail in Sect. 2.5.

NMR spectra collected under the PASADENA or ALTADENA conditions have very distinct differences with respect to the peak characteristics. PASADENA experiments are conducted with the sample in the NMR spectrometer and possess antiphase patterns for the polarised signals. A $\pi/4$ pulse is used to interrogate the magnetic states optimally. Conversely, ALTADENA experiments are initiated at low field (typically in the region of Earth's magnetic field, 0.05 mT) before being rapidly transported to the spectrometer and the NMR spectrum being collected immediately using a $\pi/2$ pulse. The resulting NMR spectrum possesses enhanced signals for the polarised resonances which are either in absorption or emission mode. Molecules polarised using SABRE are similarly interrogated optimally using a $\pi/2$ pulse; the magnetic states are again created at low magnetic field. The application of the correct RF pulse is prudent to the interrogation of the magnetic states produced under each regime.

It is because of the increased magnitude of the NMR signal response that makes PHIP such a versatile technique from the elucidation of catalytic cycles in situ to the production of contrast agents for MRI. This chapter will explore the utilization of PHIP in the fields of NMR and MRI and highlight the potential of this technique. It will cover both homogeneous and heterogeneous catalysis, NMR sequences used to interrogate the magnetic states created and also explore the role that PHIP has in producing MRI contrast agents.

1.2 Para*hydrogen*

Prior to detailing the use of *para*hydrogen in increasing signal sensitivity to both NMR and MRI, *para*hydrogen itself must be considered. Dihydrogen exits in two forms – *ortho*- and *para*-hydrogen. The existence of these two nuclear spin isomers derives from the symmetrization principle of quantum mechanics, which states that the overall wave function of the fermion dihydrogen must be antisymmetric with respect to exchange of nuclei. From this, it follows that antisymmetric rotational states (odd J) in dihydrogen are associated with symmetric (*ortho*) nuclear spin states. Likewise, symmetric rotational states (even J) are associated with antisymmetric (*para*) nuclear spin states. *Ortho*hydrogen is therefore threefold degenerate, and defined by the spin states $\alpha\alpha$, $\alpha\beta + \beta\alpha$ and $\beta\beta$, whereas *para*hydrogen exists solely as the $\alpha\beta - \beta\alpha$ spin isomer. α and β refer to the alignment of the spin with an applied magnetic field; α denotes a parallel alignment whereas β denotes an antiparallel alignment. The normalized spin states for *ortho*hydrogen can be written as

$$|T_{+1}\rangle = |\alpha\alpha\rangle$$

$$|T_{-1}\rangle = |\beta\beta\rangle$$

$$|T_0\rangle = \frac{1}{\sqrt{2}}|\alpha\beta + \beta\alpha\rangle$$

and for *para*hydrogen as

$$|S_0\rangle = \frac{1}{\sqrt{2}}|\alpha\beta - \beta\alpha\rangle$$

*Ortho*hydrogen can therefore be termed as having a triplet state whereas *para*hydrogen has a singlet state. Because these two forms have different rotational energies, their populations are temperature dependent but more importantly their interconversion is forbidden according to the selection rules. Using the right conditions it is possible to prepare exclusively the *para*hydrogen isomer.

1.3 Para*hydrogen Production*

Production of *para*hydrogen requires a spin conversion catalyst to relax the selection rules. If pure *para*hydrogen is desired, running hydrogen gas through a copper block containing silica/$FeCl_3$ catalyst at 20 K is sufficient to produce an almost quantitative yield of *para*hydrogen [12]. It must be noted that there is a wide range of materials available for catalyzing the conversion; a comprehensive list was compiled in the 1930s by Taylor and Diamond [13]. Apparatus for *para*hydrogen generation have been detailed [12, 14]. In modern times, the process of producing *para*hydrogen has been automated to enable batch delivery; details of a pulsed injection *para*hydrogen generator have recently been published [15]. Furthermore, equipment for the hyperpolarisation of ^{13}C biomolecules via PASADENA has been detailed [16], as well as assurance procedures for the quality of hyperpolarised materials [17].

1.4 *Introduction of* Para*hydrogen to a Sample*

Once formed and removed from the catalyst, *para*hydrogen is relatively stable to spin-equilibration. Practically, this means that there is plenty of time for the involvement of *para*hydrogen in a chemical reaction where the aim is to generate and then monitor reaction products by hyperpolarised NMR methods. For most spectroscopic based investigations, Young's capped NMR tubes are typically used as they, and the contained sample, can be easily degassed and then *para*hydrogen can be subsequently introduced to the sample. Shaking the sample prior to transfer to the spectrometer ensures dissolution of hydrogen in solution. Most ALTADENA based measurements involve rapid transport of the sample into the spectrometer after polarisation transfer in low magnetic field; this normally means that the lift is not used to lower the sample into the magnet proper. PASADENA, on the other hand, requires the polarisation to be transferred via weak coupling and thus polarisation transfer will only begin at high fields. Therefore rapid transport is not an issue, but more the concentration of *para*hydrogen in solution over time as it becomes consumed via hydrogenation reactions. To ensure a reasonable concentration of *para*hydrogen in solution, the use of a capillary to deliver hydrogen directly into the solution whilst it is in the spectrometer has been designed and implemented [18, 19].

There are an increasing number of reports, however, which are using polarizers to introduce *para*hydrogen and then deliver the hyperpolarised sample to the desired point of measurement. Apparatus of this type have been used in spectroscopic and MRI based investigations. For example, Duckett and co-workers have reported on the use of a polarizer to study the longitudinal magnetisation generated for the *para* proton resonance of pyridine after polarisation transfer over the field range of -150 to $+10 \times 10^{-5}$ T [20]. The process of hyperpolarisation transfer, transfer to the measurement field, data acquisition and the returning of the sample to the sample preparation chamber is reported as taking 20 s, thus enabling rapid spectral collection coupled with the ability to signal average. Chekmenev and co-workers have recently reported on the use of a centrally controlled, automated *para*hydrogen-based polarizer with in situ detection capability [21]. Using pulsed-polarisation transfer, a 5,000,000-fold signal enhancement at 48 mT was demonstrated for the labeled carbon-13 nucleus of 2-hydroxyethyl-^{13}C-propionate-$d_{2,3,3}$.

Not just restricted to NMR spectroscopic investigations, previous work by Ross and co-workers has emphasized the use of automated equipment coupled to a polarizer to produce hyperpolarised ^{13}C biomolecules via PASADENA for use in the collection of MR images in vivo [16]. The process of hydrogenating the substrate, hyperpolarisation transfer (via RF spin order transfer), delivery to the detection system and the sending of the trigger signal takes 52 s in total. The reproducibility and experimental ease that polarizers offer will see them become ever more prevalent in conducting PHIP based NMR and MRI experiments in the future.

2 Applications of PHIP to NMR

PHIP has been employed in the study of many different reactions, such as hydrogenation and hydroformylation (Sect. 2.3), and to elucidate the chemical structures of intermediates which are formed as part of the catalytic cycle. Most of these reactions have been conducted using homogeneous catalysts (Sect. 2.1), but the use of heterogeneous catalysis is slowly increasing (Sect. 2.4). The use of PHIP in conjunction with photochemical techniques to study unstable species is also detailed (Sect. 2.2). This section will detail the involvement of PHIP in advancing these arenas of chemistry, and present a detailed overview of SABRE (Sect. 2.5), an emerging methodology for polarizing molecules using *para*hydrogen without its incorporation into the analyte.

2.1 *Homogeneously Catalyzed Hydrogenation Reactions Involving* Para*hydrogen*

Homogeneously catalysed hydrogenation reactions employing *para*hydrogen are commonplace in the investigation of many systems. The sample homogeneity offered by such systems facilitates artefact free NMR spectra which possess good line shapes so that peaks are well resolved. The use of PHIP as a structural elucidation tool, either to obtain directly the chemical identity of molecules or as a mechanistic probe, has been the focus of many investigations. A number of reviews have been dedicated to this subject [12, 22]. Measurements can be conducted rapidly as hydrogen addition occurs quickly, thus entailing a substantial magnetic reservoir which can be tapped into without the need to consider relaxation times. Used in conjunction with 1D and 2D methods (see Sect. 3), PHIP provides key information to enable identification of reaction intermediates. Structural elucidation is key to ascertaining reaction cycles and the role played by the intermediates which can often only be observed using PHIP. A few examples will now be detailed.

The use of PHIP to elucidate reaction intermediates has been highlighted in a Pt catalysed alkyne hydrogenation, in which the choice of solvent could affect the alkene produced [23]. The identification by use of PHIP combined with GC-MS showed that in the presence of alcoholic based solvents, alkoxy alkenes could be produced. This observation enabled an already established catalytic cycle to be extended to show that alcohols such as methanol could act as a proton source for the hydrogenation of the alkyne [24].

New intermediates in the hydrogenation of various olefins using Wilkinson's catalyst, $RhCl(PPh_3)_3$, has been shown by PHIP [25]. Previously undetected dihydride species were observed such as the binuclear complex $(H)_2Rh(PPh_3)_2(\mu\text{-}Cl)_2Rh(PPh_3)$ (olefin) and $Rh(H)_2$(olefin)$(PPh_3)_2(Cl)$. The binuclear complex is not directly part of the catalytic cycle, and its observation suggested reduced activity of the system through the formation of a less active binuclear species. $Rh(H)_2$(olefin)$(PPh_3)_2(Cl)$, however, is an intermediate in hydrogenation catalysis and, from the PHIP spectra, the phosphines were found to be *cis* to one another as were the two hydrides. These observations enabled previously established reaction cycles to be expanded as well as alluding to the reduced effectiveness of the catalysts over time.

Conclusive evidence for the mechanism of hydrogen addition to $Pt(Ph_2CH_2CH(Me)OPPh_2)$ has been obtained through the use of PHIP [26]. Hydrogen addition was found to occur in a concerted pairwise manner by monitoring the hyperpolarised signals generated when *para*hydrogen was reacted with the unsymmetrical platinum complex. The resulting dihydride species, in which the hydrides are magnetically inequivalent, possess antiphase peaks which are observed as doublets of doublets due to coupling with ^{31}P and ^{195}Pt. The 2D spectroscopy of this dihydride complex is discussed in more detail in Sect. 3.4.

As well as structural elucidation, PHIP has been used in conjunction with other chemical methods to obtain novel hyperpolarised products. A recent contribution by Bargon has highlighted how molecules that are not normally polarisable can become hyperpolarised by utilising chemical processes other than hydrogenation [27]. Their study focuses on the hydrogenation of vinyl acetate forming ethyl acetate. The use of vinyl acetate instead of ethenol (which tautomerises to give predominantly ethanal) results in the double bond being in the desired position for hydrogenation. Subsequent basic hydrolysis yields acetic acid and ethanol, the latter of which possesses an absorption peak (CH_2 protons) and an emission peak (CH_3 protons). The signal enhancement is low (~fourfold) but the process of hydrogenation, hydrolysis and transport into the measurement field is greater than 20 s so there is significant relaxation.

In Sects. 2.2 and 2.3 the application of PHIP to two specific examples involving hydrogenation will be discussed: its use in conjunction with photochemical methods and its employment for studying hydroformylation reactions.

2.2 *In Situ Photochemistry in Combination with PHIP*

The unstable nature of some complexes means that most characterisation methods using NMR are limited due to the way the sample is manipulated prior to data acquisition. Groups have used UV light to circumvent this; in particular the use of modified probes which enable the photolysis of samples within the bore of the magnet allow for the characterisation of intermediates formed by conventional methods.

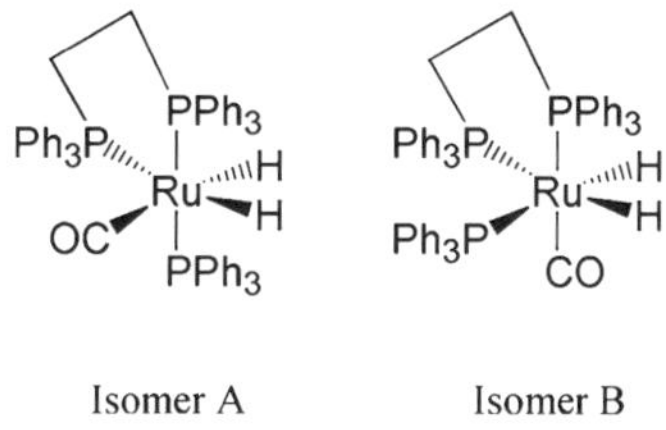

Fig. 1 The two isomers produced after the irradiation of $Ru(CO)_2(PPh_3)(dppe)$ with concurrent exposure to *para*hydrogen

Isomer A Isomer B

One of the main contributions in this area was in 2005 by the group of Duckett. Their investigations into the laser initiated (either pulsed or continuous-wave) reaction between complexes of the type $Ru(CO)_3(L_2)$ (where L_2 = dppe (1,2-bis(diphenylphosphino)ethane) or dpae (1,2-bis(diphenylarsino)ethane) and *para*hydrogen led to the generation of a pure nuclear spin state with respect to the hydride ligands [28]. The two complexes formed by reaction with *para*hydrogen ($Ru(CO)_2(L_2)(H)_2$) after laser flash photolysis possess hydrides with spin states of purity 89.8 ± 2.6% (dppe) and 106 ± 4% (dpae). The dppe system was studied using continuous wave and flash laser irradiation. The use of flash irradiation was far superior to continuous wave as it prevented dephasing and relaxation effects. This work demonstrated the applicability of such a system in producing a defined pure magnetic state which would be suitable for employment in quantum computing.

A related system, $Ru(CO)_2(PPh_3)(dppe)$, was also probed using *para*hydrogen based NMR studies [29]. $Ru(CO)_2(PPh_3)(dppe)$ was shown to react differently when probed using either thermal or photochemical methods. Under thermal conditions, PPh_3 is lost from $Ru(CO)_2(PPh_3)(dppe)$, enabling the oxidative addition of hydrogen to form $Ru(CO)_2(dppe)(H)_2$. The formation of this product was confirmed by the observation of enhanced hydride resonances using *para*hydrogen which were quenched upon the addition of an excess of PPh_3. Photochemically, loss of CO dominates and thus an alternative dihydride is produced $Ru(CO)(PPh_3)(dppe)(H)_2$ (Isomer A in Fig. 1). The exposure of $Ru(CO)_2(PPh_3)(dppe)$ to *para*hydrogen with concurrent UV laser photolysis resulted in the formation of a new set of signals; these were shown to belong to Isomer B (see Fig. 1). An alternative photochemical pathway was identified by the irradiation of $Ru(CO)_2(PPh_3)(dppe)$ in the presence of pyridine, which led to the observation of $Ru(H)_2(CO)(dppe)(pyridine)$. This pathway involves CO and PPh_3 loss. The presence of two isomers of $Ru(H)_2(CO)$ (dppe)(pyridine) was again confirmed by the use of *para*hydrogen. This report highlights the importance of *para*hydrogen in conjunction with photochemical studies so that all the generated species can be detected.

Further ruthenium complexes of the type $Ru(CO)_3(L)_2$ where L = PPh_3, PMe_3, PCy_3 and $P(p\text{-tolyl})_3$, have been studied using photochemical methods in conjunction with *para*hydrogen [30]. Two competing processes were identified; one in which CO loss occurs prior to hydrogen addition to form the *cis-cis-trans*-L isomer of Ru $(CO)_2(L)_2(H)_2$ and another in which a single photon induces loss of CO and L, thus leading to the formation of the *cis-cis-cis* isomer. A summary of the products formed after reaction, via both pathways, is shown in Fig. 2. The pathway which involves just

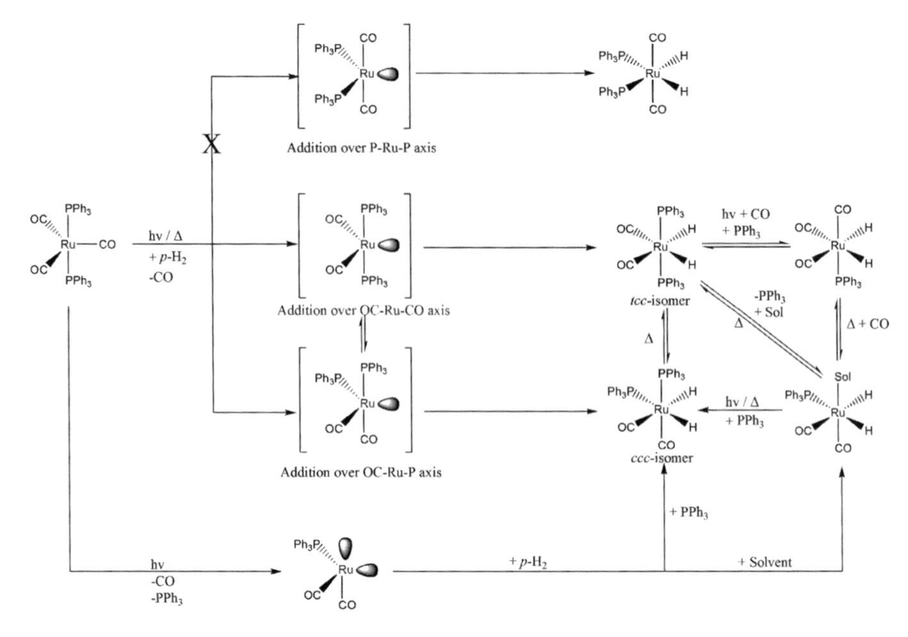

Fig. 2 Summary of *para*hydrogen addition to $Ru(CO)_2(PPh_3)_2$ and $Ru(CO)_2(PPh_3)$ fragments and the subsequent structures of the products formed

CO loss involves the 16-electron complex $Ru(CO)_2(L)_2$. The utilisation of PHIP in conjunction with theoretical calculations elucidated that hydrogen addition occurs over the OC–Ru–CO axis.

Iridium complexes have also been studied using photochemical techniques and PHIP. The rapidity of NMR characterisation of thermally and photochemically generated complexes of the general formula $[Ir(H)_3(CO)_{3-x}(PPh_3)_x]$ (x = 1–3) has been reported [31]. The hydride region of the ^{1}H NMR spectrum is very diagnostic as to the complex formed, and the ability to complete these experiments using ^{31}P decoupling and the employment of ^{13}CO facilitated the complete assignment of all the hydrides observed. The detection of these unstable hydride complexes was achieved without the need for high pressures of hydrogen gas and, more importantly, requiring very small amounts of the initial complex (~1 mg). Without the increase in signal intensity offered by PHIP these observations would not have been possible.

2.3 The Study of Hydroformylation Systems

Hydroformylation is a key industrial process for the synthesis of aldehydes using metal complexes. Understanding the process by which the aldehyde is formed can lead to the development of better catalysts. PHIP has been employed to achieve a greater understanding in these systems by enabling the direct observation of intermediate species.

Cobalt and rhodium systems containing phosphines are typically used as industrial catalysts. Iridium catalysts are generally slower because of reduced rates of migratory insertion, but this may be advantageous in terms of NMR measurements in that it can be observed on an NMR time scale. An investigation of a hydroformylation reaction catalysed by an iridium containing catalyst has been described, in which *para*-hydrogen is used to gain insight into the reaction mechanism [32]. Ir(CO)$(PPh_3)_2(\eta^3\text{-}C_3H_5)$ when reacted with *para*hydrogen at 273 K forms Ir(CO)$(PPh_3)(\eta^3\text{-}C_3H_5)(H)_3$ and two square based pyramidal isomers of the formula Ir(CO)$(PPh_3)(\eta^3\text{-}C_3H_5)(H)_2$. The isomers were only detected when using *para*hydrogen, and were not detectable under thermal conditions. By warming to 295 K, propene and propane were detected via their PHIP enhanced ^{1}H NMR signature. The liberation of propene and propane enabled the subsequent formation of *mer* and *fac* isomers of Ir(CO)$(PPh_3)_2(H)_3$. In the presence of CO and *para*hydrogen, hydrogenation was suppressed and carbonylation occurs. In this mixed atmosphere, two novel dihydride products *cis,cis*-$Ir(CO)_2(PPh_3)(COCH_2CHCH_2)(H)_2$ and *cis,cis*- $Ir(CO)_2(PPh_3)(\eta^1\text{-}CH_2CH{=}CH_2)(H)_2$ were detected. The former goes on to eliminate propene, whereas the latter yields the hydroformylation product. The use of ^{13}CO in both instances enabled the geometry to be assigned due to the observation of only a single hydride resonance.

A xantphos containing iridium complex has also been used to study hydroformylation [33]. The use of PHIP in this report was to allow the observation of Ir(xantphos)(CO)$(H)_2$(COEt), the precursory complex involved in the formation of ethanol after subsequent reaction with a further *para*hydrogen molecule. PHIP was also employed advantageously in conjunction with ^{1}H and ^{31}P NMR spectroscopy to elucidate fully the structures of the isomers of Ir(xantphos)(CO)$(H)_2$(X). Interestingly, the observation of polarisation (net emission) in the aldehyde product formed upon the reaction of Ir(xantphos)(CO)$(H)_2$(COEt) with *para*hydrogen originates from one-H PHIP. This effect, so named because only one proton from *para*hydrogen is transferred to a product, has been reported previously by Permin and Eisenberg [34] for the analogous complex Ir(dppe)$(CO)_2$(COEt). Permin and Eisenberg, who were also studying hydroformylation reactions, had postulated that the overlapping hydride resonances of the intermediate acyl dihydride complex had resulted in a specific condition causing a second-order effect. However, the observation of one-H PHIP when using Ir(xantphos)(CO)$(H)_2$(COEt) suggests that this is not the case, as the hydride resonances are well separated and thus there is no second-order effect. Upon re-examination of Ir(dppe)$(CO)_2$(COEt) at 500 MHz instead of 400 MHz, the overlap in the hydride region is now non-existent, yet one-H PHIP was still observed. This set of experiments highlighted that one-H PHIP is not as specific as previously thought, and may actually be more commonplace. Indeed a further report of one-H PHIP has been observed in the hydrogenation of diphenylacetylene catalyzed by a platinum(II) bis-phosphine triflate complex [23]. When the reaction was conducted in methanol-d_4 at 314 K, the hydrogen atom of the Z isomer of the vinylether produced was shown to exhibit polarised signals due to one-H PHIP.

The PHIP effect has been employed to map fully the hydroformylation reaction using $Co(\eta^3\text{-}C_3H_5)(CO)_2(PCy_3)$ [35]. The detection and characterisation of reaction intermediates which did not contain hydride ligands was made possible by the use of *para*hydrogen which resulted in anti-phase signals being observed in the ^1H NMR spectrum. These complexes were identified as $Co(COCH(CH_3)_2)(CO)_3(PCy_3)$ and $Co(COCH_2CH_2CH_3)(CO)_3(PCy_3)$.The composition of these intermediates, in that they both contain PCy_3, led to the conclusion that the reaction proceeded initially with H_2 via CO loss. Other enhanced resonances were observed for $HCOCHCH_2CH_3$ and $HCOCH(CH_3)_2$, the two corresponding hydroformylation products. In a further report, analogous complexes of the type $Co(\eta^3\text{-}C_3H_5)(CO)_2(PR_2R')$ (R, R' = Ph, Me; R, R' = Me, Ph; R = R' Ph, Cy, CH_2Ph) were used to study the same hydroformylation reaction [36]. Reactions of these allyl complexes led to the observation of PHIP in both liberated propane and propene. The detection of branched and linear acyl containing complexes was again observed. Additionally, when the complex containing PPh_2Me was used, additional PHIP enhanced signals for $Co(COCH_2CH_2CH_3)(CO)_2(PPh_2Me)$(propene) and $Co(COCH(CH_3)_2)(CO)_2(PPh_2Me)$(propene) were also detected. These arise from the trapping of the 16 electron fragments $Co(COCH_2CH_2CH_3)(CO)_2(PPh_2Me)$ and $Co(COCH(CH_3)_2)(CO)_2(PPh_2Me)$ by propene, which is PHIP enhanced, rather than CO.

2.4 *Heterogeneously Catalyzed Hydrogenation Reactions Involving* Para*hydrogen*

PHIP based reactions are typically performed using homogeneous catalysts as this ensures good sample homogeneity for spectroscopic based studies. There is, however, a growing need to be able to separate the catalyst from the hyperpolarised material produced. This is because, for in vivo measurements, the biocompatibility of the catalyst cannot always be assured. It also means that more biologically compatible media can be used without cause for concern with respect to catalyst solubility. It is for these two reasons that attention is being directed to design and create systems that can be used to polarise effectively a range of suitable substrates. The first example of the PASADENA effect at a solid surface was demonstrated in 2001 by Weitekamp and co-workers [37]. Reversible hydrogen adsorption onto the surface of zinc(II) oxide was studied using normal and *para*-enriched hydrogen. The anti-phase nature of PASADENA derived signals was observed in the presence of *para*hydrogen, which contrasted with a single broad signal observed in the presence of normal hydrogen. The observation of this polarised signal facilitated an investigation into hydrogen adsorption modes occurring on the surface of the metal oxide.

The first report of heterogeneously catalyzed liquid-phase hydrogenation of alkenes and alkynes was by Duckett and co-workers [38]. A range of platinum, silver, iron, palladium and rhodium solid supported catalysts were investigated for the hydrogenation of methyl propiolate. There was no clear correlation between average particle size and the intensity of the polarised signals. Of the materials tested, the Pt/silica material gave the most intense PASADENA derived signal, despite possessing a low metal content (0.5 wt%) compared with other materials tested. The observed signal intensity, although small, clearly possessed anti-phase character associated with a PASADENA type derived signal.

Koptyug et al. provided the first ALTADENA based study using heterogeneous catalysis [39]. Initially they showed that Rh/TiO_2 and Rh/AlO(OH) could hydrogenate propylene in the gas phase. This was then extended to a liquid phase based study, in which the propylene gas was dissolved in either toluene or acetone; solvent choice was dependent on the catalyst used. ALTADENA polarisation patterns were observed in the ^1H NMR spectrum, thus highlighting the fact that heterogenous based catalysts could be used for the production of continuously flowed polarised liquids. The hydrogenation of styrene using the Rh/TiO_2 supported metal catalysts quite clearly differentiates the phase character of the signals produced via either an ALTADENA or a PASADENA route, the former being either in emission (CH_2) or absorption (CH_3) mode and the latter having anti-phase character.

2.5 SABRE

The requirement of unsaturated bonds in molecules so they can be used in conjunction with PHIP does mean that the scope of the technique is limited to the chemical nature of the analyte. A new technique which uses *para*hydrogen to improve the sensitivity of the NMR experiment is being pioneered at the University of York by the Duckett group. This new technique does not require the incorporation of *para*hydrogen into the analyte. It is the *J*-coupling which modulates the transfer of polarisation from *para*hydrogen derived hydrides to the analyte of interest which are brought into contact at a metal surface [40]. This new approach, referred to as signal amplification by reversible exchange (SABRE) [11], involves the creation of magnetic states at low magnetic field followed by rapid transport to the NMR spectrometer and acquisition of NMR data. Like ALTADENA, magnetisation is optimally read out using a $\pi/2$ pulse.

The effect was first reported when complexes of the type $[Ir(COD)(PR_3)_2]BF_4$ (R = Ph, (*p*-Tol) or *p*-C_6H_4OMe) in the presence of pyridine was exposed to an atmosphere of *para*hydrogen [41]. The use of ^{15}N-labeled pyridine led to the observation of anti-phase peaks in the ^1H NMR spectrum, which arises because the hydride ligands are magnetically inequivalent as they belong to an AA"XX" spin system. The ^{15}N label was sensitized by a factor of 120 times using a PH-INEPT (detailed in Sect. 3.1) pulse sequence compared to a 1,024 scan ^{15}N NMR control spectrum. This result was achieved without incorporation of *para*hydrogen into the analyte.

Subsequent studies showed that the sensitization of nuclear spins could be fine tuned by modification of the coordination sphere of the metal complex. A further complex, $[Ir(COD)(PCy_3)(py)][BF_4]$, was also found to be a capable polarisation transfer catalyst [42]. Upon reaction with *para*hydrogen in the presence of pyridine, the complex *fac,cis*-$[Ir(PCy_3)(py)_3(H)_2]BF_4$ results. After polarisation in a field of 0.5×10^{-4} T, the three ^1H resonances of pyridine exhibit enhanced signals in either an absorption (*meta* resonance) or emission (*ortho* and *para* resonances) mode. Enhancements for the *ortho*, *para* and *meta* ^1H resonances were found to be −144-, −47- and +220-fold. Sensitization of ^{15}N and ^{13}C resonances was also demonstrated. When using ^{15}N-labeled pyridine a 128-fold increase was observed whilst ^{13}C enhancements of 15:5:8 (*ortho*:*meta*:*para*) was obtained when using a refocused $^{13}C\{^1H\}$ pulse sequence. The identity of the phosphine was shown to have an effect on the absolute signal strengths of pyridine after polarisation transfer at 0.5×10^{-4} T. Of those tested, it was the most electron rich and sterically demanding ligand, PCy_3Ph, which proved most beneficial in polarisation transfer to the hydrogen atoms in pyridine.

A further report using $Ir(COD)(PCy_3)(py)]BF_4$ to sensitize ^1H spins in conjunction with NMR and MRI has been published by Duckett and co-workers [11]. This report highlights not only the application of SABRE to NMR but also to MRI. The sensitization of analytes at low concentrations was demonstrated. A 6-nmol sample of pyridine was successfully polarised using $[Ir(H)_2(PCy_3)(pyridine)_3][BF_4]$ in a field of 2×10^{-2} T. Even when the thermal, non-hyperpolarised, trace is expanded by 128-fold, the hyperpolarised signals (which all have emission character) are far larger. A sample containing 50 μmol of nicotinamide was polarised similarly at a field of 0.5×10^{-4} T, after which all four ^1H resonances were shown to be hyperpolarised in the ^1H NMR spectrum. Again the resonances display emission character aside from the *meta* ^1H resonance which is in absorption mode. A refocused $^{13}C\{^1H\}$ spectrum was also recorded for this sample in a single transient. The authors note that 3-fluoropyridine, nicotine, pyridazine, quinoline, quinazoline, quinoxaline and dibenzothiophene can also be polarised using this method because they all weakly associate with the metal complex through their basic donor sites. ^1H True-FISP (fast imaging with steady state precession) MRI images of an 8-mm sample tube containing glass cylinders of internal diameter 1 mm were presented, which showed a 160-fold increase in signal strength when the sample is hyperpolarised using SABRE compared to its thermal equilibrium magnetisation.

The use of SABRE as a trace analysis tool has been explored by Blümich and co-workers [43]. They report that 12 nmol of pyridine could be detected in a 0.4-mL NMR sample using just a single scan after polarisation transfer via SABRE followed by acquisition of ^1H NMR data in a low field spectrometer. The polarisation transfer catalyst $[Ir(COD)(PCy_3)(py)][BF_4]$ was used. The application of SABRE to sensitize the signals of pyridine was compared against a thermal prepolarisation method using a 2-T Halbach magnet. The limit of detection where the $S/N = 1$ was estimated to be 1 μL using the thermal prepolarisation method and 1 nL using SABRE. Therefore, compared with the thermal prepolarisation method, SABRE provides a 1,000-fold improvement in terms of the detection limit.

SABRE has been implicated in the selective detection of drug molecules in a low field spectrometer [44]. Harmine, nicotine and morphine were all polarised using SABRE at 3.9 mT using $[Ir(COD)(PCy_3)(py)][PF_6]$ as the polarisation transfer catalyst. The ^1H NMR spectra for all three drug molecules were compared with those obtained using thermal prepolarisation at 2 T. The ^1H NMR spectra obtained following polarisation using SABRE showed considerable increase in the amplitude of the signal relative to that obtained using prepolarisation at 2 T. In fact, it was found that a few milligrams of each drug molecule could be readily detected after polarisation transfer using the SABRE conditions for magnetisation transfer, even though the hydrogen gas used for the study contained only 50% *para*hydrogen. The same group also showed that it was possible to hyperpolarize small amount of amino acids and peptide chains and detect them in low magnetic fields (~0.25 mT) [45]. A similar, yet distinct, approach had been reported prior to this by Duckett and co-workers [46]. Use of $IrCl(CO)(PPh_3)_2$ in the presence of adenine or purine led to the observation of multiple PHIP enhanced hydride resonances which are indicative of the nitrogen involved in binding and its chemical surroundings upon ligation.

Modification of the coordination sphere surrounding the iridium metal centre has been found to improve the sensitization of nuclei of interest [20]. Exchanging the phosphine for an N-heterocyclic carbene (NHC) led to the observation of increased signal intensity in the ^1H NMR spectrum. For example, when using the complex [Ir(IMes)(COD)Cl] to polarise pyridine, a 266-fold increase in signal intensity was observed for the *meta* protons after polarisation transfer at 0.5×10^{-4} T followed by interrogation by RF pulses in a 9.4 T magnet. This compares with an 18-fold increase when using $[Ir(COD)(PCy_3)(py)][BF_4]$. This result is therefore suggestive that the chemical make-up of the metal complex directly influences polarisation transfer, and thus means that the ligand sphere around the metal ion may be "tuned" to increase effectively magnetisation transfer.

3 Pulse Sequences for Increasing Sensitivity Using PHIP

Due to the naturally low abundance of heteronuclei, sensitivity gains from using PHIP are still relatively small. A number of pulse sequences in which hyperpolarisation can be transferred from ^1H nuclei to heteronuclei efficiently have been developed to tackle this problem. These sequences have been based on well-established 1D and 2D methods such as INEPT. Other pulse sequences have been designed to detect hyperpolarised signals selectively and suppress thermally derived signals to aid in spectral assignment and to probe the nature of hyperpolarised peaks. This section will look at both aspects of this arena of research. Reports on this subject have been published previously, in particular on the transfer of *para*hydrogen-induced hyperpolarisation to heteronuclei by Kuhn and Bargon [47].

3.1 INEPT

INEPT (insensitive nuclei enhanced by polarisation transfer) was originally developed to increase the signal strength of nuclides which are low in natural abundance and possess a low gyromagnetic ratio. By transferring magnetisation from protons to coupled heteronuclei of interest, a gain in sensitivity is achieved. Initial studies had viewed the use of INEPT from a chemical standpoint. The observation of anti-phase peaks in the ^{31}P spectrum of $[Ir(H)_2Br(CO)(dppb)]$ just 1 min after thawing from −196 °C in the presence of *para*hydrogen compared almost exactly with that of an INEPT spectrum with room-temperature equilibrated spin states [48]. However, this experiment just exemplified that spin overpopulations could be generated via the oxidative addition of *para*hydrogen to a metal centre followed by subsequent polarisation transfer to ^{31}P nuclei. Using the INEPT+ sequence developed by Sorensen and Ernst [49], Eisenberg and co-workers were able to record ^{13}C spectra of labelled and unlabelled $[Ir(H)_2Br(CO)dppe]$ [50]. Using labelled ^{13}CO and *para*hydrogen, a 158-fold enhancement in signal strength was observed for a 32-scan spectrum compared with a 256-transient collected using thermal equilibrated hydrogen gas. This enhancement is quite phenomenal in that 1 scan collected using PHIP is effectively equivalent to 25,000 scans collected using normal conditions. When natural abundance CO was used, the resulting spectrum was 25-fold stronger in S/N than that produced at thermal equilibrium.

The INEPT sequence has been adapted for use with PHIP by Bargon and co-workers, who have developed three sequences based on the INEPT building block (Fig. 3); PH-INEPT, PH-INEPT+ and PH-INEPT(+π/4) [51]. All of these sequences require the two protons of *para*hydrogen to add to an unsaturated molecule in chemically inequivalent positions. Additionally, it is worthy of note that the PH-INEPT sequence does not transfer any proton magnetisation stemming from systems at thermal equilibrium. Consequently, the corresponding spectrum only displays signals originating from products recently formed by reaction with *para*hydrogen.

The need to develop more than one pulse programme for conducting INEPT experiments using *para*hydrogen, arose from difficulties in probing systems in which the *J*-coupling between the protons derived from *para*hydrogen is small or non-existent. The PH-INEPT(+π/4) successfully circumvents this problem by converting trilinear linear unobservable terms of the type $I_zI_yS_x$ into observable terms by application of a 45° pulse on proton to yield observable $I_zI_zS_x$. Thus the coupling between the two *para*hydrogen derived protons is irrelevant. Compared with the INEPT + sequence, which yields anti-phase signals in *para*hydrogen systems and in-phase signals in systems at thermal equilibrium, the PH-INEPT (+π/4) was found to perform much better.

3.2 SEPP

Most sequences discussed so far have revolved around the direct modification of the pulse sequence by changing an initial π/2 for a π/4 pulse in order to interrogate magnetisation transfer conducted under the PASADENA regime. A simple selective excitation and detection scheme for observing enhanced signals via PHIP has been published [52]. This scheme has been named SEPP (selective excitation of polarisation using PASADENA); see Fig. 4 for pulse sequence. SEPP converts the

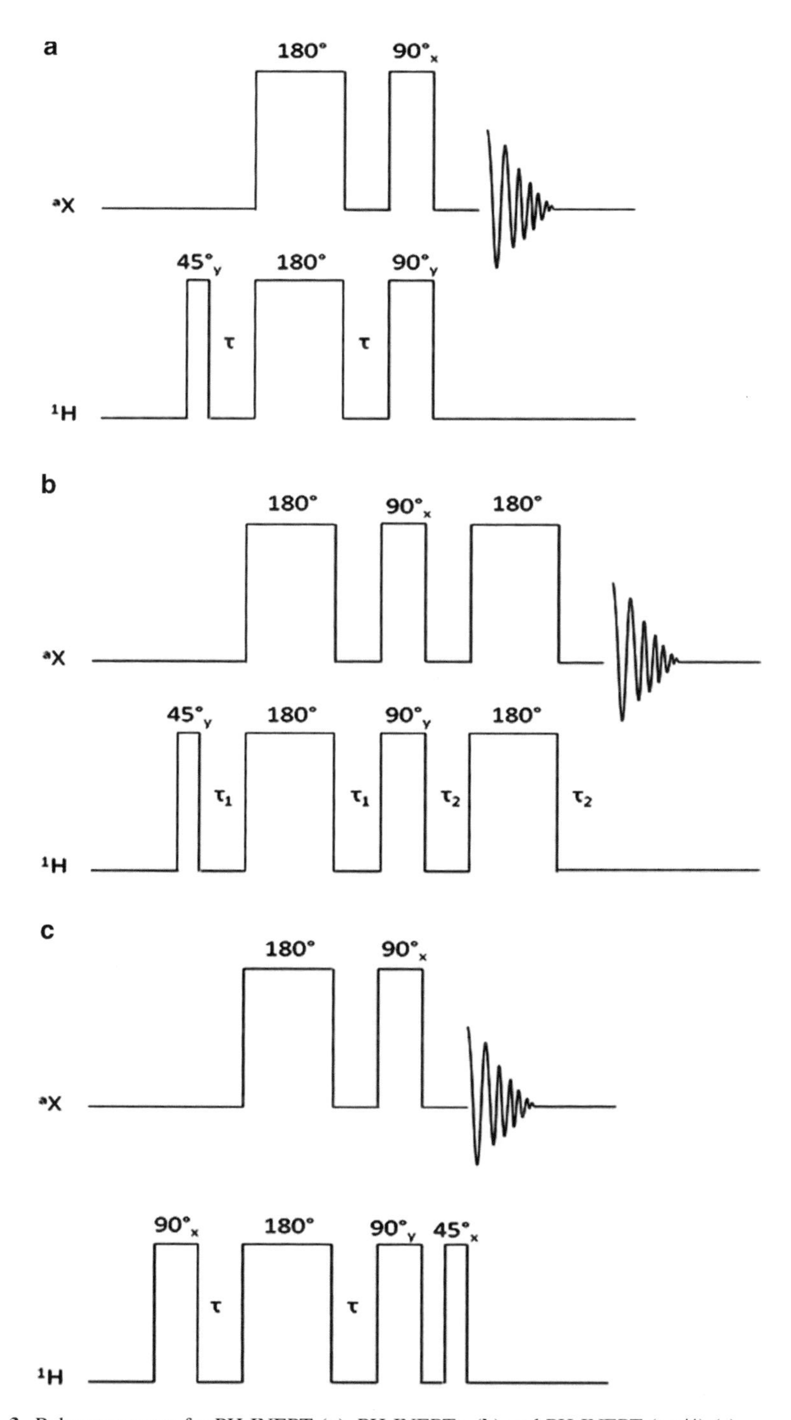

Fig. 3 Pulse sequences for PH-INEPT (**a**), PH-INEPT+ (**b**) and PH-INEPT (+π/4) (**c**)

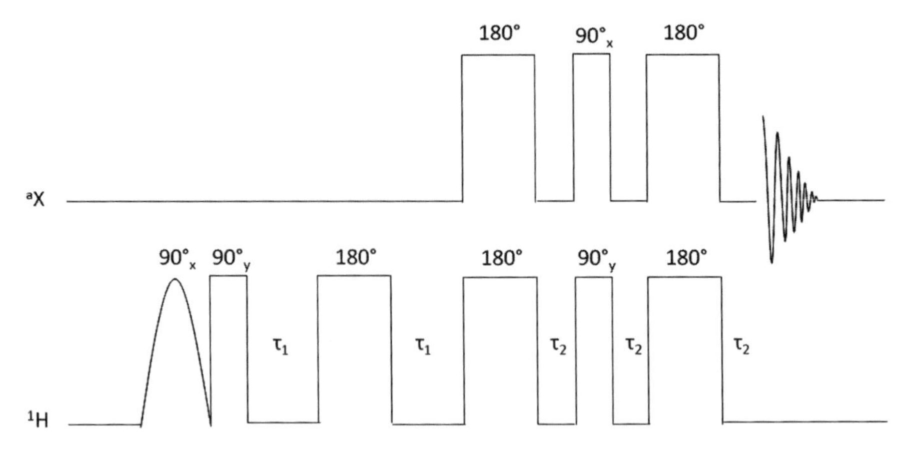

Fig. 4 Pulse sequence for SEPP-INEPT

longitudinal two-spin order term I_zI_z into in-phase magnetisation on one of the protons originating from *para*hydrogen. Given that the conversion of longitudinal two spin order is only 50% using a $\pi/4$ pulse, SEPP allows for a more efficient and versatile handling of PASADENA signals.

SEPP has been combined with a number of different pulse sequences in order to investigate *para*hydrogen based systems. Bargon and co-workers have reported sequences for SEPP-INEPT, SEPP-INEPT+ and SEPP-DEPT and used them successfully to monitor the hydrogenation of 1-hexyne and phenylacetylene catalysed by [Rh[norbornadiene](PPh_3)$_2$] [53].

Sengstschmid et al. have demonstrated the use of 1D TOCSY and doubly selective HOHAHA transfer in combination with the SEPP sequence [52]. A 1D-SEPP-HOHAHA is presented showing a single step transfer from the proton located on the 2 position of 1-hexyne to the adjacent methylene group whilst hydrogenation occurs to form 1-hexyne. Also presented is a multistep transfer which demonstrates that four stepwise transfers is possible using this technique, whilst still giving sufficient S/N after eight scans. Eight scans were also adequate to enable all ^{1}H NMR resonances to be detected using the 1D SEPP-TOCSY.

3.3 *INADEQUATE*

*Para*hydrogen polarisation can be transferred to pairs of hetero nuclei using a modified INADEQUATE sequence as reported by Bargon and co-workers [54]. The pulse sequence used is shown in Fig. 5. Using this pulse sequence, ^{13}C-^{13}C couplings were obtained for the hydrogenation product 1,4-diphenylbut-1-en-3-yne from 16 scans under the PASADENA condition. Employment of the conventional INADEQUATE sequence was not sufficient to detect ^{13}C-^{13}C couplings after 15 h of measurement time. Modulation of the evolution times τ_1 and τ_2 enabled the

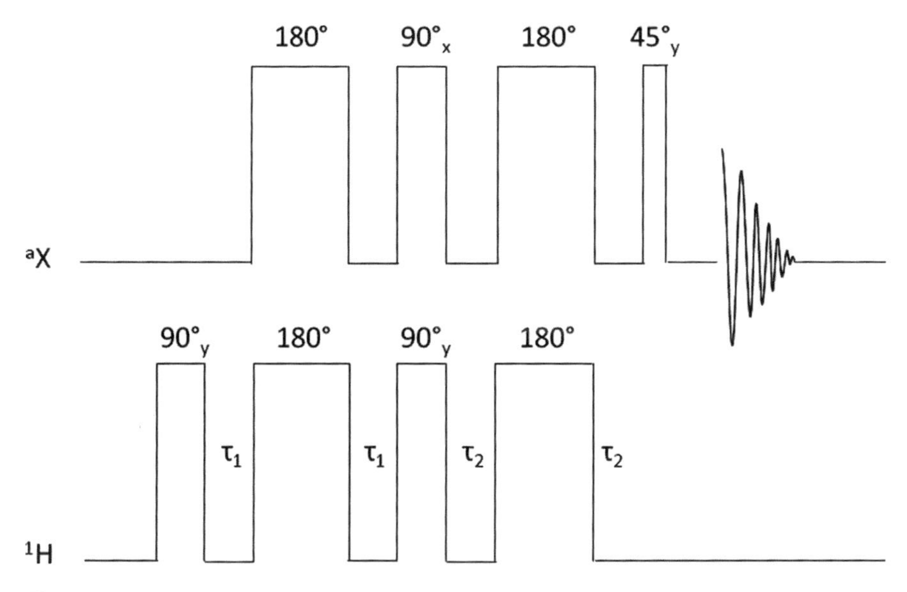

Fig. 5 Pulse sequence for PH-INADEQUATE.

polarisation to be transferred exclusively to selected pairs. When τ_1 and τ_2 were both set to 0.0032 s polarisation was transferred to the two olefinic carbon nuclei. Use of a longer time of 0.028 s reduced the selectivity of the transfer and, as such, three pairs of carbon nuclei were detected.

Cowley et al. have reported on the use of INADEQUATE in conjunction with SABRE [20]. The traditional INADEQUATE sequence rather than PH-INADEQUATE was employed. The *ortho* and *meta* ^{13}C NMR resonances of 4-picoline were detected with a common $J_{13C-13C}$ splitting of 54 Hz. This was obtained in just four scans using an automated polarizer system to prepare and transfer the hyperpolarised material.

3.4 HMQC

The HMQC sequence has been modified for use with *para*hydrogen derived polarisation transfer experiments under the high field regime by changing the sequence such that the initial $\pi/2$ pulse is replaced by a $\pi/4$ pulse. Duckett and co-workers have used such a sequence to probe the reaction chemistry of [Rh(PMe$_3$)$_3$Cl] and [Rh(PMe$_3$)$_4$]Cl with hydrogen [55]. Using the modified HMQC sequence in conjunction with *para*hydrogen, the ^{31}P and ^{103}Rh nuclei of the addition products were detected indirectly due to their coupling to ^{1}H. This demonstrates how PHIP can be used to detect, with respect to ^{103}Rh, nuclei that have low magnetogyric ratios and would therefore not be routinely probed. Couplings to ^{195}Pt have been detected via the use of this modified HMQC sequence [26]. Even when less than 1 mg of the ^{195}Pt

Fig. 6 *tcc*- and *ccc*-forms of $[Ru(CO)_2(PR_3)_2(H)_2]$

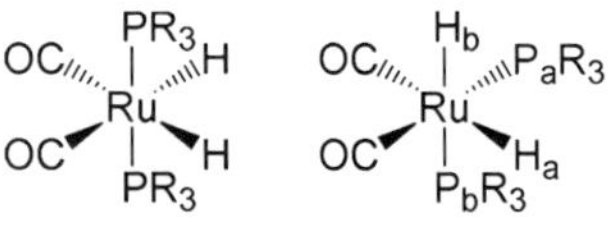

tcc-form *ccc*-form

containing complex is used, $[Pt((dpp)_2mop)(Cl)_2]$ ($((dpp)_2mop)$ is $Ph_2PCH_2CH(Me)OPPh_2$), 2D spectra were obtained in 2–5 min. From this, a pair of doublets of doublets was observed due to coupling to the magnetically inequivalent ^{31}P nuclei of the ligand.

This modified HMQC sequence has also been used to probe directly the ligand sphere of a Ru(II) dihydride [56]. Examination of the spectra produced highlighted that, whilst H_b in the *ccc*-form of the complex (see Fig. 6) is *trans* to P_b and *cis* to P_a, H_a is *cis* to both P_a and P_b and is, therefore, *trans* to CO. The presence of an additional cross peak connects the hydride and phosphorus resonances of the *tcc*-form of the complex. Using normal hydrogen, the *ccc*-form is not observable, thus highlighting the importance of using *para*hydrogen in these studies, especially in conjunction with 2D methods.

3.5 OPSY

An NMR method which efficiently removes signals derived from nuclei with thermally equilibrated spin state populations whilst leaving those signals derived from PHIP has been reported. Gradient assisted coherence selection is used to achieve the discrimination required. This technique, known as OPSY (only *para*hydrogen spectroscopy) [57], relies on the generation of the $2I_xS_x$ term produced for *para*hydrogen derived nuclei after the first $\pi/2$ pulse. Subsequent precession during gradient pulse and evolution by the application of a second $\pi/2$ pulse and a further gradient pulse generates terms of the type I_zS_y which are observable (see Fig. 7 for pulse sequence). The application of this technique was validated by monitoring the hydrogenation reaction of 4-vinylcyclohexene to 4-ethylcyclochexene catalysed by [Rh(COD)(dppb)][BF_4]. Comparison of the OPSY-DQ filtered spectrum with that of the ^1H PHIP NMR spectrum showed very clearly the polarised signals of the methyl group of the hydrogenated product and the ring protons of the alternative hydrogenation product vinylcyclohexane. The most important observation in the OPSY-DQ filtered spectrum was, however, the observation of the polarised signal belonging to the methylene resonance of 4-ethylcyclohexene and a further enhanced ring proton peak of vinylcyclohexane. Both of these signals were obscured by thermal signals in the ^1H PHIP NMR spectrum. The OPSY filter was also applied to a COSY pulse sequence. The 2D spectrum obtained showed strong peaks that arise from the major hydrogenation product, 4-ethylcyclohexene, and weaker peaks for the minor product vinylcyclohexane.

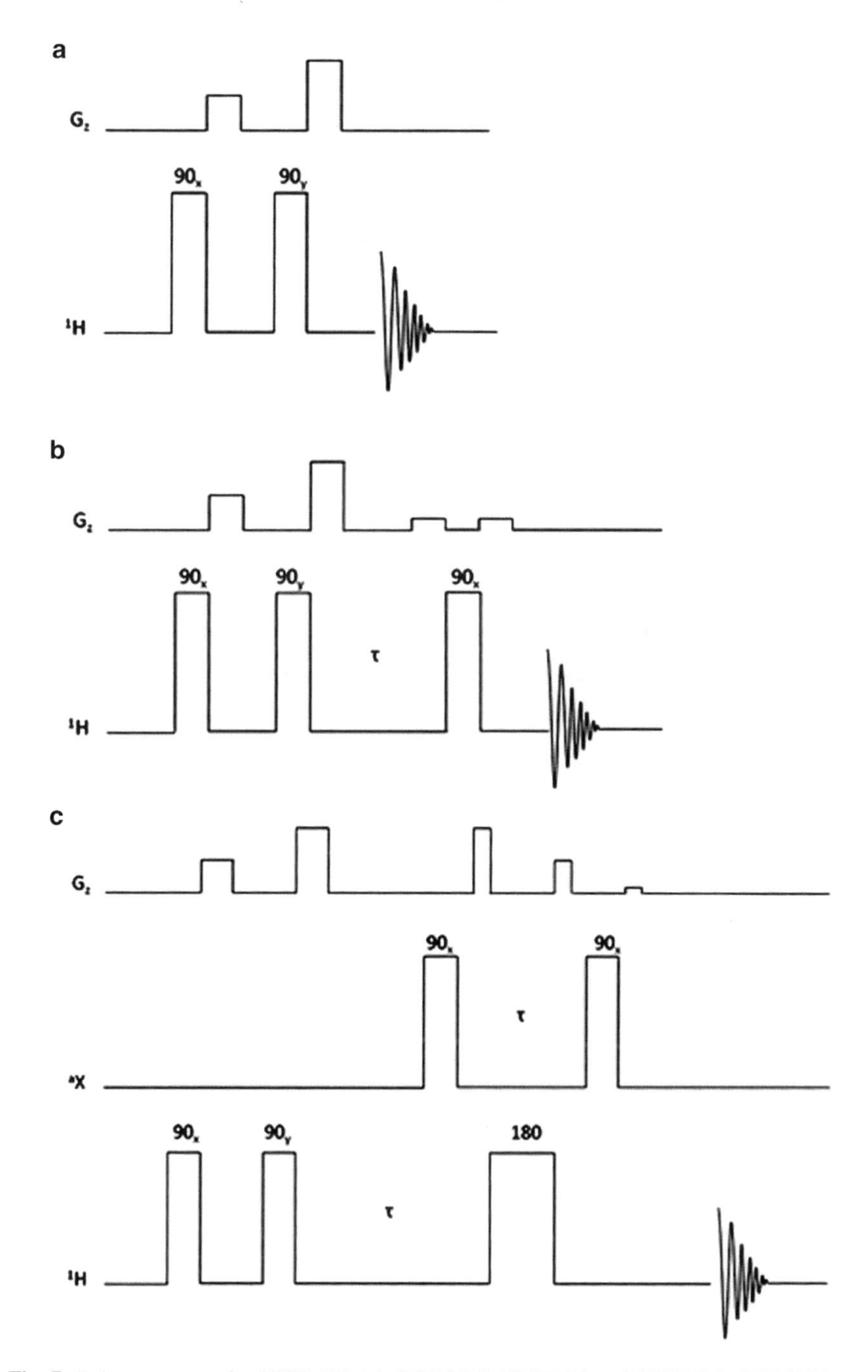

Fig. 7 Pulse sequences for OPSY-DQ (**a**), OPSY-DQ COSY (**b**) and OPSY-DQ HMBC (**c**)

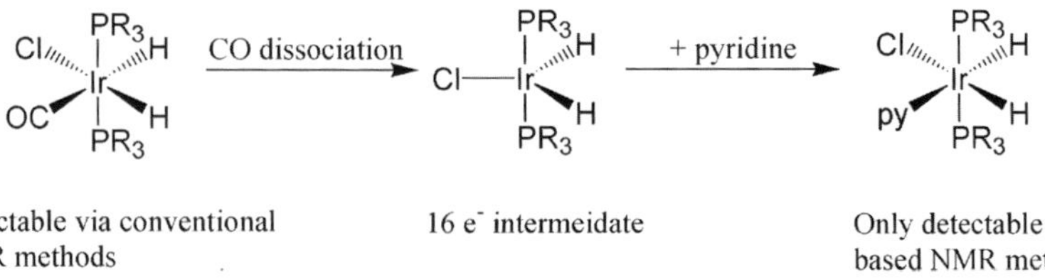

CO dissociation + pyridine

Detectable via conventional NMR methods

16 e⁻ intermeidate

Only detectable via PHIP based NMR methods

Fig. 8 Scheme showing the formation of [Ir(py)(P(*p*-$MeC_6H_4)_3)_2(H)_2Cl$] via CO dissociation and subsequent trapping with pyridine (R = *p*-MeC_6H_4)

Tang et al. report on the employment of the OPSY sequence to select only those signals belonging to methyl 2-acetamidopropanote (MAP), the hydrogenation product of methyl 2-acetamidoacrylate (MAA) [58]. The rhodium catalysed reaction was conducted in sodium dodecyl sulphate (SDS) and D_2O. The OPSY sequence enabled the suppression of those signals derived from the solvent and SDS, thus entailing the observation of CH_2 and CH_3 resonances of MAP. This pair of signals was interrogated using the OPSY sequence repeatedly in order to ascertain the longevity of the polarised signal. Lifetimes exceeding that of T_1 were measured, with signal still being observed after 147 s. The slow continuous diffusion of hydrogen through the ionic liquid maintains a steady state of hydrogenated product being produced by PASADENA, thus prolonging the lifetime of the hyperpolarised signal.

The pulse sequence family utilising the OPSY filter has been recently expanded to include HMBC [59]. In a comparison of a conventional $\pi/4$ ^1H-^{31}P HMBC with that of an OPSY-ZQ HMBC for a sample containing [Ir(CO)(P(*p*-$MeC_6H_4)_3)_2Cl$] and pyridine in d_6-benzene, initially both spectra exhibit two sets of enhanced signals; one set is for *trans*-[Ir(CO)(P(*p*-$MeC_6H_4)_3)_2(H)_2Cl$], the other for [Ir(py)(P(*p*-$MeC_6H_4)_3)_2(H)_2Cl$]. The former complex is dominant whereas the latter is only detectable via the PHIP effect. After the reaction has been allowed to progress, the presence of both hyperpolarised and non-hyperpolarised signals are detected in the $\pi/4$ ^1H-^{31}P HMBC experiment, thus making identification of PHIP derived signals difficult and, furthermore, making it difficult to ascertain whether the reaction is still occurring. Conversely, only one set of signals are detected in the OPSY-ZQ HMBC experiment – those belonging to [Ir(py)(P(*p*-$MeC_6H_4)_3)_2(H)_2Cl$] formed by CO dissociation from *trans*-[Ir(CO)(P(*p*-$MeC_6H_4)_3)_2(H)_2Cl$] and subsequent trapping with pyridine (Fig. 8).

4 Applications of PHIP to MRI

Applications of PHIP have not just been restricted to NMR applications. The inherent ability of *para*hydrogen to magnify the signal response generated by an analyte of interest has found widespread utility in MRI. Here, PHIP has been used to create contrast agents which are themselves hyperpolarised in order to monitor their biodistribution and to gain significant increases in signal intensity. It has also been implemented in the study of hydrogenation reactions within the gas phase.

4.1 Production of Hyperpolarised Contrast Agents

One of the main advantages of using PHIP is that the resulting hyperpolarised molecules have relatively long T_1 values. This allows for the injection of a hyperpolarised bolus into a subject in order to study the biolocalisation. Predominantly the use of ^{13}C has been the nucleus of choice, the advantageous nature of its long T_1 values making it highly applicable to MRI applications. However, there is still much interest in the use of ^{1}H as the detected nucleus, as clinics worldwide routinely collect such images. There is also interest in ^{19}F as it provides a background-free active marker in images whilst also being a component in many drug molecules, e.g. FDG which is used in PET (positron emission tomography).

The use of ^{19}F as a detectable spin nucleus in MRI has been explored by Bommerich and co-workers [60]. Manipulation of the spin states at sub-Earth fields via field cycling enabled a strong coupling to be formed between ^{19}F and ^{1}H of 3-fluorophenylacetylene, thus entailing the propagation of polarisation from ^{1}H to ^{19}F. ^{19}F imaging experiments, performed in a 4.7-T magnet using FLASH and a flip angle of 5°, showed relatively good intensity up until ~30 s. After 1 min, thermal equilibrium had been restored. However, to obtain the same S/N as possessed by the first image in the series (acquired after 9 s) would require 2,500 scans. Thus a substantial improvement in the signal intensity is obtained by the use of PHIP in conjunction with field cycling.

Using apparatus described in Sect. 1.3 and elsewhere [16], Ross and co-workers have reported on the successful collection of 3D hyperpolarised ^{13}C spectra in phantoms and mice models [61]. An in vivo ^{13}C sub-second image was presented of the rat brain. The image was collected 9 s after close-arterial injection of 1 mL of 25 mM of hyperpolarised succinate. Prior to this, a series of 16 experiments had demonstrated that the polarisation obtained using their apparatus was very stable; the average polarisation measured over the 16 runs was 15.3 $\pm$ 1.9%.

The potential that SABRE offers in terms of MRI is an exciting recent development. As SABRE does not chemically modify the substrate, there is the potential to polarise approved drug molecules or molecules which are found naturally in the human body. This circumvents the need to find unsaturated precursors as would be required for normal PHIP type experiments in which the molecule needs to be hydrogenated in order to polarise it. Furthermore, the scope for the types of molecules that could be used as biological tracers is widened. Thus far, True-FISP MR images of an 8-mm tube containing 1-mm internal diameter cylinders have been successfully imaged which have a 160-fold increase in signal intensity over those collected using thermal equilibrium magnetisation [11]. In reality, the signal intensity is far greater than 160-fold when the slice thickness is accounted for; the slice thickness of the hyperpolarised image is 40 times less than that of the scan obtained using thermal magnetisation.

Methods have been developed to overcome the efficient relaxation processes that exist in liquids after polarisation transfer. Most NMR measurements are performed using a "batch process" in that magnetisation is read out shortly after

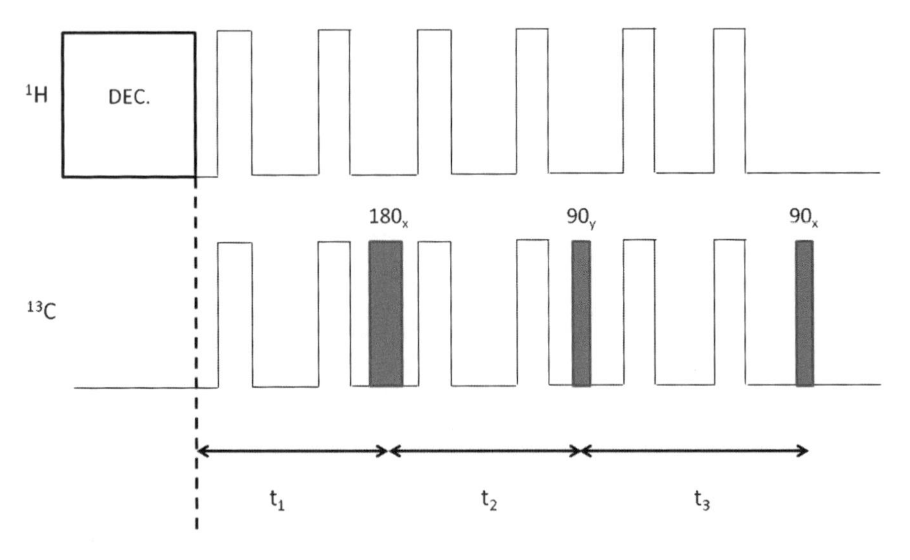

Fig. 9 Pulse sequence for polarizing hydroxyethylpropionate

preparation. Use of small flip angles can maximise the longevity of the polarised state, especially for 2D methods. Roth et al. have reported on the use of hollow fibre membranes (HFM) to introduce *para*hydrogen to a solution [62]. The use of HFM also circumvents the normally small liquid-gas interface of the hydrogen bubbles. Using a circular system enables the hydrogenation process to be maintained in conjunction with a continuous sampling of the magnetisation generated, using either 1D or 2D spectroscopic techniques. An exemplar of this method enabled a hyperpolarised ^{1}H-^{1}H COSY to be collected in ~7 min compared with ~58 min for the analogous thermal spectrum.

The advent of ^{13}C imaging in MR, using PHIP, can be approached from the inclusion of labelled ^{13}C nuclei in the prospective contrast agent and/ or implementation of field cycling to transfer spin polarisation from ^{1}H to ^{13}C. Goldman et al. discuss the use of RF, either CW or pulses, to achieve the same order transfer [63]. When considering hydroxyethylpropionate, a pulse sequence is described for the production of a polarisation level of 83.7% (Fig. 9). In reality, the maximum polarisation was about 49–50%; the difference is due to the departure of the RF pulses from ideal conditions. However, compared with field cycling in which the theoretical maximum is 28% (and observed was 21–25%), the use of pulses offers a distinct advantage in this respect.

The design of the unsaturated molecule which is to be hydrogenated can be tailored to increase its resulting bio-applicability. The incorporation of glucose is one design feature which has been evaluated by Reineri et al. [64]. Several glucose and glucosamine derivatives were synthesised which possessed alkyne moieties as suitable sites for hydrogenation. Interestingly, the presence of an amide functionality was found to have a negative impact on both the hydrogenation process and the subsequent attainment of PHIP effects in the NMR spectrum. Ester derivatives

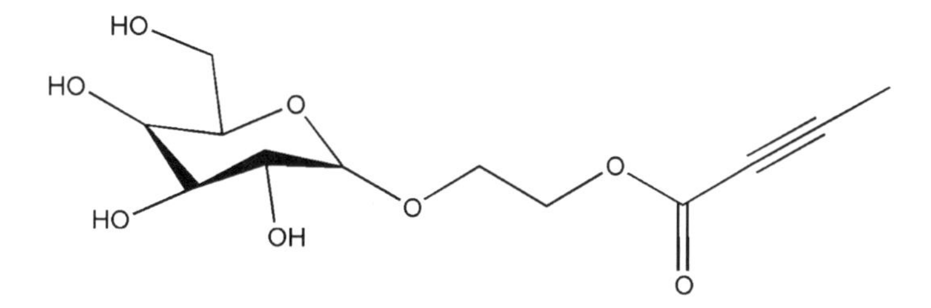

Fig. 10 Chemical structure of the glucose alkyne derivative incorporating an ethylene spacer

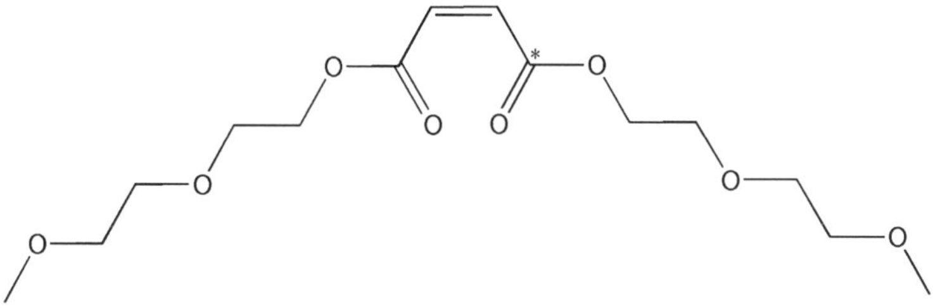

Fig. 11 Structure of the oligooxyethylenic alkyne; *asterisk* denotes the location of the ^{13}C atom

conversely possessed good *para*hydrogenation yields (an example of a glucose alkyne derivative hydrogenated is shown in Fig. 10) and ^{13}C enhancement after field cycling was employed. T_1 measurements showed that the incorporation of a short ethyleneglycol spacer improved the T_1 value of ^{13}C carbonyl to 21 s at 9.4 T. This is significant, as it means that there is the opportunity for the molecule to localise within tissues during in vivo experiments, prior to data acquisition.

In a similar vein, oligooxyethylenic alkynes have been synthesized and investigated via MRI [65]. The solubility of the alkyne shown in Fig. 11 enabled MRI measurements to be concluded in aqueous media. In the presence of oxygen, the T_1 of the hydrogenated species was 33 s in a field of 0.5 T for the ^{13}C labelled atom of the carbonyl functional group in the symmetrical alkene. Field cycling was again employed to produce magnetisation of the I_z type predominantly on carbon. RARE images of a 130 mM hyperpolarised solution in acetone gave good S/N. The corresponding hyperpolarised image in water gave considerably lower S/N, but did involve a fast distillation procedure to remove the acetone prior to acquisition, and thus relaxation was far more important.

A more recent contribution to this area has focussed on the production of hyperpolarised 1-^{13}C-phospholactate [66]. This report investigated the feasibility of hydrogenating phospholactate using a rhodium water soluble catalyst, the T_1 of the resulting ^{13}C hyperpolarisation and the characterization of the homo- and heteronuclear spin–spin couplings. The ^{13}C labelled nucleus was found to have a T_1 of 36 $\pm$ 2 s at 0.0475 T after polarisation via PASADENA. In the ^{13}C NMR spectrum, after hyperpolarisation transfer, a 4,000-fold gain in sensitivity was achieved at 3 T; this corresponded to a 1% polarisation level. An exploration of the couplings within ^{13}C-phospholactate suggested that perdeuteration of the precursor ^{13}C-phosphoenolpyruvate would be a more viable contrast agent but more challenging to synthesise. Deuteration would result in an increased T_1 and, as the authors suggest, reduce the spin system to only three spins which participate in polarisation transfer dynamics. This three spin system is perfectly suited for use with the polarisation transfer described by Goldman and Johannesson [67], which converts the proton pair *para* order derived from *para*hydrogen into ^{13}C polarisation using RF pulses. Thus, the increase in sensitivity would be greater, therefore enabling this molecule to be tracked more easily in vivo.

4.2 Imaging of Hydrogenation Reactions

Pines and co-workers have shown how PHIP can be used to image catalytic hydrogenation in micro-reactors [68]. The conversion of propylene to propane catalysed by Wilkinson's catalyst immobilized on a tightly packed bed of silica allowed flow maps, active regions in the catalyst bed and controlled transport of polarisation to be resolved. These experiments highlight the applicability of PHIP to gain an insight into this type of reaction, and show that the MR aspect of this field is not restricted to advancing clinical diagnostics. Furthermore, if coupled with methods to transfer polarisation to heteronuclei, additional insight could be obtained using the methodology described.

Prior to this report, Pines and co-workers had presented cross sectional images of hyperpolarised propylene gas through a bed of glass capillaries [69]. Hyperpolarised propylene was produced under the ALTADENA condition, which led to an enhancement factor of 300 relative to thermally polarised gas. The random packing of the capillaries was used to demonstrate that gas phase MRI could be visualised through a porous medium.

5 Conclusion

The utilisation of PHIP in both NMR and MRI based investigations facilitates a greater understanding about reaction mechanisms, allows for better structural characterisation and produces much improved signal intensity, thus curtailing the number of scans required to obtain the same spectrum or image. Reactions can be conducted in either homogeneous or heterogeneous media. The ease with which *para*hydrogen may be introduced to the sample of interest, coupled with the increased technological advances in polarizers for preparing and delivering the hyperpolarised material, means that PHIP has an ever-increasing utility.

In terms of the future, there is much promise for PHIP, especially when considering recent developments such as SABRE and the development of hyperpolarised contrast agents. Pulse sequence design will also have a major role to play in the development of new techniques and increasing the applicability of existing ones in NMR and MRI. Hyperpolarised contrast agent design was discussed along with pulse sequences for optimising the observable signal obtainable from clinical images. SABRE was highlighted as an exciting prospect as it offers a route to hyperpolarise molecules without the need for hydrogenating them, thus increasing enormously the range of analytes which can be investigated.

Acknowledgments The authors would like to thank the EPSRC (grant number EP/G009546/1) and the University of York for their financial support.

References

1. Brewer W, Kopp M (1976) Hyperfine Interact 2:299
2. Ardenkjaer-Larsen JH, Fridlund B, Gram A, Hansson G, Hansson L, Lerche MH, Servin R, Thaning M, Golman K (2003) PNAS 100:10158
3. Eisenschmid TC, Kirss RU, Deutsch PP, Hommeltoft SI, Eisenberg R (1987) J Am Chem Soc 109:8089
4. Eisenberg R (1991) Acc Chem Res 24:110
5. Natterer J, Bargon J (1997) Prog Nucl Magn Reson Spectrosc 31:293
6. Duckett SB, Mewis RE (2012) Acc Chem Res. http://dx.doi.org/10.1021/ar2003094
7. Green RA, Adams RW, Duckett SB, Mewis RE, Williamson DC, Green GGR (2012) Prog Nucl Magn Reson Spectrosc. http://dx.doi.org/10.1016/j.pnmrs.2012.03.011
8. Bowers CR, Weitekamp DP (1986) Phys Rev Lett 57:2645
9. Bowers CR, Weitekamp DP (1987) J Am Chem Soc 109:5541
10. Pravica MG, Weitekamp DP (1988) Chem Phys Lett 145:255
11. Adams RW, Aguilar JA, Atkinson KD, Cowley MJ, Elliott PIP, Duckett SB, Green GGR, Khazal IG, Lopez-Serrano J, Williamson DC (2009) Science 323:1708
12. Duckett SB, Wood NJ (2008) Coord Chem Rev 252:2278
13. Taylor HS, Diamond H (1935) J Am Chem Soc 57:1251
14. Gamliel A, Allouche-Arnon H, Nalbandian R, Barzilay CM, Gomori JM, Katz-Brull R (2010) Appl Magn Reson 39:329
15. Feng B, Coffey AM, Colon RD, Chekmenev EY, Waddell KW (2012) J Magn Reson 214:258
16. Hovener JB, Chekmenev EY, Harris KC, Perman WH, Robertson LW, Ross BD, Bhattacharya P (2009) Magn Reson Mater Phys Biol Med 22:111
17. Hovener JB, Chekmenev EY, Harris KC, Perman WH, Tran TT, Ross BD, Bhattacharya P (2009) Magn Reson Mater Phys Biol Med 22:123
18. Giernoth R, Huebler P, Bargon J (1998) Angew Chem Int Ed 37:2473
19. Hubler P, Giernoth R, Kummerle G, Bargon J (1999) J Am Chem Soc 121:5311
20. Cowley MJ, Adams RW, Atkinson KD, Cockett MCR, Duckett SB, Green GGR, Lohman JAB, Kerssebaum R, Kilgour D, Mewis RE (2011) J Am Chem Soc 133:6134
21. Waddell KW, Coffey AM, Chekmenev EY (2010) J Am Chem Soc 133:97
22. Duckett SB, Blazina D (2003) Eur J Inorg Chem 2003:2901
23. Boutain M, Duckett SB, Dunne JP, Godard C, Hernandez JM, Holmes AJ, Khazal IG, Lopez-Serrano J (2010) Dalton Trans 39:3495

24. Lopez-Serrano J, Duckett SB, Dunne JP, Godard C, Whitwood AC (2008) Dalton Trans 4270
25. Duckett SB, Newell CL, Eisenberg R (1994) J Am Chem Soc 116:10548
26. Jang M, Duckett SB, Eisenberg R (1996) Organometallics 15:2863
27. Trantzschel T, Bernarding J, Plaumann M, Lego D, Gutmann T, Ratajczyk T, Dillenberger S, Buntkowsky G, Bargon J, Bommerich U (2012) PCCP, 14:5601
28. Blazina D, Duckett SB, Halstead TK, Kozak CM, Taylor RJK, Anwar MS, Jones JA, Carteret HA (2005) Magn Reson Chem 43:200
29. Blazina D, Dunne JP, Aiken S, Duckett SB, Elkington C, McGrady JE, Poli R, Walton SJ, Anwar MS, Jones JA, Carteret HA (2006) Dalton Trans 2072
30. Dunne JP, Blazina D, Aiken S, Carteret HA, Duckett SB, Jones JA, Poli R, Whitwood AC (2004) Dalton Trans 3616
31. Hasnip S, Duckett SB, Taylor DR, Taylor MJ (1998) Chem Commun 923
32. Godard C, Duckett SB, Henry C, Polas S, Toose R, Whitwood AC (2004) Chem Commun 1826
33. Fox DJ, Duckett SB, Flaschenriem C, Brennessel WW, Schneider J, Gunay A, Eisenberg R (2006) Inorg Chem 45:7197
34. Permin AB, Eisenberg R (2002) J Am Chem Soc 124:12406
35. Godard C, Duckett SB, Polas S, Tooze R, Whitwood AC (2005) J Am Chem Soc 127:4994
36. Godard C, Duckett SB, Polas S, Tooze R, Whitwood AC (2009) Dalton Trans 2496
37. Carson PJ, Bowers CR, Weitekamp DP (2001) J Am Chem Soc 123:11821
38. Balu AM, Duckett SB, Luque R (2009) Dalton Trans 5074
39. Koptyug IV, Zhivonitko VV, Kovtunov KV (2010) Chem. Phys. Chem. 11:3086
40. Adams RW, Duckett SB, Green RA, Williamson DC, Green GGR (2009) J Chem Phys 131
41. Atkinson KD, Cowley MJ, Duckett SB, Elliott PIP, Green GGR, Lopez-Serrano J, Khazal IG, Whitwood AC (2009) Inorg Chem 48:663
42. Atkinson KD, Cowley MJ, Elliott PIP, Duckett SB, Green GGR, Lopez-Serrano J, Whitwood AC (2009) J Am Chem Soc 131:13362
43. Gong Q, Gordji-Nejad A, Blümich B, Appelt S (2010) Anal Chem 82:7078
44. Glöggler S, Emondts M, Colell J, Müller R, Blümich B, Appelt S (2011) Analyst 136:1566
45. Glöggler S, Müller R, Colell J, Emondts M, Dabrowski M, Blümich B, Appelt S (2011) PCCP 13:13759
46. Wood NJ, Brannigan JA, Duckett SB, Heath SL, Wagstafft J (2007) J Am Chem Soc 129:11012
47. Kuhn LT, Bargon J (2007) Top Curr Chem 276:25
48. Eisenschmid TC, McDonald J, Eisenberg R, Lawler RG (1989) J Am Chem Soc 111:7267
49. Sørensen OW, Ernst RR (1983) J Magn Reson (1969) 51:477
50. Duckett SB, Newell CL, Eisenberg R (1993) J Am Chem Soc 115:1156
51. Haake M, Natterer J, Bargon J (1996) J Am Chem Soc 118:8688
52. Sengstschmid H, Freeman R, Barkemeyer J, Bargon J (1996) J Magn Reson, Ser A 120:249
53. Barkemeyer J, Bargon J, Sengstschmid H, Freeman R (1996) J Magn Reson, Ser A 120:129
54. Natterer J, Barkemeyer J, Bargon J (1996) J Magn Reson, Ser A 123:253
55. Duckett SB, Barlow GK, Partridge MG, Messerle BA (1995) J Chem Soc-Dalton Trans 3427
56. Duckett SB, Mawby RJ, Partridge MG (1996) Chem Commun 383
57. Aguilar JA, Elliott PIP, Lopez-Serrano J, Adams RW, Duckett SB (2007) Chem Commun 1183
58. Tang JA, Gruppi F, Fleysher R, Sodickson DK, Canary JW, Jerschow A (2011) Chem Commun 47:958
59. Aguilar JA, Adams RW, Duckett SB, Green GGR, Kandiah R (2011) J Magn Reson 208:49
60. Bommerich U, Trantzschel T, Mulla-Osman S, Buntkowsky G, Bargon J, Bernarding J (2010) PCCP 12:10309
61. Bhattacharya P, Harris K, Lin AP, Mansson M, Norton VA, Perman WH, Weitekamp DP, Ross BD (2005) Magn Reson Mater Phys Biol Med 18:245
62. Roth M, Kindervater P, Raich H-P, Bargon J, Spiess HW, Münnemann K (2010) Angew Chem Int Ed 49:8358
63. Goldman M, Johannesson H, Axelsson O, Karlsson M (2006) C R Chim 9:357

64. Reineri F, Santelia D, Viale A, Cerutti E, Poggi L, Tichy T, Premkumar SSD, Gobetto R, Aime S (2010) J Am Chem Soc 132:7186
65. Reineri F, Viale A, Giovenzana G, Santelia D, Dastru W, Gobetto R, Aime S (2008) J Am Chem Soc 130:15047
66. Shchepin RV, Coffey AM, Waddell KW, Chekmenev EY (2012) J Am Chem Soc 134:3957
67. Goldman M, Johannesson H (2005) C R Phys 6:575
68. Bouchard L-S, Burt SR, Anwar MS, Kovtunov KV, Koptyug IV, Pines A (2008) Science 319:442
69. Bouchard L-S, Kovtunov KV, Burt SR, Anwar MS, Koptyug IV, Sagdeev RZ, Pines A (2007) Angew Chem Int Ed 46:4064

Top Curr Chem (2013) 338: 105–122
DOI: 10.1007/128_2012_357

Published online: 6 September 2012

The Solid-State Photo-CIDNP Effect and Its Analytical Application

Photo-CIDNP MAS NMR to Study Radical Pairs

Bela E. Bode, Smitha Surendran Thamarath, Karthick Babu Sai Sankar Gupta, A. Alia, Gunnar Jeschke, and Jörg Matysik

Abstract Photochemically induced dynamic nuclear polarization (photo-CIDNP) is an effect that produces non-Boltzmann nuclear spin polarization which can be observed as modification of signal intensity in NMR spectroscopy. The effect is well known in liquid-state NMR where it is explained most generally by the classical radical pair mechanism (RPM). In the solid-state, other mechanisms are operative in the spin-dynamics of radical pairs such as three-spin mixing (TSM) and differential decay (DD). Initially the solid-state photo-CIDNP effect has been solely observed on natural photosynthetic reaction centers (RCs). Therefore the analytical capacity of the method has been explored in experiments on reaction centers (RCs) of the purple bacterium of *Rhodobacter* (*R.*) *sphaeroides*. Here we will provide an account on phenomenology, theory, and analytical capacity of the solid-state photo-CIDNP effect.

Keywords Spin hyperpolarisation · Photo-CIDNP · Photosynthesis · Photochemistry · Spin-chemistry

B.E. Bode
Leiden Institute of Chemistry, Einsteinweg 55, 2300 RA Leiden, The Netherlands

EaStCHEM School of Chemistry and Biomedical Sciences Research Complex, University of St Andrews, St Andrews, KY16 9ST Fife, Scotland, UK

S.S. Thamarath, K.B.S.S. Gupta, A. Alia, and J. Matysik (✉)
Leiden Institute of Chemistry, Einsteinweg 55, 2300 RA Leiden, The Netherlands
e-mail: j.matysik@chem.leidenuniv.nl

G. Jeschke
Institut für Physikalische Chemie, Eidgenössische Technische Hochschule, Zürich, Switzerland

Contents

1 The Solid-State Photochemically Induced Dynamic Nuclear Polarization Effect

The solid-state photochemically induced dynamic nuclear polarization (photo-CIDNP) effect, discovered by Zysmilich and McDermott in 1994 [1], allows for signal enhancement by factors of several 10,000 s for ^{13}C magic angle spinning (MAS) NMR at a magnetic field of 4.7 T (i.e., 200 MHz ^{1}H frequency) in reaction centers (RCs) of the purple bacteria *Rhodobacter* (*R.*) *sphaeroides* wildtype (WT) [2] and the carotenoid-less mutant R26 [3]. Such strong signal enhancement allows, for example, selectively observing photosynthetic cofactors to form radical-pairs at nanomolar concentrations in membranes, cells [3, 4] and entire plants (G.J. Janssen et al., unpublished). Due to the long ^{13}C relaxation time in solids, the nuclear polarization of subsequent photocycles can be accumulated in continuous illumination experiments, making photo-CIDNP MAS NMR a sensitive analytical tool for studying radical pairs [5–8]. The effect has been observed in all natural photosynthetic RCs studied so far [9], and for a long time, despite great efforts, experiments on other systems have failed. Recently we observed the effect in a blue-light photoreceptor, the phototropin mutant LOV1-C57S [10]. The short lifetime of the radical pair in natural photosynthesis (some 10 ns) is leading to broad matching windows which facilitate the observation of the effect. On the other hand, blue-light photoreceptors with radical-pair lifetimes in the microsecond range show sharp matching windows (Fig. 1) and are therefore more difficult to approach with standard NMR equipment.

From RCs of *R. sphaeroides*, the best investigated photosynthetic RC (for review see [11, 12]), the field-dependence of the amplitude of the solid-state photo-CIDNP effect has been studied in the range from 1.4 to 17.6 T in WT and R26, respectively (Figs. 2, 3, 4, and 5) [13]. Within the entire field regime, the intensity patterns are different between spectra of WT and R26 RCs. While the first are entirely emissive spectra, in the latter both emissive and absorptive lines occur. The optimum for the spectral resolution is reached at about 4.7 T since at 2.4 T the spectral dispersion becomes too poor, while at higher field artificial line-broadening is required to compensate for lower signal-to-noise ratio. WT RCs show the maximum enhancement around 2.4 T (Figs. 2 and 4) corresponding to the high-field matching window of the solid-state photo-CIDNP effect [by the TSM and DD mechanisms; see Sect. 2]. On the other hand, for R26 RCs the maximum enhancement is not reached at 1.4 T (Figs. 3 and 5) and the ratio between positive and

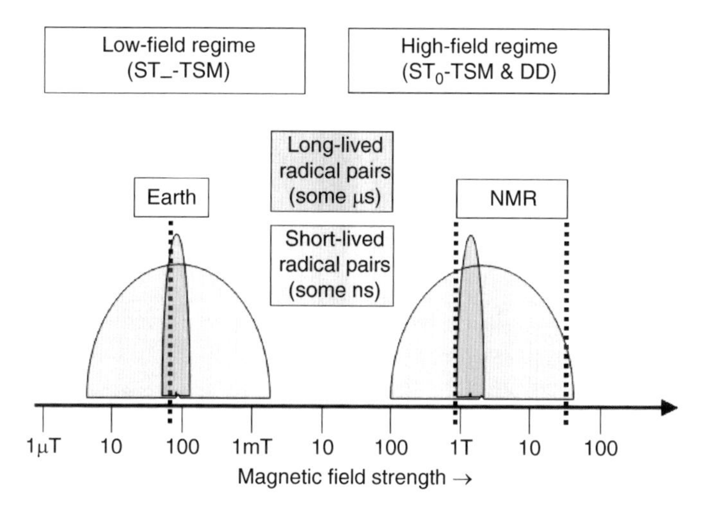

Fig. 1 Signal enhancement (*vertical axis*) obtained by the solid-state photo-CIDNP effect due to three-spin mixing (TSM) depending on the strength of the magnetic field (*horizontal axis*) and the lifetime of the radical pair (broad vs thin semispheres)

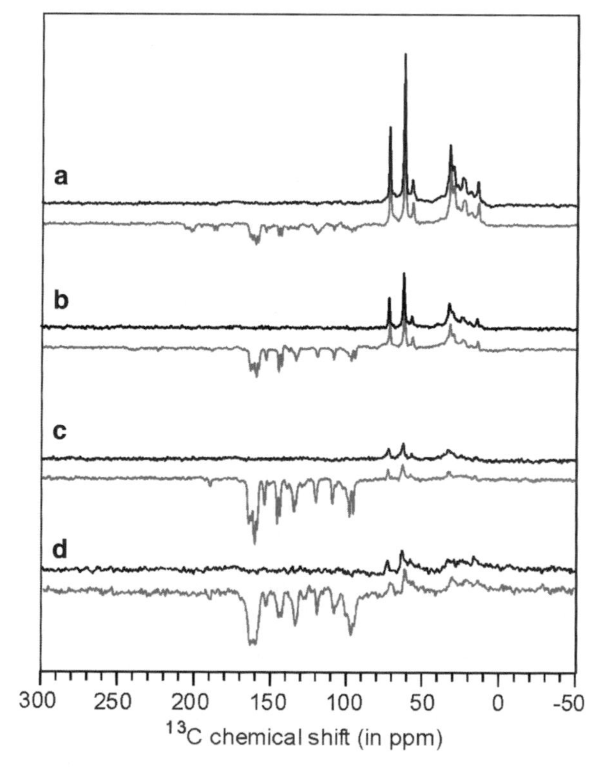

Fig. 2 ^{13}C MAS NMR spectra of quinone depleted RCs of *R. sphaeroides* WT in the dark (*black*) and under illumination (*gray*) at 17.6 T (**a**), 9.4 T (**b**), 4.7 T (**c**), and 2.4 T (**d**). The spectra were obtained at 235 K under an MAS frequency of 8 kHz under ^{1}H decoupling

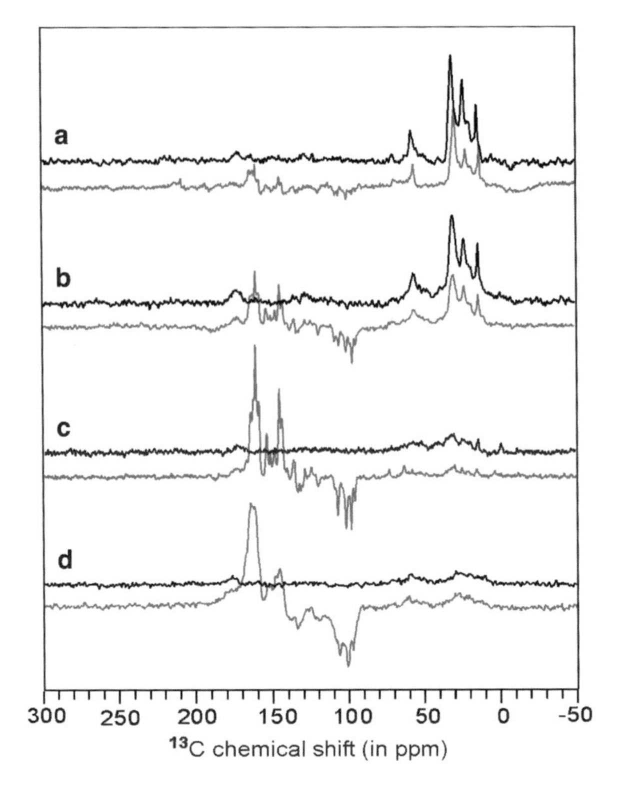

Fig. 3 ^{13}C MAS NMR spectra of quinone depleted RCs of *R. sphaeroides* R26 in the dark (*black*) and under illumination (*gray*) at 17.6 T (**a**), 9.4 T (**b**), 4.7 T (**c**), and 2.4 T (**d**). The spectra were obtained at 235 K under an MAS frequency of 8 kHz under ^{1}H decoupling

negative signals is changed strongly in favor of the first. For 1.4 T we estimate an enhancement factor of at least 80,000 due to the DR mechanism (see Sect. 2). Experiments at even lower fields would require a field cycling system to avoid further loss of resolution.

2 The Solid State Photo-CIDNP Effect: Theory

The cyclic spin-chemical processes producing such high nuclear polarizations are now understood for a spin-correlated radical pair interacting with a single nuclear spin in quinone depleted RCs of *R. sphaeroides* [11, 13]. Under illumination, RCs (Fig. 6) form radical pairs with the primary electron donor P, the so-called "special pair" of two bacteriochlorophylls (BChl) as radical cation, and the primary electron acceptor Φ, a bacteriopheophytin (BPhe), as radical anion (Scheme 1).

The radical pair mechanism (RPM), well established in liquid-state photo-CIDNP, is active in spin-sorting, i.e., enriching one nuclear spin state in one of

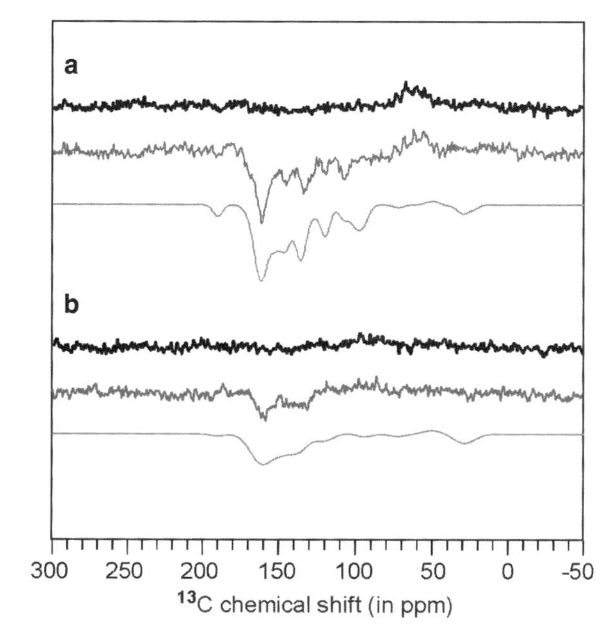

Fig. 4 ^{13}C MAS NMR spectra of quinone depleted RCs of *R. sphaeroides* WT in the dark (*black*) and under illumination (*gray*) at 2.4 T (**a**) and 1.4 T (**b**). The spectra were obtained at 235 K under an MAS frequency of 8 kHz without ^{1}H decoupling

the two decay channels of the radical pair and depleting it in the other. Since the product state in Scheme 1 is identical for both branches of the radical-pair decay, in steady-state experiments using continuous illumination the process of spin sorting does not lead to enhanced nuclear polarization. The polarizations arising from the two channels exactly cancel. Time-resolved experiments (Fig. 7, spectra C), however, are able to observe the spin-sorting by the RPM. Figure 7 shows the evolution of signal intensity on the microsecond timescale. Initially positive (absorptive) transient nuclear polarization occurs (light gray) and is visible up to 10 μs. This initial phase is due to the RPM and shows selectively the enriched nuclear polarization on the singlet decay pathway. This polarization is only transiently visible, in this case because the nuclear polarization occurring on the triplet decay pathway is shifted and broadened beyond detection by the nearby paramagnetic carotenoid triplet [14]. After the decay of the transient nuclear polarization from the singlet decay channel, a new pattern occurs, showing an entirely negative (emissive) envelope on the 100-μs timescale. Equilibration of the polarization by spin diffusion on the millisecond timescale leads to the all-emissive steady-state intensity pattern (Fig. 7, spectrum B) [15].

The all-emissive steady-state pattern is caused by two solid-state mechanisms called the three-spin mixing (TSM) [16, 17] and the differential decay (DD) [18]. These mechanisms transfer the initial electron spin zero-quantum coherence, which is created upon birth of the radical pair in the S state in the S–T_0 manifold of states, into

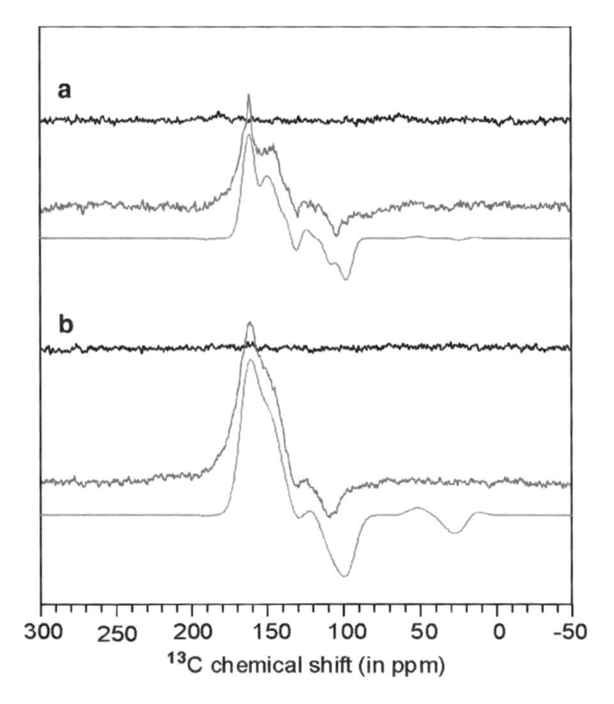

Fig. 5 ^{13}C MAS NMR spectra of quinone depleted RCs of *R. sphaeroides* R26 in the dark (*black*) and under illumination (*gray*) at 2.4 T (**a**) and 1.4 T (**b**). The spectra were obtained at 235 K under an MAS frequency of 8 kHz without 1H decoupling

net nuclear polarization. In the electron–electron-nuclear TSM mechanism, the symmetry of the coherent spin evolution in the correlated radical pair is broken by state mixing due to electron–electron coupling and pseudosecular hyperfine coupling (HFC). State mixing is maximized at the double matching condition $2|\Delta\Omega| = 2|\omega_I| = |A|$, i.e., the difference between the electron Zeeman frequencies $\Delta\Omega$, the nuclear Zeeman frequency ω_I, and the secular part of the hyperfine interaction A must match (Fig. 8). Such magnetization transfer near an avoided level crossing is reminiscent of optical nuclear polarization (ONP) in molecular triplet states. The difference between ONP and solid-state photo-CIDNP lies in the facts that the latter requires a double matching and that the matched interactions are different, so that solid-state photo-CIDNP occurs at the high magnetic fields used in high-resolution solid-state NMR work. In contrast, ONP is insignificant at such fields, where typical zero-field splittings of molecular triplets are much smaller than the electron Zeeman interaction.

In the DD mechanism, the symmetry between the singlet and triplet decay pathways is broken by different lifetimes of the S and of the T_0 states of the radical pair and by pseudosecular HFC. In this case, only a single matching of interactions $2|\omega_I| = |A|$ is required and the difference between singlet and triplet radical pair lifetimes must be of the order of the inverse HFC. During the radical pair evolution the TSM and DD mechanisms in RCs of *R. sphaeroides* WT lead to a set of entirely

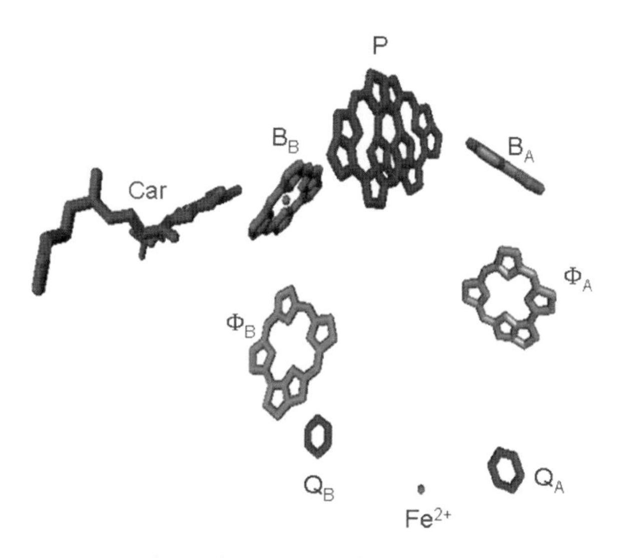

Fig. 6 Cofactor arrangement in reaction centers (RCs) of *R. sphaeroides* wildtype (WT). The primary electron donor *P*, the special pair, transfers an electron selectively into the A branch of cofactors. Reduction of bacteriopheophytin Φ_A leads to the formation of the primary radical pair. Upon removal of the quinones, cyclic electron transfer (Scheme 1) is induced by transient formation of the primary radical pair and subsequent back-transfer. Electron back-transfer leads either to the singlet ground state or the donor triplet state. The lifetime of the donor triplet state depends on presence (WT) or absence of the nearby carotenoid cofactor C

emissive (negative) signals, whose relative intensity encodes information on spin density distribution in the radical pair state [19]. Since $\Delta\Omega$ and ω_I depend on the magnetic field, while *A* does not, both the TSM and DD mechanisms create maximum absolute nuclear polarization at a matching field. Experimentally, the maximum of enhancement is found around 2.4 T (Figs. 2 and 4) which agrees well with the calculated value [20]. The all-emissive spectra of WT RCs are caused by a predominance of the TSM over the DD mechanism, for which the sign of the signal depends on the signs of the secular HFC and of the **g** tensor difference [20].

In RCs having a long donor triplet lifetime as in the carotenoid-less mutant R26 of *R. sphaeroides*, contributions from a third mechanism have been observed [21]. Here the polarization generated by RPM, which has the same amplitude and opposite signs in the singlet and triplet decay branches (and thus usually cancels in steady state experiments), is partially maintained [22, 23] due to different longitudinal nuclear relaxation in the two branches (for review see [24]). In the solid state this has been termed the differential relaxation (DR) mechanism to emphasize that RPM polarization is modified according to the different relaxation rates for different nuclei [20]. This mechanism explains the differences between photo-CIDNP MAS NMR spectra of RCs of *R. sphaeroides* WT and R26 [13]. The DR mechanism relies on enhanced nuclear relaxation in the triplet branch, which is in turn caused by fluctuations of the anisotropic HFCs of these nuclei to the donor triplet (^3P) state. Therefore, relative line intensity in this case also encodes

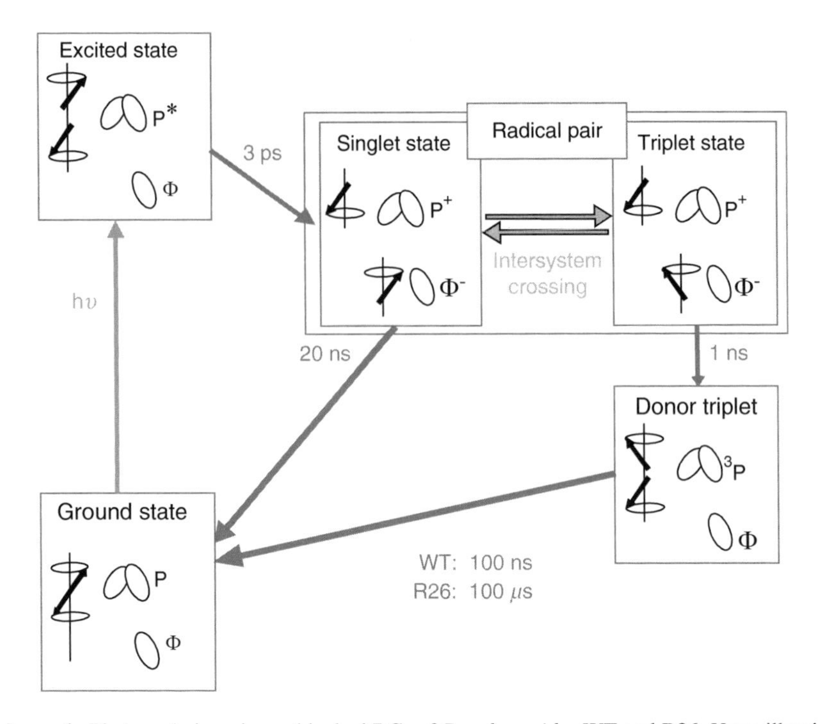

Scheme 1 Photocycle in quinone-blocked RCs of *R. sphaeroides* WT and R26. Upon illumination and fast electron transfer from an excited singlet state, a radical pair is formed in a pure singlet state having electron two-spin order of unity. The radical pair is formed by a radical cation at the two donor BChls (special pair, *P*) and a radical anion on the BPhe acceptor cofactor (Φ) of the active branch. The chemical fate of the radical pair depends on its electronic spin state: while the singlet state is allowed to recombine, for the triplet state a direct recombination to the ground state is spin-forbidden and a donor triplet (^3P) is formed instead. The lifetime of ^3P depends on the relaxation channels provided by the environment. Therefore it is short in WT RCs having a nearby carotenoid and significantly longer in the carotenoid-less mutant R26

information on the electron spin density distribution in the ^3P state [13] (see Sect. 4). The DR polarization dominates at sufficiently low fields and can be experimentally separated from TSM and DD contributions (Fig. 5; 1.4 T for RCs of *R. sphaeroides*). Simulations, using an RPM polarization pattern experimentally obtained from time-resolved experiments (see next paragraph) [25] and DFT-computed HFC values, suggest that the maximum of the DR enhancement is also around 1.5 T. The DR mechanism in the solid state and "cyclic reaction" mechanism [22] in the liquid state are identical except for the fact that they rely on different relaxation mechanisms.

The above discussion is valid for spin dynamics at fields of more than 25 mT, significantly higher than Earth's magnetic field (~50 μT). At fields below 25 mT, an analog coherent low-field TSM has been proposed for cyclic photoreactions due to mixing in the $S–T_-$ or $S–T_+$ manifold leading to significant nuclear polarization up to 10%, almost nine orders of magnitudes larger than the Boltzmann polarization at thermal equilibrium [26]. If the magnitudes of the electron Zeeman interaction, half

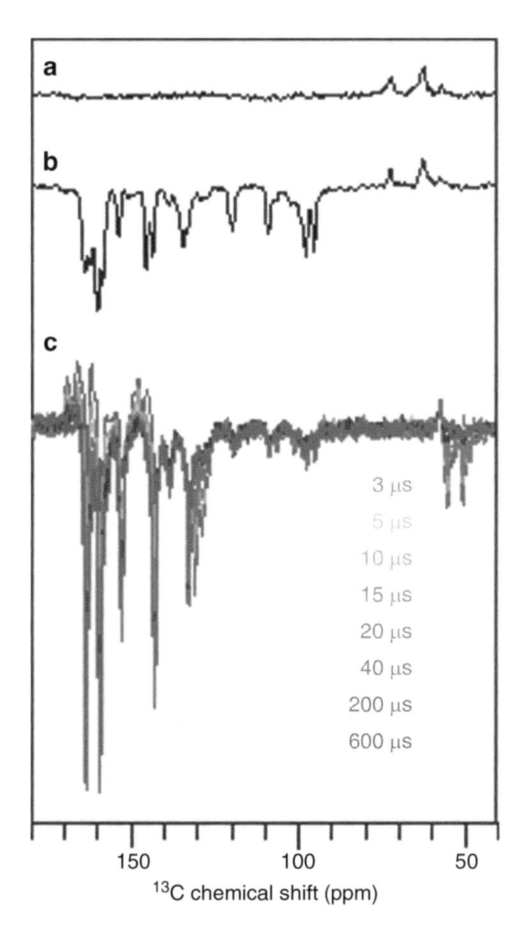

Fig. 7 ^{13}C MAS NMR spectra of WT RCs measured in the dark (**a**), under continuous illumination (**b**), and after a nanosecond-laser flash (**c**). The sample is selectively ^{13}C isotope labeled at the tetrapyrrole cofactors (4-Ala label pattern). The laser pulse length is ~10 ns and the wavelength 532 nm

the electron–electron coupling, and a quarter of the HFC match, three of the eight levels become degenerate. The pseudosecular contributions of the *isotropic* HFC and of the electron–electron coupling are then sufficient for level mixing and thus for transfer of electron two-spin order to nuclear polarization. The dominant part of the electron–electron coupling in a matching situation at the Earth's field is usually the dipole–dipole coupling since, at distances where this coupling is reduced to about 2.4 MHz, exchange coupling is negligible unless there is a continuous conjugated pathway or conducting material between the two electron spins. At Earth's field, the effect is maximal at a distance of about 30 Å between the two radicals, which roughly coincides with the separation between the donor and secondary acceptor in RCs (Fig. 9). Numerical computations show that many nuclei in the chromophores and their vicinity are likely to become polarized. Theory predicts that only modest HFC of a few hundred kilohertz is required to generate polarization of more than 1% for radical–radical distances between 20 and 50 Å, that is, for a large number of radical pairs in electron-transfer proteins. This suggests that CIDNP effects at Earth field may be common in biological systems [26].

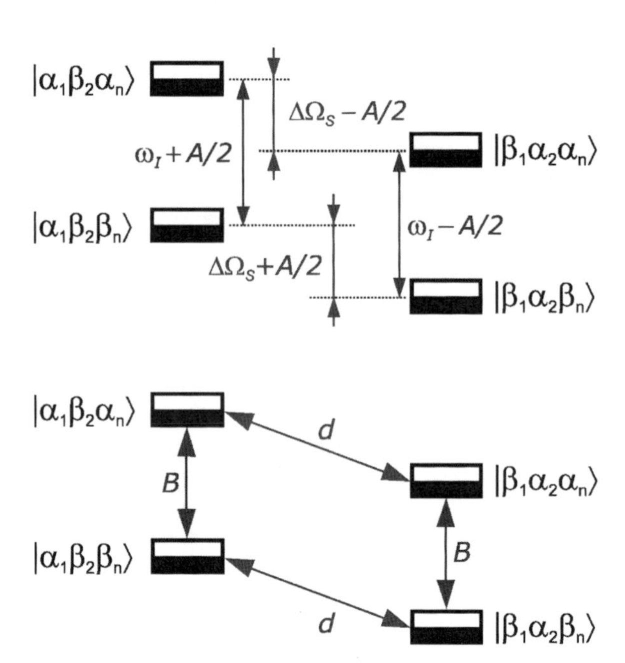

Fig. 8 Energy gaps (*top*) and mixing terms (*bottom*) required for three-spin mixing (TSM) in the high-field limit. At lower fields, the matching conditions are easier to fulfil ($\Delta\Omega_S$ = difference in electron Zeeman frequency, ω_I = nuclear Zeeman frequency, A = secular part of the hyperfine interaction, d = effective electron–electron coupling in the S-T_0 manifold, B = pseudosecular part of the hyperfine interaction)

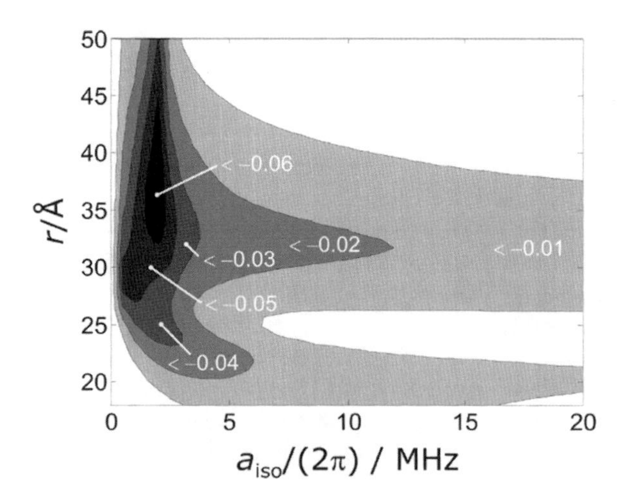

Fig. 9 Contour plot of the dependence of nuclear polarization on isotropic HFC $a_{iso}/2\pi$ and distance r between the two electron spins. *Shades of gray* correspond to levels of nuclear polarization as indicated by the labels. The electron Zeeman frequency $\omega_S/2\pi$ = 1.4 MHz corresponds to the Earth field

We have seen that the TSM and DD mechanisms acting in RCs of *R. sphaeroides* show a broad maximum at high fields, which is around 2 T (Figs. 2 and 4), as expected for a matching mechanism. The low-field theory predicts the occurrence of a second broad maximum (Fig. 9). The size of the matching window is determined by the radical-pair lifetime (Fig. 1). Natural photosynthetic RCs – and presumably also the phototropin mutant LOV1-C57S – having a short radical-pair lifetime of a few tens of nanoseconds, have broad enhancement functions making the solid-state photo-CIDNP effect easily detectable. On the other hand, at much longer radical-pair lifetimes, lifetime broadening is not significant and the matching conditions become narrow, making the effect more difficult to detect in many blue-light photoreceptors and artificial RCs. Therefore, for exploring the effect in such systems, the availability of various fields is the key factor. There is, however, an opposite effect in blue-light photoreceptors simplifying the observability of the effect: due to the comparably small size of the radicals and their asymmetric structure, the range of hf factors of individual nuclei is larger than in photosynthetic systems, providing the chance to observe a small number of nuclei at a certain field (see Sect. 4).

3 Photo-CIDNP MAS NMR: Analytical Applications

Classical solid-state NMR experiments provide a wealth of information on the sample, in particular:

1. Chemical shift (Fig. 10a, a′), chemical shift anisotropy, and *J*-couplings reflect local ground-state electronic properties of the system.
2. The linewidth is related to order and dynamics. Specifically dedicated NMR experiments can recognize dynamics and separate it from disorder.
3. The intensity of crosspeaks provides information on interactions and distances and allows for unequivocal signal assignments.

Exactly such type of information can also be obtained from photo-CIDNP MAS NMR experiments, even with improved sensitivity and selectivity. To that end, we have modified pulse-sequences for classical two-dimensional MAS NMR experiments for photo-CIDNP, mostly by changing the initial cross-polarization step to a direct carbon pulse. For example, we used modified RFDR and DARR [27] pulse sequences to obtain exact chemical shift assignments of selectively ^{13}C labeled cofactors [25, 28]. This straightforward NMR analysis has proven that the special pair is already differing between its two cofactors in the electron ground state (Fig. 10a, a′). In particular, P_L is the special cofactor of the special pair, while P_M is rather similar to a BChl cofactor in chloroform and to the accessory BChl cofactors [25, 28].

Solid-state photo-CIDNP experiments also provide information not available in standard NMR, for example, due to the induction of non-equilibrium distribution of polarization. Recently we measured the effect of ^{13}C–^{13}C spin diffusion on a selectively isotope labeled special pair [27]. In this experiment, the equilibration of the high photo-CIDNP polarization is observed. Since spin diffusion depends on

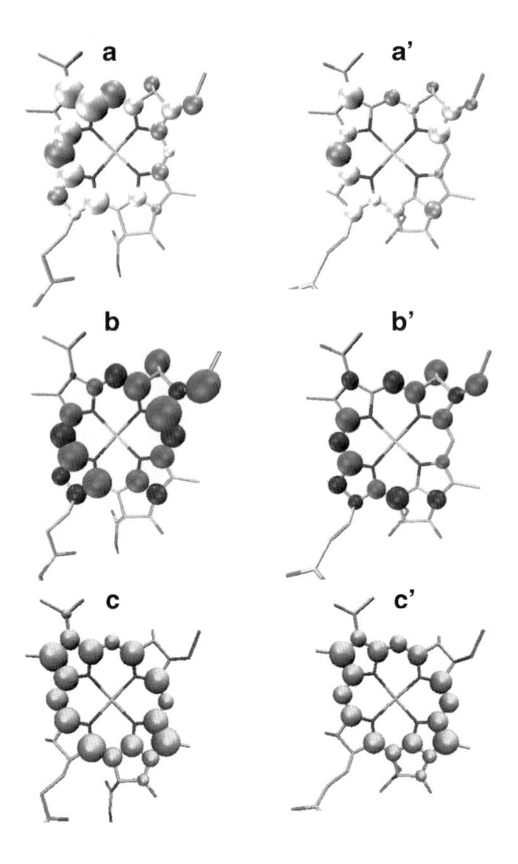

Fig. 10 Experimentally determined electronic structures of the special pair in different states showing the two halves P_L (*left*) and P_M (*right*). Relative electron densities in the electronic ground state (**a**, **a**′), s-spin densities of the radical cation (**b**, **b**′), and p_Z electron spin density in the donor triplet state (**c**, **c**′)

local mobility, the local dynamics of the special pair were reconstructed. From this data we conclude that, although the entire special pair is rather rigid, there is a gradient in rigidity from the soft end of P_M to the hard end of P_L. It might be that this gradient of dynamics is relevant for the symmetry break of electron transfer in *R. sphaeroides*.

Another form of information obtained from photo-CIDNP experiments is due to the fact that photo-CIDNP intensities are related to local electron spin densities. There are three approaches to obtain electron spin densities from photo-CIDNP intensities:

1. Both TSM and DD contribute to the photo-CIDNP build-up by unbalancing the ratio of α- to β-nuclear spins in the two decay channels. Both mechanisms require anisotropy of the hyperfine interaction (ΔA). According to theoretical considerations and numerical simulations, the polarization arising at the end of the photocycle in *R. sphaeroides* WT from these two mechanisms is roughly proportional to ΔA^2, and thus to the square of *p* spin density on the carbon atoms [19].
2. Time-resolved photo-CIDNP MAS NMR allows for observation of transient nuclear polarization originating selectively from the electronic S state. In such experiments, the light pulse for excitation (~8 ns) and the NMR pulse for

detection (~4 μs) are much shorter than the lifetime of the *P* molecular triplet (^3P) of R26 RCs (~100 μs). Thus, the transient nuclear polarization of the S state can be detected by a population of the electronic ground state that arises primarily from the singlet decay pathway, whereas nuclear spins in the triplet branch are still invisible because of paramagnetic shift and broadening [15]. Thus, effects attributable to isotropic coupling between electrons and nuclei become observable, similar to RPM-based photo-CIDNP in liquids. The sign rules are the same as for cage products from a singlet-born pair in RPM-based photo-CIDNP [29]. Therefore, the time-resolved experiment can provide estimates of the isotropic hyperfine interaction a_{iso} (Fig. 10b, b′) [25].
3. The intensities caused by the DR mechanism, which can be isolated in low-field experiments [13], are related to the local electron spin densities in the donor triplet state ^3P. Such analysis shows that the electron spin density in the donor triplet state of the special pair is almost equally spread over the two macrocycles (Fig. 10c, c′) [13]. Assuming that ^3P is a combination of electron spin density of a HOMO and a LUMO, and also approximating that the isotropic hyperfine interactions in the radical cation state provide a picture of the HOMO, the reconstruction of the excited state from NMR data might be possible [13].

In addition to standard NMR information, local dynamics, and electronic structures, there are three more important parameters which can be determined from photo-CIDNP experiments:

1. Time-resolved photo-CIDNP MAS NMR experiments provide kinetic information, for example on the lifetime of the triplet state of the donor or a nearby carotenoid.
2. From field-dependence of the effect in the low-field range, the distance of the cofactors forming the radical pair can be estimated, since the low-field matching condition is set by electron–electron coupling. The width of the matching condition is related to the radical-pair lifetime. Therefore precise and comparable field-dependent experiments, for example with a shuttle system, are required.
3. The intensity pattern contains the information as to whether the radical pair has been formed from a singlet or triplet excited state precursor [20]. In particular for blue-light photo-receptors, it often is not clear whether the electron is transferred from an excited singlet or triple state. Here a definite answer can be found.

Another analytically relevant aspect, which we presently explore [27], is the transfer of the high photo-CIDNP polarization to the neighborhood in "spin-torch" experiments [27] allowing, for example, study of the protein pocket around the cofactor. While the enhanced polarization dissipates rapidly in CHHC type experiments into the proton pool, ^{13}C–^{13}C spin torch experiments allow for exploration in a 6–7 Å radius in which the flow of polarization can be observed on a millisecond timescale. Via a chain of ^{13}C labels ("relay stations") the nuclear polarization can flow up to 40 Å [27]. Hence, spin-torch experiments allow study of the environment of the radical pair in great detail. Such experiment might be useful to probe putative light-induced changes relevant for signaling.

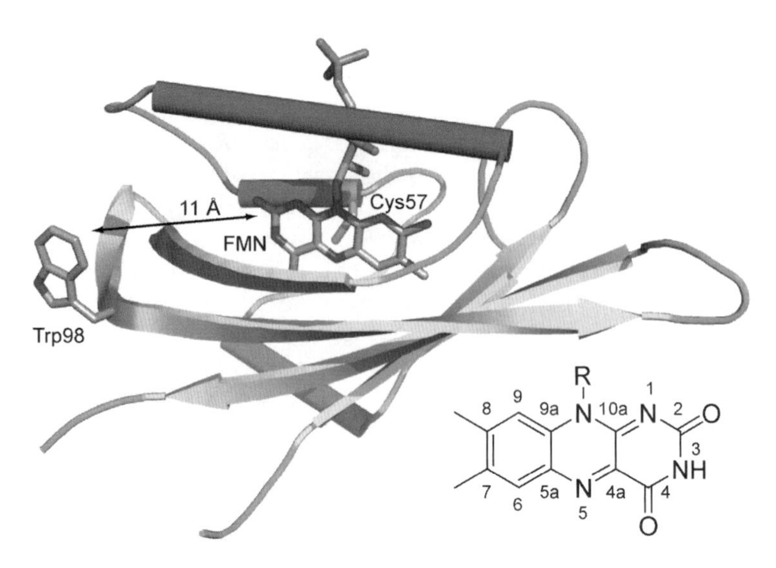

Fig. 11 Structure of the blue-light photoreceptor phototropin LOV1-C57S. The flavin cofactor is shown at the *bottom*

4 Photo-CIDNP in Blue-Light Photoreceptors

In liquid-state photo-CIDNP, flavin compounds are frequently used as dyes to produce triplets which act as single electron acceptors and thus produce radical pairs. Classical reaction partners in such solution systems are aromatic amino acids such as tyrosines and tryptophanes. The method has been brought to perfection by Kaptein and co-workers by probing surfaces of proteins with liquid-state photo-CIDNP [30]. The effect has been well explained by the classical RPM and its "cyclic reaction" modification. Observation by Weber et al. of photo-CIDNP in the C450A-mutant of the phototropin LOV2-domain of *Avena sativa* [31, 32] by ^{13}C liquid-state NMR, however, cannot be explained by the classical RPM. Also an explanation based on the "cyclic-reaction" scheme is difficult because signals from both electron donor and acceptor occur.

Phototropins, blue-light photoreceptors related to cryptochromes, regulate key responses of plants to light, such as phototropic movement and chloroplast relocation. Upon illumination, the triplet excited state of the flavin reacts with a nearby cysteine residue to form a covalent adduct as the signaling state [33]. Mutation of the reactive cysteine to serine or alanine abolishes this adduct formation. Instead, a less efficient competing pathway of electron transfer from a tryptophan leads to transient accumulation of a flavin anion radical on illumination [31, 32] and finally to formation of a flavin neutral radical [34]. The radical is re-oxidized by oxygen.

The mutant C57S of the phototropin-LOV1 domain from the green alga *Chlamydomonas reinhardtii* (Fig. 11) was the first non-photosynthetic system showing the solid-state photo-CIDNP effect (Figs. 12 and 13) [10]. This experiment

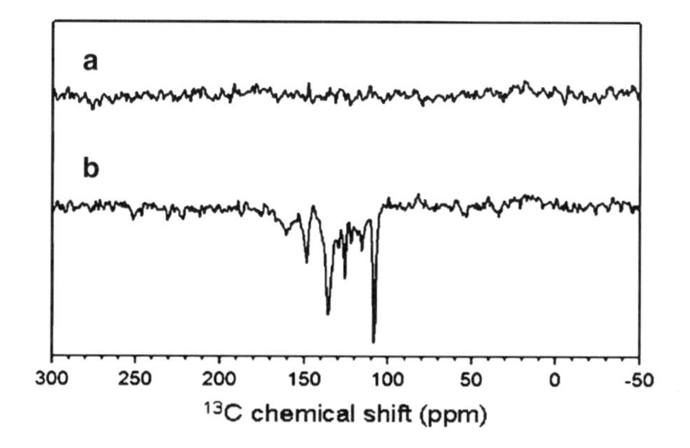

Fig. 12 ^{13}C MAS NMR spectra of phototropin LOV1-C57S obtained with 8-kHz MAS at a magnetic field of 2.3 T in the dark (**a**) and under continuous illumination with white light (**b**)

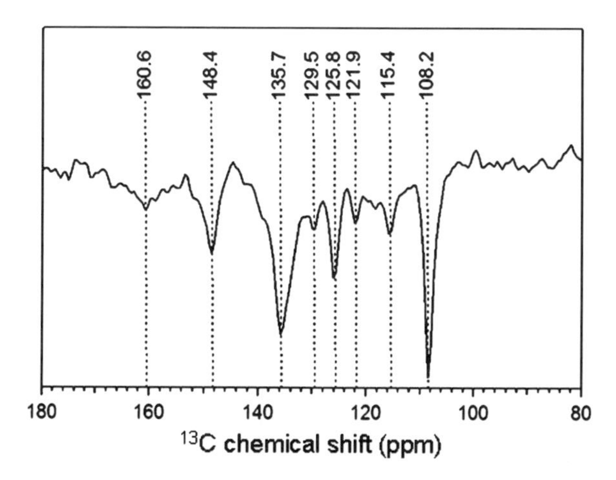

Fig. 13 Expanded view on the aromatic region of the ^{13}C MAS NMR spectrum of phototropin LOV1-C57S showing the solid-state photo-CIDNP effect (spectrum 2B)

was done at 2.4 T under continuous illumination. While natural RCs have a radical pair lifetime of a few tens of nanoseconds, this mutant presumably has a radical pair lifetime of a few hundreds of nanoseconds, still allowing for a broad enhancement function and the detection of the effect at various fields. Chemical shift assignments demonstrate that the radical pair is formed by the FMN cofactor and tryptophan residue Trp-98, the only Trp in the protein. An edge-to-edge distance of about 11 Å between the FMN and Trp-98 is suitable for electron transfer. An assignment to a histidine or tyrosine would be difficult to reconcile with the resonance at 108.2 ppm (Cγ of Trp). These data demonstrate that the solid-state photo-CIDNP effect is not limited to natural photosynthetic systems, providing the possibility to develop a more generally applicable method for signal enhancement.

References

1. Zysmilich MG, McDermott AE (1994) Photochemically induced dynamic nuclear polarization in the solid-state ^{15}N spectra of reaction centers from photosynthetic bacteria *Rhodobacter sphaeroides* R-26. J Am Chem Soc 116:8362–8363
2. Prakash S, Alia, Gast P, de Groot HJM, Jeschke G, Matysik J (2005) Magnetic field dependence of photo-CIDNP MAS NMR on photosynthetic reaction centers of *Rhodobacter sphaeroides* WT. J Am Chem Soc 127:14290–14298
3. Prakash S, Alia, Gast P, de Groot HJM, Matysik J, Jeschke G (2006) Photo-CIDNP MAS NMR in intact cells of *Rhodobacter sphaeroides* R26: molecular and atomic resolution at nanomolar concentration. J Am Chem Soc 128:12794–12799
4. Janssen GJ, Daviso E, van Son M, de Groot H, Alia A, Matysik J (2010) Observation of the solid-state photo-CIDNP effect in entire cells of cyanobacteria *Synechocystis*. Photosynth Res 104:275–282
5. Roy E, Alia A, Gast P, van Gorkom HJ, de Groot HJM, Jeschke G, Matysik J (2007) Photochemically induced dynamic nuclear polarisation observed in the reaction center of the green sulphur bacteria *Chlorobium tepidum* by ^{13}C MAS NMR. Biochem Biophys Acta 1767:610–615
6. Diller A, Roy E, Gast P, van Gorkom HJ, de Groot HJM, Glaubitz C, Jeschke G, Matysik J, Alia A (2007) ^{15}N-photo-CIDNP MAS NMR analysis of the electron donor of photosystem II. Proc Natl Acad Sci USA 104:12843–12848
7. Prakash S, Alia A, Gast P, de Groot HJM, Jeschke G, Matysik J (2007) ^{13}C chemical shift map of the active cofactors in photosynthetic reaction centers of *Rhodobacter sphaeroides* revealed by photo-CIDNP MAS NMR. Biochemistry 46:8953–8960
8. Roy E, Rohmer T, Gast P, Jeschke G, Alia A, Matysik J (2008) Characterization of the primary electron pair in reaction centers of *Heliobacillus mobilis* by ^{13}C photo-CIDNP MAS NMR. Biochemistry 47:4629–4635
9. Matysik J, Diller A, Roy E, Alia A (2009) The solid-state photo-CIDNP effect. Photosynth Res 102:427–435
10. Thamarath SS, Heberle J, Hore P, Kottke T, Matysik J (2010) Solid-state photo-CIDNP effect observed in phototropin LOV1-C57S by ^{13}C magic-angle spinning NMR spectroscopy. J Am Chem Soc 132:15542–15543
11. Hoff AJ, Deisenhofer J (1997) Photophysics of photosynthesis. Structure and spectroscopy of reaction centers of purple bacteria. Phys Rep 287:2–247
12. Hunter CN, Daldal F, Thurnauer MC, Beatty JT (2008) The phototropic purple bacteria. Springer, Dordrecht, The Netherlands
13. Thamarath SS, Bode BE, Prakash S, Sai Sankar Gupta KB, Alia A, Jeschke G, Matysik J (2012) Electron spin density distribution in the special pair triplet of *Rhodobacter sphaeroides* R26 revealed by magnetic field dependence of the solid-state photo-CIDNP effect. J Am Chem Soc 134:5921–5930
14. Wirtz AC, van Hemert MC, Lugtenburg J, Frank HA, Groenen EJJ (2007) Two stereoisomers of spheroidene in the *Rhodobacter sphaeroides* R26 reaction center: a DFT analysis of resonance Raman spectra. Biophys J 93:981–991
15. Daviso E, Alia A, Prakash S, Diller A, Gast P, Lugtenburg J, Matysik J, Jeschke G (2009) Electron-nuclear spin dynamics in a bacterial photosynthetic reaction center. J Phys Chem C 113:10269–10278
16. Jeschke G (1997) Electron–electron-nuclear three-spin mixing in spin-correlated radical pairs. J Chem Phys 106:10072–10086
17. Jeschke G (1998) A new mechanism for chemically induced dynamic nuclear polarization in the solid state. J Am Chem Soc 120:4425–4429
18. Polenova T, McDermott AE (1999) A coherent mixing mechanism explains the photoinduced nuclear polarization in photosynthetic reaction centers. J Phys Chem B 103:535–548

19. Diller A, Prakash S, Alia, Gast P, Matysik J, Jeschke G (2007) Signals in solid-state photochemically induced dynamic nuclear polarization recover faster than with the longitudinal relaxation time. J Phys Chem B 111:10606–10614
20. Jeschke G, Matysik J (2003) A reassessment of the origin of photochemically induced dynamic nuclear polarization effects in solids. Chem Phys 294:239–255
21. McDermott A, Zysmilich MG, Polenova T (1998) Solid state NMR studies of photoinduced polarization in photosynthetic reaction centers: mechanism and simulations. Sol State Nuc Magn Reson 11:21–47
22. Closs GL (1975) On the overhauser mechanism of chemically induced nuclear polarization as suggested by Adrian. Chem Phys Lett 32:277–278
23. Goldstein RA, Boxer SG (1987) Effects of nuclear spin polarization on reaction dynamics in photosynthetic bacterial reaction centers. Biophys J 51:937–946
24. Hore PJ, Broadhurst RW (1993) Photo-CIDNP of biopolymers. Prog Nucl Magn Reson Spectrosc 25:345–402
25. Daviso E, Prakash S, Alia A, Gast P, Neugebauer P, Jeschke G, Matysik J (2009) The electronic structure of the primary electron donor of reaction centers of purple bacteria at the atomic resolution as observed by photo-CIDNP ^{13}C MAS NMR. Proc Natl Acad Sci USA 106:22281–22286
26. Jeschke G, Anger BC, Bode BE, Matysik J (2011) Theory of solid-state photo-CIDNP in the Earth's magnetic field. J Phys Chem A 115:9919–9928
27. Sai Sankar Gupta KB (2011) Spin-torch experiments on reaction centers of *Rhodobacter sphaeroides*. PhD Thesis, Leiden University
28. Schulten EAM, Matysik J, Alia, Kiihne S, Raap J, Lugtenburg J, Gast P, Hoff AJ, de Groot HJM (2002) ^{13}C MAS NMR and photo-CIDNP reveal a pronounced asymmetry in the electronic ground state of the special pair of *Rhodobacter sphaeroides* reaction centres. Biochemistry 41:8708–8717
29. Kaptein R (1971) Simple rules for chemically induced dynamic nuclear polarization II: relation with anomalous ESR spectra. Chem Commun 732–733
30. Kaptein R, Dijkstra K, Nicolay K (1978) Laser photo-CIDNP as a surface probe for proteins in solution. Nature 274:293–294
31. Richter G, Weber S, Römisch W, Bacher A, Fischer M, Eisenreich W (2005) Photochemically induced dynamic nuclear polarization in a C450A mutant of the LOV2 domain of the *Avena sativa* blue-light receptor phototropin. J Am Chem Soc 127:17245–17252
32. Eisenreich W, Joshi M, Weber S, Bacher A, Fischer MJ (2008) Natural abundance solution ^{13}C NMR studies of a phototropin with photoinduced polarization. J Am Chem Soc 130:13544–13545
33. Salomon M, Christie JM, Knieb E, Lempert U, Briggs WR (2000) Photochemical and mutational analysis of the FMN-binding domains of the plant blue light receptor, phototropin. Biochemistry 39:9401–9410
34. Kottke T, Dick B, Fedorov R, Schlichting I, Deutzmann R, Hegemann P (2003) Irreversible photoreduction of flavin in a mutated phot-LOV1 domain. Biochemistry 42:9854–9862

Top Curr Chem (2013) 338: 123–180
DOI: 10.1007/128_2012_371

Published online: 25 October 2012

Parahydrogen-Induced Polarization in Heterogeneous Catalytic Processes

Kirill V. Kovtunov, Vladimir V. Zhivonitko, Ivan V. Skovpin, Danila A. Barskiy, and Igor V. Koptyug

Abstract Parahydrogen-induced polarization of nuclear spins provides enhancements of NMR signals for various nuclei of up to four to five orders of magnitude in magnetic fields of modern NMR spectrometers and even higher enhancements in low and ultra-low magnetic fields. It is based on the use of parahydrogen in catalytic hydrogenation reactions which, upon pairwise addition of the two H atoms of parahydrogen, can strongly enhance the NMR signals of reaction intermediates and products in solution. A recent advance in this field is the demonstration that PHIP can be observed not only in homogeneous hydrogenations but also in heterogeneous catalytic reactions. The use of heterogeneous catalysts for generating PHIP provides a number of significant advantages over the homogeneous processes, including the possibility to produce hyperpolarized gases, better control over the hydrogenation process, and the ease of separation of hyperpolarized fluids from the catalyst. The latter advantage is of paramount importance in light of the recent tendency toward utilization of hyperpolarized substances in in vivo spectroscopic and imaging applications of NMR. In addition, PHIP demonstrates the potential to become a useful tool for studying mechanisms of heterogeneous catalytic processes and for in situ studies of operating catalytic reactors. Here, the known examples of PHIP observations in heterogeneous reactions over immobilized transition metal complexes, supported metals, and some other types of heterogeneous catalysts are discussed and the applications of the technique for hypersensitive NMR imaging studies are presented.

Keywords Heterogeneous hydrogenation · Hyperpolarized gases · Immobilized metal complexes · Nuclear magnetic resonance · Parahydrogen-induced polarization · Supported metal catalysts

---

K.V. Kovtunov, V.V. Zhivonitko, I.V. Skovpin, D.A. Barskiy, and I.V. Koptyug (✉)
International Tomography Center, SB RAS, 3A Institutskaya St, Novosibirsk 630090, Russian Federation

Novosibirsk State University, 2 Pirogova St, Novosibirsk 630090, Russian Federation
e-mail: koptyug@tomo.nsc.ru

Contents

Abbreviations

ALTADENA	Adiabatic longitudinal transport after dissociation engenders nuclear alignment
BMPY	*N*-butyl-4-methylpyridinium
COD	1,4-Cyclooctadiene
DPPB	1,4-Bis(diphenylphosphino)butane
HET-PHIP	Parahydrogen-induced polarization in heterogeneous processes
ID	Inside diameter
IL	Ionic liquid
MOF	Metal-organic framework
MRI	Magnetic resonance imaging
NMR	Nuclear magnetic resonance
PASADENA	Parahydrogen and synthesis allow dramatically enhanced nuclear alignment
PCy_3	Tricyclohexylphosphine
PHIP	Parahydrogen-induced polarization
Py	Pyridine
RD	Remote detection
RF	Radiofrequency
RT	Room temperature
SABRE	Signal amplification by reversible exchange
SILP	Supported ionic liquid phase
SNR	Signal-to-noise ratio
Tf_2N	Bis(trifluoromethylsulfonyl)amide
TOF	Turnover frequency
XPS	X-ray photoelectron spectroscopy

1 Introduction

The entire field of research related to hyperpolarization in magnetic resonance is demonstrating rapid and exciting progress lately, and parahydrogen-induced polarization (PHIP) is no exception in this respect. Observed and misinterpreted [1], predicted theoretically [2], and then re-discovered experimentally [3], PHIP (or PASADENA, Parahydrogen And Synthesis Allow Dramatic Enhancement of Nuclear Alignment) was a curiosity at first, which gradually evolved into a hypersensitive tool for research in the field of homogeneous catalysis [4–8]. The PHIP phenomenon is based on the conversion of the correlated state of nuclear spins of parahydrogen (or, in general, the *ortho* or *para* spin isomers of H_2 or D_2) into the significant enhancements of the NMR signals of molecular species by means of a catalytic process (see, e.g., Chap. 2 of this book). Any gain in signal enhancement in NMR and MRI is always welcome, but usually one can only dream about an enhancement as large as four to five orders of magnitude and more. Signal enhancements provided by PHIP and other members of the family of hyperpolarization techniques described throughout this book are a very rare example of the dream come true.

It is obvious at this point that one of the major driving forces for the development of the field of hyperpolarization in magnetic resonance is its potential application in human MRI research and possibly in medical diagnostics. However, most of the spectroscopic and imaging PHIP studies reported to date are performed using homogeneous catalysts – transition metal complexes dissolved in a liquid phase along with the substrate that undergoes hydrogenation [2, 6–14]. The presence of the dissolved catalyst is one of the major obstacles en route toward human studies as it has to be somehow removed from the hyperpolarized liquid before the latter can be administered into the body. In this respect, SABRE [10, 15] (see also Chap. 3 of this book) is no different from the classical PHIP as it also requires a dissolved catalyst and in fact represents a catalytic process, notwithstanding the fact that the product and the substrate are chemically identical and differ only in the states of their nuclear spins.

Removal of a catalyst from the reaction products is also one of the key issues in industrial catalysis which is heavily employed in the technologies of the modern chemical industry (e.g., petrochemical, pharmaceutical, fine chemical, food industry, etc.). Over 90% of all commercial chemical processes make use of heterogeneous catalysts [16], i.e., catalysts that constitute a phase (e.g., solid) which can easily be separated from the reaction mixture. Quite often the ease of catalyst separation associated with heterogeneous catalysts by far outweighs the superior catalytic properties of homogeneous catalysts. There are other advantages in using heterogeneous catalysts as well, for instance the possibility to flow reactants continuously through a bed of catalyst particles instead of performing a reaction in a batch mode/reactor. Similarly, utilization of heterogeneous catalysts in hydrogenation processes could be a viable strategy for developing novel PHIP techniques and applications that would benefit from easy removal of the catalysts from the hyperpolarized fluids and the possibility of their continuous production.

Unfortunately, one cannot just utilize the hydrogenation catalysts developed for industrial applications in order to advance PHIP technology. The reason is that those extremely efficient catalysts were developed for a somewhat different purpose, namely to maximize the efficiency of production of the desired product of a hydrogenation process, i.e., to maximize both the conversion of the substrate into the products and the selectivity toward the desired product. However, for PHIP studies it is not sufficient to produce a product in large quantities; this product should also exhibit polarization of nuclear spins. For the conventional PHIP effects to be observed, the initial correlation of the two nuclear spins of the parahydrogen molecule has to be converted into the polarization of the nuclear spins of the product molecule formed in the hydrogenation reaction. This usually implies that the two hydrogen atoms originally belonging to the same parahydrogen molecule should not lose each other throughout the catalytic cycle. While there are known examples when only one hydrogen atom ends up in a product molecule and yet exhibits hyperpolarization of its nuclear spin [17, 18], this is rather an exception. It still requires that the two hydrogen atoms remain close to each other through some part of the catalytic cycle, e.g., in a reaction intermediate where spin dynamics can convert their correlation into the observable hyperpolarization before the two hydrogen atoms go their separate ways. In the majority of PHIP examples, however, the two hydrogen atoms need to stay together all the way from the parahydrogen molecule through a number of reaction intermediates and finally into the same product molecule. Therefore, this pairwise addition of the two hydrogen atoms from the same parahydrogen molecule to the same substrate molecule can be considered as a prerequisite for the observation of PHIP effects in the product molecule formed upon hydrogenation of the substrate. In contrast, supported metals, some of the most efficient industrial-type hydrogenation catalysts, do not sustain this pairwise addition, but instead tend to add random hydrogen atoms to double and triple bonds of unsaturated substrates. As a result, for many years the use of heterogeneous catalysts was not even considered in the context of PHIP research.

Nevertheless, the possibility to observe PHIP in heterogeneous reactions (HET-PHIP) has been demonstrated successfully and conclusively for various types of heterogeneous catalysts [19–23] including the supported metal catalysts expected to have the "wrong" reaction mechanism. This makes it possible to move on and formulate a number of new objectives in this research field that need to be addressed.

Two related but separate directions can be outlined. One direction is the further expansion and development of the scope of the PHIP phenomenon and methodology. In particular, it is important to advance our understanding of the HET-PHIP phenomenon by testing a broader range of heterogeneous catalysts and identifying those types of catalysts that can produce HET-PHIP most efficiently. Comparison with their homogeneous counterparts is essential and can be quite informative. For some heterogeneous catalysts (e.g., supported metals), the reaction mechanisms are expected to be very different from those of homogeneous hydrogenation processes, so the question is what mechanism is responsible for PHIP formation? However, even in those cases when the mechanisms for heterogeneous catalysts are expected to be similar to those of their homogeneous analogs (e.g., immobilized vs. homogeneous metal complexes), the presence of a different phase (e.g., the solid support) can still affect the formation of PHIP. Therefore, for all heterogeneous catalysts it is important to identify the essential differences with the homogeneous case in terms of the reaction mechanism and to establish what factors can influence the magnitude and other characteristics of PHIP. In certain cases it may be necessary to go back and re-examine the results published previously to clarify the nature of the actual catalyst that is responsible for the observed signal enhancements. For instance, observation of PHIP during hydrogenation of phenylacetylene into styrene and ethylbenzene in acetone-d6 with the use of colloidal $Pd_x[N(octyl)_4Cl]_y$ catalyst [24] was taken as the evidence in favor of homogeneous hydrogenation mechanism, obviously because at that time the possibility of HET-PHIP observation was not even considered.

Another direction encompasses the development of novel spectroscopic and imaging applications of the PHIP phenomenon. Two major sub-domains can be identified here. One is the development of PHIP technology as a source of catalyst-free hyperpolarized liquids and gases for advanced hypersensitive spectroscopic and imaging applications, from the studies of HET-PHIP formation and temporal and spatial evolution in beds of heterogeneous catalysts to the advanced biomedical MRI and MRS studies, and more. Another sub-domain is the development of HET-PHIP as a highly sensitive tool for the in situ and *operando* studies of the mechanisms of heterogeneous catalytic processes. This has already been demonstrated in homogeneous hydrogenations where the strong signal enhancement offered by PHIP provides high sensitivity essential for the detection of short-lived reaction intermediates and the detailed analysis of reaction mechanisms and kinetics [4–8]. Further progress in the fields of HET-PHIP phenomena and practice could make it possible to develop a similar hypersensitive tool for in situ and *operando* studies of the mechanisms of heterogeneous catalytic processes.

Before we proceed to the discussion of HET-PHIP, some comments on the interrelation between homogeneous and heterogeneous catalysis are in order. In fact, classifying catalysts as "homogeneous" and "heterogeneous" is a more complicated issue than it may seem. For the purpose of this discussion, homogeneous catalysts are complexes of transition metals such as the well-known Wilkinson's catalyst $RhCl(PPh_3)_3$, Vaska's catalyst $Ir(CO)Cl(PPh_3)_2$, Crabtree's catalyst $[(COD)Ir(PCy_3)(Py)]^+PF_6^-$ (COD = 1,4-cyclooctadiene, PCy_3 = tricyclohexylphosphine, Py = pyridine), and other molecular complexes including those used in many conventional PHIP and SABRE studies. To carry out a homogeneous catalytic reaction using these catalysts, they are dissolved in an appropriate solvent and are present in solution along with the substrate to be hydrogenated and/or hyperpolarized. In contrast, heterogeneous catalysts are macroscopic particles of a complex nature that are not dissolved in the liquid and thus constitute a separate solid phase which can be removed relatively easily from the liquid or gas phase containing the substrate and the product(s). The catalytic reaction in this case takes place at the surface of the solid catalyst upon adsorption of the reactants, and the product desorbs from the surface into the bulk liquid or gas phase. There are many different types of heterogeneous catalysts, including, but not limited to, transition metal complexes (for instance, those mentioned above) immobilized on (attached to) macroscopic solid supports, supported metal catalysts comprising metal nanoparticles residing on the surface of solid materials, and unsupported particles or films of metals and metal oxides. Such a working definition of heterogeneous catalysts is in fact a dramatic oversimplification which is intended to streamline the following discussion of the modern status of HET-PHIP research. It is likely, however, that as the field of HET-PHIP research develops further, it will be necessary to refine and expand this definition to include other types of catalysts currently employed in modern heterogeneous catalysis.

Homogeneous catalysts often have a catalytically active center with a structure which is well-defined on the molecular level and thus can be synthesized reproducibly. In contrast, heterogeneous catalysts such as supported metal nanoparticles are often known to possess several different types of active sites with significantly different catalytic behavior which can operate in parallel. Even slight (and often uncontrollable) modifications of the preparation procedure can markedly affect their catalytic behavior such as activity and selectivity. Further complications arise because catalytic substances introduced in the reaction mixture are often precatalysts rather than the actual catalysts. For both homogeneous and heterogeneous catalysts, some kind of activation is often required to observe catalytic activity, during which the (pre)catalyst can undergo significant structural changes. Furthermore, an initially homogeneous precatalyst can produce heterogeneous catalytic species (e.g., transformation of a metal salt or complex into metal clusters and nanoparticles), and for a supposedly heterogeneous catalyst some fraction of the metal can cross over to the homogeneous phase (e.g., leaching of an immobilized complex, partial dissolution of metal particles under reactive conditions), leading to the homogeneous nature of a supposedly heterogeneous reaction.

In many catalytic studies reported in the literature, the nature of the actual catalyst is merely assumed from the nature of the precatalyst employed [25]. It is not surprising, therefore, that there are numerous examples in catalytic research when the nature of the catalytically active species has been misidentified. Examples related to PHIP are already known and will be mentioned below. Therefore, while characterization of the catalyst prior to the reaction is very important, it is no less important to check the catalyst integrity or transformation after the reaction. However, the most reliable results can be achieved only if the nature of the catalytically active species is addressed in the *operando* mode, i.e., during the reaction. Still, the problem cannot be solved by simply identifying the nature of the majority of species which potentially can have catalytic activity. Indeed, the predominant species can be catalytically inactive, whereas minor species (e.g., metal clusters or nanoparticles) can have an exceptionally high catalytic activity. This "homeopathic effect" can lead to false conclusions about the nature of the catalytically active species even for studies performed under reactive conditions [25]. While HET-PHIP can potentially become a powerful tool for mechanistic studies in heterogeneous catalysis, it should be developed and applied with care, as the homeopathic effect of minor species or even impurities possessing an exceptionally high catalytic activity multiplied by the signal enhancement provided by PHIP can lead to "hyperhomeopathic effect" and data misinterpretation.

2 HET-PHIP with Immobilized Metal Complexes

As mentioned above, to produce PHIP the catalyst has to be able to add an H_2 molecule to an unsaturated substrate in a pairwise manner. Transition metal complexes used as homogeneous catalysts are known to perform this pairwise addition, as confirmed by numerous observations of PHIP in homogeneous hydrogenations with parahydrogen [3–9] (see also Chap. 2 of this book). Therefore, heterogenization of such complexes by means of their immobilization on a suitable solid support was likely the most promising route toward the first observation of HET-PHIP. Indeed, the use of immobilized transition metal complexes in catalytic hydrogenations of unsaturated substrates with parahydrogen constituted the first instance of HET-PHIP observation in heterogeneous reaction processes [19].

In catalysis, many different strategies to heterogenize (immobilize) homogeneous catalysts have been developed [26–30], including covalent binding, ionic binding, hydrogen bonding, physisorption, entrapment or encapsulation, the use of supported liquid (e.g., aqueous) phase, supported ionic liquid phase, and more. However, basically all these strategies are not reliable enough to be used in industrial catalysis [28]. Nevertheless, utilization of immobilized noble metal complexes at this stage of HET-PHIP research is quite important since, similar to their homogeneous analogs, it allows tailored preparation of catalysts with a single and well-defined type of active site. This approach provides the uniformity of active sites essential for the elucidation of mechanisms and structure-function relationships.

However, the apparent simplicity of the concept of HET-PHIP observation using immobilization of transition metal complexes is deceptive. In fact, immobilization of a metal complex on a solid support introduces a large number of new factors that can affect (or even prevent) the formation and/or observation of PHIP effects. In particular, solid support can be considered as a bulky ligand which can modify the catalytic behavior of the metal center by introducing various steric and electronic effects. These effects will depend on the mode of immobilization and on the length and mobility of the tether that links the active center to the support. Alteration of molecular mobility and the lifetime of reaction intermediates caused by immobilization may also significantly affect the degree of relaxation of nuclear spin coherence/polarization before the reaction product is formed. Finally, even if the product molecule is produced with hyperpolarized nuclear spins, the lifetime of this hyperpolarization can be reduced drastically because of the enhanced nuclear spin relaxation due to the presence of the solid phase, i.e., the catalyst support.

In general, the preparation and use of the immobilized catalysts requires extraordinary attention to the experimental conditions. In particular, immobilization should be performed in an oxygen-free atmosphere. In addition, particular attention should be paid to reaction operation conditions, since immobilized complexes are apt to reduce readily in the presence of hydrogen at elevated temperatures, resulting in the formation of metal particles. Furthermore, many heterogenized catalysts tend to leach from the support into solution during the reaction, which contaminates the liquid phase with the metal and also raises the question as to whether the heterogeneous or the re-dissolved species are responsible for the experimental observations. All these complications led to the fact that the research in the field of unambiguous and reliable application of immobilized complexes in HET-PHIP is still at an early stage.

In this section we consider all reported instances of the use of immobilized complexes in HET-PHIP and discuss the valuable practical and theoretical points related to such applications. We note that in all HET-PHIP experiments described in this chapter, "parahydrogen" refers to H_2 with the *ortho:para* ratio of ca. 1:1 since conversion of H_2 to parahydrogen in all studies was performed at 77 K.

2.1 Immobilized Rh Complexes

2.1.1 Immobilized Wilkinson's Complex

It appears that among all immobilized Rh complexes, Wilkinson's complex (RhCl $(PPh_3)_3$) has been studied most extensively in terms of HET-PHIP observation [19, 31, 32]. To the best of our knowledge, three types of this catalyst differing in immobilization support were examined in hydrogenations using parahydrogen under different experimental conditions. It is also rather convenient to separate the discussions of the liquid-phase and the gas-phase hydrogenations because of the basic differences in their experimental implementations.

Liquid-Phase Hydrogenations

The first demonstration of a successful use of immobilized Wilkinson's complex for producing hyperpolarized substances was reported by Koptyug et al. [19]. Two types of immobilized Wilkinson's catalysts, supported either on styrene-divinylbenzene copolymer or on diphenylphosphinoethyl-modified silica, were used in the hydrogenation of styrene as a test process. For the experiments, a 0.1 M solution of styrene in benzene-d6 was thoroughly degassed with N_2 bubbling to ensure a low concentration of dissolved O_2. Moreover, the operations such as assembling of the flow system and adding the catalysts were performed while N_2 flow was left on to purge any oxygen. Both catalysts demonstrated clear PASADENA signals in the ^{1}H NMR spectra after bubbling parahydrogen through the 10-mm NMR sample tube held at 65°C inside the NMR magnet and containing the solution and the catalyst (Fig. 1a, b). The ALTADENA spectra were also detected when bubbling of the sample with parahydrogen was performed outside the magnet before it was placed in the NMR magnet for signal detection.

Two significant differences between these two types of immobilized Wilkinson's complexes were observed in the experiments. First, the polymer-supported catalyst required a long activation time at elevated temperatures, sometimes in excess of 1 h. Presumably this was needed for the catalyst beads to swell and expose the catalytic

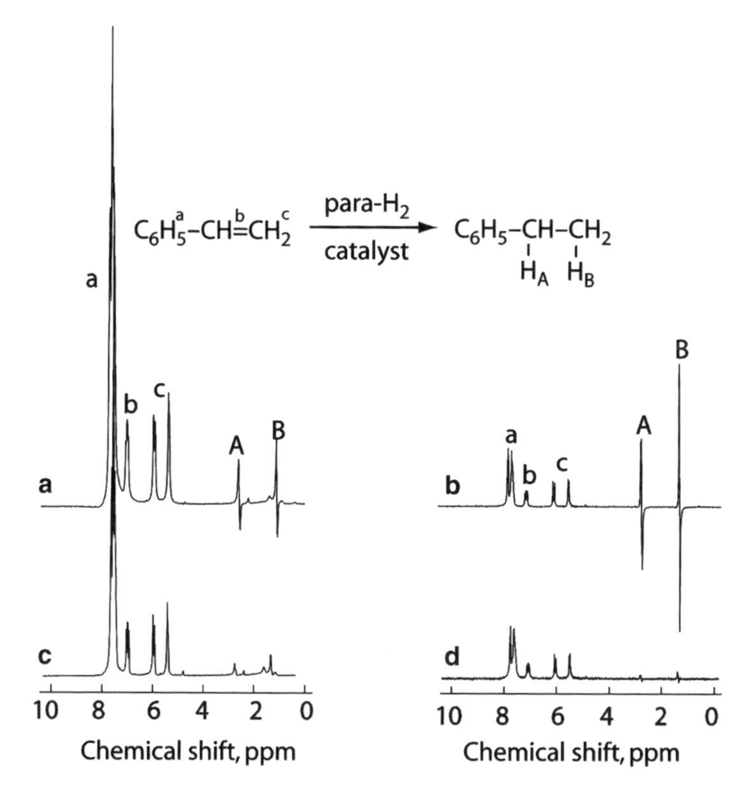

Fig. 1 ^{1}H NMR spectra detected during the in situ (PASADENA) hydrogenation of styrene in C_6D_6 at 65°C using parahydrogen. The antiphase multiplets of the polarized protons in the product ethylbenzene are labeled A and B and appear at 2.6 and 1.2 ppm, respectively. (**a**) The spectrum obtained with the Wilkinson's catalyst supported on diphenylphosphinoethyl-modified silica after hydrogenation for 8 s inside the magnet. (**c**) The spectrum acquired after the catalyst was removed and the supernatant solution was bubbled with parahydrogen for 15 min. The spectrum is plotted on the same vertical scale as in (**a**) and shows no PASADENA or product formation. (**b**) The spectrum obtained with the polymer-supported Wilkinson's catalyst after hydrogenation for 15 s. (**d**) The spectrum acquired after the catalyst was removed and the solution was bubbled with parahydrogen for 15 s. The spectrum is plotted on the same vertical scale as the one in (**b**) and shows a very small antiphase pattern. Reprinted with permission from [19]. Copyright 2007 American Chemical Society

centers to the reactants [33]. Second, higher gas flow rates and longer bubbling times (more than 1 min) were critical for a reliable observation of polarization effects in hydrogenations using the polymer-supported catalyst. On the other hand, Wilkinson's complex supported on modified silica did not require any activation period and became active as soon as the temperature was raised, and bubbling parahydrogen for a few seconds was sufficient to observe strong polarization in the ^{1}H NMR spectra with this catalyst.

In general, metal complex leaching off the support is the potential problem, which should be considered whenever an immobilized metal complex is involved in a catalytic process in a liquid phase. In terms of PHIP, this leads to the ambiguity on whether the polarization is produced by the immobilized catalyst or not, and additional tests are required in order to elucidate this issue. Separation of the reaction solution from the heterogeneous catalysts followed by bubbling of parahydrogen through the supernatant liquid confirmed that the polarization was generated almost exclusively in the heterogeneous reaction both for the polymer-supported catalyst and for the catalyst supported on the modified silica [19]. Indeed, some drastic differences in the ^{1}H NMR signals corresponding to the hydrogenation product with (Fig. 1a, b) and without the immobilized catalysts (Fig. 1c, d) were observed. In addition, the insignificance of the leaching was confirmed by detecting the ^{31}P NMR spectra of the supernatant liquid obtained after the hydrogenation using the polymer-supported catalyst [19].

At the same time, leaching can be strongly affected by the experimental conditions, especially the type of solvent used, temperature, and the duration of the reaction process. Therefore, the results described above for the Wilkinson's complex supported on the polymer and on the modified silica should be considered in conjunction with the particular experimental conditions, and a reasonable conclusion can be made that reaction in benzene at 65°C on the time scale of several hours does not lead to significant metal complex leaching off the support. However, if experimental conditions are changed significantly, the necessary tests should be performed to verify that leaching is not a problem.

The issue of leaching for the Wilkinson's complex supported on mesoporous SBA-15 material was considered in detail by Gutmann et al. [32]. The mesoporous material had 2-(diphenylphosphino)ethyl linkers attached to its surface, which were intended for the anchoring of the complex. For some reason, probably because of the lower Rh content or a substantially different structure of the support, the catalyst did not show significant activity for styrene hydrogenation reaction performed in benzene-d6 at 60°C. Moreover, no PHIP was observed after bubbling parahydrogen under these reaction conditions. This observation is in a striking contrast with the results obtained earlier for the Wilkinson's complex supported on the polymer and the modified silica [19]. At the same time, when more polar solvents (methanol-d4 and acetone-d6) were used, strong ALTADENA ^{1}H NMR signals corresponding to ethylbenzene were observed in the spectra as the reaction was performed using the SBA-15-supported catalyst with the solvents heated to their boiling points [32]. Further study of filtrate solutions showed that complex leaching off the SBA-15 support was exclusively responsible for the observed polarization. Authors also concluded that leaching was probably caused by the presence of the adsorbed Wilkinson's complex, which likely can be detached from the surface much easier as compared to the complex coordinated to the ligand-containing linker. In any case, as no leaching was observed when benzene was used as a solvent, there is clear evidence that an increase in the solvent polarity leads to more significant leaching of the complex.

The successful observation of PASADENA and ALTADENA in the hydrogenations using immobilized Wilkinson's catalyst [19] confirmed the notion that the reaction mechanism does not essentially change upon the immobilization of the transition metal complex. Indeed, observation of PHIP provides direct evidence that the hydrogenation mechanism involves a pairwise hydrogen addition step, as is the case when the original Wilkinson's complex catalyzes a homogeneous hydrogenation process. On the other hand, specific details of the hydrogen addition are not known when an immobilized complex is used, and it is rather important to look for further confirmation that the homogeneous reaction mechanism is preserved upon metal complex immobilization. Importantly, PHIP can provide such information as a spin-labeling technique which can reveal the pathways of hydrogen transfer [4, 6, 34–37]. In the study reported by Skovpin et al. [31] it was shown that propyne hydrogenation in benzene-d6 over immobilized Wilkinson's complex supported on modified silica exhibits a stereoselective *syn* addition of hydrogen molecule, which is much the same as for homogeneous hydrogenation of alkynes using dissolved Wilkinson's complex [38]. In the study, the gas mixture containing propyne and parahydrogen was bubbled for ca. 10 s through 4 mL of benzene-d6 suspension of the immobilized Wilkinson's complex in a 10-mm sample tube held at 70°C, followed by the acquisition of ^1H NMR spectra. Under these conditions, the antiphase PASADENA patterns were observed only for the two resonances corresponding to vicinal vinyl protons of product propene which are in the *cis* position relative to each other [31]. The absence of PASADENA polarization for the *trans* molecular configuration in propene provided the evidence that propyne hydrogenation over the immobilized catalyst proceeded stereoselectively via *syn* addition of H_2, which in turn serves as an additional confirmation for the preservation of homogeneous reaction mechanism upon immobilization of the metal complex.

Gas-Phase Hydrogenations

In the context of PHIP research, hyperpolarization of gases is important from the mechanistic, methodological, and practical points of view. As the concentration of nuclear spins in gases is roughly three orders of magnitude lower than in liquids, the NMR signals of gases particularly suffer from low signal-to-noise ratio (SNR). The importance of hyperpolarization for the practical applications of gas-phase NMR and MRI is thus difficult to overestimate. A very limited range of gases (^{129}Xe, ^3He, ^{83}Kr) can be hyperpolarized using optical pumping techniques [39–43]. There are no reports on hyperpolarized gases produced using homogeneous PHIP as this would be, at best, very difficult to achieve. At the same time, HET-PHIP is naturally suited for gas-phase hydrogenations with parahydrogen, thus significantly extending the list of gases that can be hyperpolarized. Gas-phase HET-PHIP applications are easy to implement because gaseous substrates such as low-mass unsaturated hydrocarbons are easy to premix with parahydrogen gas. Next, one can accurately control the concentrations of individual gases in the gaseous mixture including the substrate and H_2 during the hydrogenation process which, in contrast to the liquid-phase hydrogenations, is not limited by the dissolution of H_2. All this makes it easier to get an understanding of the processes on a quantitative level.

However, in the first HET-PHIP experiments yet another aspect of heterogeneous gas-phase hydrogenation was of paramount importance, namely the fact that leaching of an immobilized complex is impossible in the absence of the liquid phase (solvent). Therefore, any observation of polarization under these conditions must stem from a heterogeneous catalytic reaction. This is the ultimate proof that in the early studies [19, 21] the observed PHIP was indeed HET-PHIP, i.e., that it was generated in a heterogeneous catalytic reaction. At the same time, the absence of a solvent may have a dramatic influence on the mechanism of hydrogenation in the gas-phase as compared to liquid-phase hydrogenations, and therefore one should expect some differences if an immobilized catalyst is used for gas-phase hydrogenations.

The first demonstration of HET-PHIP production in gas-phase hydrogenations was reported by Koptyug et al. [19] for the Wilkinson's catalyst immobilized on diphenylphosphinoethyl-modified silica. The experimental procedure consisted of preparing the gas mixture of parahydrogen and propene in a gas cylinder, and passing the mixture through the catalyst layer. PASADENA experiments were performed with the catalyst in a 10-mm NMR tube positioned in a high magnetic field (7 T) and a sample temperature of 80°C. The spectrum was acquired while the gas mixture was flowing through the NMR tube. In ALTADENA experiments the hydrogenation was performed outside the NMR magnet in an S-shaped reaction cell made of 1/8″ copper tubing which was packed with 0.1 g of the catalyst and held at 70–150°C while the mixture flowed through it and then to the NMR magnet for spectrum acquisition.

Both types of experiments showed the presence of enhanced PHIP signals in the ^{1}H NMR spectra (Fig. 2). At the same time, apart from the different polarization patterns, the ALTADENA spectrum revealed stronger polarized signals as compared to the PASADENA experiment. The likely reason for this is a lower reaction yield due to a different reaction temperature and possibly the mutual cancellation of the inner components of the antiphase multiplets for the PASADENA polarization pattern. The results of this work were further exploited in the subsequent utilization of HET-PHIP for gas-phase NMR imaging [44–46], and the signal enhancements of up to 300 compared to thermal polarization of nuclear spins were achieved using the immobilized Wilkinson's catalyst.

We note, however, that at that early stage of research the nature of the catalytic centers responsible for the hydrogenation process was not considered in detail, mostly, it seems, because at that time it was strongly believed that supported metal catalysts described later in this chapter were not expected to produce any PHIP. At the same time, the catalytic cycle for the Wilkinson's complex in solution implies a more or less active involvement of a solvent, which is absent in the case of gas-phase hydrogenations. In order to gain a deeper understanding of this issue, Skovpin et al. [31] studied the *cis–trans* stereoselectivity of hydrogenation over the immobilized Wilkinson's complex under gas-phase conditions. Propyne hydrogenation was used in the study to allow a direct comparison to the liquid-phase stereoselectivity experiment described in Sect. 2.1.1, subsection Liquid-Phase Hydrogenations. Three stages of the catalyst evolution were observed in the experiments. During the first stage when the reaction was performed at moderate temperatures (≈70°C), stereoselective *syn* addition of hydrogen to a propyne molecule was observed, as confirmed by the observation of PASADENA polarization for the proton resonances corresponding to

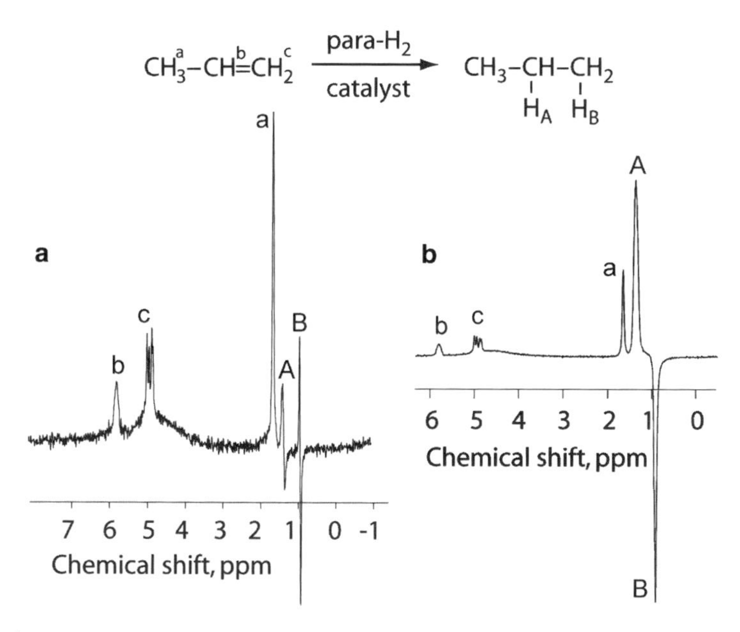

Fig. 2 ^{1}H NMR spectra from the gas-phase hydrogenation of propene with parahydrogen catalyzed by immobilized Wilkinson's catalyst supported on diphenylphosphinoethyl-modified silica. (**a**) PASADENA experiment; the sample temperature was 80°C. (**b**) ALTADENA experiment; propene was hydrogenated to propane by passing the propene/parahydrogen mixture through the catalyst packed in a cell maintained at 150°C. Subsequently, the resulting gaseous reaction mixture flowed into the high field of the NMR magnet at a flow rate of 100 standard cubic centimeters per minute (sccm). In both spectra the signals of hyperpolarized propane appear at 1.4 and 0.96 ppm and are labeled as A and B. The remaining peaks are from the unreacted propene, and the hump between 4 and 5 ppm is from *ortho*-H_2. Adapted with permission from [19]. Copyright 2007 American Chemical Society

cis molecular configuration of the product propene (Fig. 3a). This stage lasted for a short period of time on the order of 1 min. Thereafter the catalyst became inactive, and no polarization and/or product formation was detected for tens of minutes during this second stage until the temperature was elevated significantly. The third stage was observed while the catalyst was held at an elevated temperature (≈110°C) for several minutes, and was characterized by an essential increase in activity for the hydrogenation process, which was accompanied by the appearance of PASADENA-type polarization of the ^{1}H NMR signals (Fig. 3b). Moreover, there was no stereoselectivity of hydrogen addition during the third stage since polarization was observed for both *trans* and *cis* propene configurations. In addition, inspection of the catalyst showed that its color has changed from orange to dark gray. The obvious change in the nature of active catalytic centers during the catalyst evolution was eventually associated with partial reduction of the metal complex at elevated temperatures, and was confirmed by XPS analysis [31].

Therefore, the possibility of the reduction of metal complexes should be critically considered if an immobilized metal complex is used in hydrogenations performed at elevated temperatures. As for HET-PHIP in the gas-phase hydrogenations of

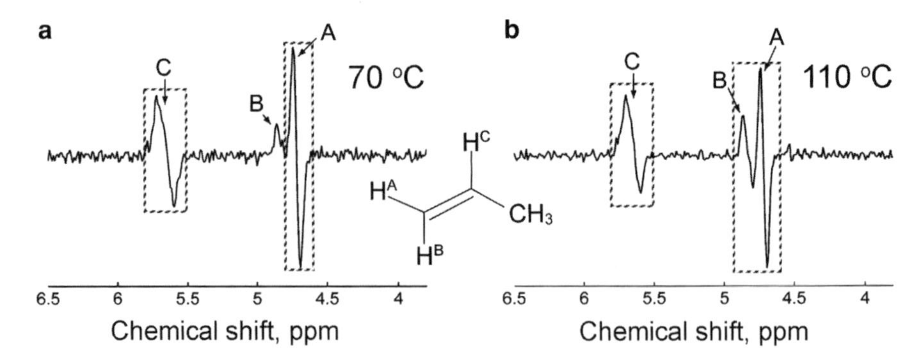

Fig. 3 ^{1}H NMR PASADENA signal patterns detected for vinyl protons of propene molecule in gas-phase propyne hydrogenation with parahydrogen. The immobilized Wilkinson's complex supported on diphenylphosphinoethyl-modified silica was used as the catalyst. (**a**) Hydrogenation performed at 70°C resulted in the observation of antiphase PASADENA signals corresponding to vinyl protons in *cis* positions relative to each other (H^A and H^C). The spectrum was recorded in the first moments after reaction initiation. (**b**) After the temperature was increased to 110°C, antiphase PASADENA signals for all vinyl protons of product propene (H^A, H^B and H^C) were observed, indicating the loss of reaction stereoselectivity (see text). The *dashed rectangles* indicate the PASADENA polarization pattern positions

propene over immobilized Wilkinson's complex [19, 44–46], the polarization observed at high temperatures (≈100°C, depending on the time of exposure to H_2 atmosphere) most probably should be ascribed to the partially reduced catalyst. On the other hand, it seems that during the first stage of the evolution of the immobilized Wilkinson's complex, the gas-phase hydrogenation proceeds similarly to homogeneous hydrogenations since stereoselective hydrogenation is observed. The rapid deactivation of the catalyst indicates the instability of catalytically active species in the absence of a solvent along with the existence of side processes such as formation of significantly less active dimeric Rh complexes [31], which in the end lead to the disappearance of catalytic activity. It should be stressed, however, that these findings regarding the possible transformations of the immobilized complexes during the reaction do not cast any doubt on the conclusions made earlier regarding the heterogeneous nature of the reaction process.

2.1.2 Other Types of Immobilized Rh Complexes

In addition to the immobilized Wilkinson's catalyst, several other immobilized Rh complexes were studied in the context of HET-PHIP research, even though the information available for these catalysts is not as detailed as for the immobilized Wilkinson's complex.

Cationic Rh complexes of different structure were utilized in homogeneous hydrogenations using parahydrogen in many reported studies [4, 6, 36, 47, 48], and demonstrate even higher activities compared to the neutral Rh complexes including Wilkinson's catalyst. Cationic complexes are less prone to dimerization,

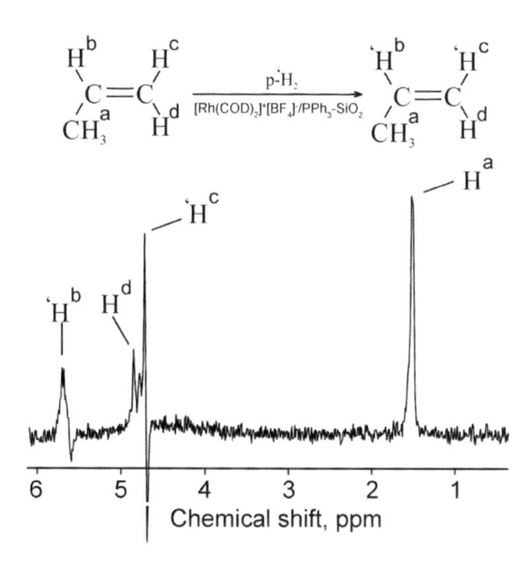

Fig. 4 ^{1}H NMR PASADENA spectrum acquired during the gas-phase reaction of parahydrogen and propene catalyzed by the immobilized $[Rh(COD)_2]^+[BF_4]^-$ complex supported on diphenylphosphinoethyl-modified silica. The reaction was performed at 25°C. The PASADENA polarization signals correspond to the protons of the CH ($'H^b$) and CH_2 ($'H^c$) groups of propene. The signal from the CH_3 group of propene is labeled as H^a. The signal of the hydrogen atom H^d of the vinyl fragment of propene shows no polarization. Reprinted with kind permission from Springer-Verlag: [31], Fig. 7

which is likely the main route of the deactivation of the immobilized Wilkinson's complex under gas-phase reaction conditions [31] (see Sect. 2.1.1, subsection Gas-Phase Hydrogenations). Thus it is rather interesting to test the heterogenized analogues of cationic complexes for generating HET-PHIP. In the work published by Skovpin et al. [31], $[Rh(COD)_2]^+[BF_4]^-$ (COD = cycloocta-1,5-diene) was used as a precursor of an immobilized cationic catalyst, which was synthesized by supporting the complex onto diphenylphosphinoethyl-modified silica. It was found that as the mixture of propene and parahydrogen started to flow through the catalyst layer in a 10-mm NMR tube positioned in a high magnetic field, the polarized signals appeared in the ^{1}H NMR spectrum even at room temperature (Fig. 4).

Note that the reagent molecule, propene, exhibits polarization which is revealed as the PASADENA signals corresponding to the vicinal protons in *cis* positions with respect to each other ($'H^c$ and $'H^b$ in Fig. 4). This may indicate the presence of an exchange process leading to a mutual pairwise exchange between the two hydrogen atoms of propene and the parahydrogen molecule. Earlier PHIP experiments performed under homogeneous hydrogenation conditions revealed the existence of hydrogen exchange processes for several cationic complexes [4, 37, 47]. It appears, therefore, that such exchange may be a feature of cationic complexes even after their immobilization. Unfortunately, the immobilized catalyst proved to be unstable under gas-phase reaction conditions, and even very moderate heating (70°C) in the presence of H_2 resulted in the reduction of the complex, which was accompanied by an almost complete disappearance of polarization.

Silica-immobilized tridentate Rh complex [Rh(COD)(sulfos)]–SiO_2 (sulfos = $^{-}O_3S(C_6H_4)CH_2C(CH_2PPh_2)_3$) is another example of a cationic substance used for generating HET-PHIP via heterogeneous hydrogenations at 70–150°C [19]. Further investigations showed that most likely the immobilized complex served as a precursor of the catalytically active material, since partial reduction of the complex occurs at the employed operating conditions. This catalyst was utilized in gas-phase hydrogenation of propene and demonstrated a significant hyperpolarization of the product propane at elevated temperatures both in PASADENA and ALTADENA experiments. Additionally, it was used for boosting sensitivity in microfluidic gas-flow imaging experiments described in Sect. 4.2 [46].

Catalytic reactions can be carried out using ionic liquids (IL) as a reaction medium. Observation of PHIP effects in ionic liquids has been reported by Gutmann et al. [49]. However, PHIP effects in that study could not be observed in the (organic liquid/ionic liquid) biphasic systems, but were successfully observed only in homogeneous solutions where IL containing a $[Tf_2N]^-$ anion simply enhanced the solubility and thus activity of a cationic rhodium complex $[Rh(COD)(DPPB)][BF_4]$ (DPPB = 1,4-bis(diphenylphosphino)butane) used as the catalyst for homogeneous hydrogenation of ethyl acrylate.

In many catalytic reactions, including hydrogenation, supported ionic liquid phase (SILP) catalysts have been successfully used [50, 51]. Usually, heterogeneous SILP catalysts comprise a thin film of an ionic liquid supported on the surface of a (porous) solid material with a homogeneous catalyst dissolved in the ionic liquid. HET-PHIP in the hydrogenation of propene using rhodium complex $[Rh(COD)(PPh_3)_2][BF_4]$ dissolved in the ionic liquid $[BMPY][Tf_2N]$ (BMPY = *N*-butyl-4-methylpyridinium, Tf_2N = bis(trifluoromethylsulfonyl)amide) supported on silica gel was studied by Gong et al. [52]. PASADENA experiments were performed with the mixture of propene and parahydrogen flowing continuously through the catalyst put at the bottom of a U-shaped cell. The cell was positioned in such a way that the catalyst layer was residing below the RF coil. At low flow rates (~2 mL/min) no PHIP was observed despite the substantial levels of conversion of propene into propane. This is likely because, due to the low flow rate, nuclear spin relaxation destroyed the polarization before the gas could leave the catalyst and reach the detection zone. When the flow rate of the gas mixture was increased to 50 mL/min (hydrogen: propene ratio 3:2), the conversion of propene to propane decreased significantly, but at the same time the strongly enhanced signals of propane were observed (the maximum enhancement factors were evaluated as 200–400). The signal enhancement was strongest in the first 10 min of reaction and dropped by more than a factor of 20 at later times, while at the same time the catalytic activity in terms of propane yield was growing for the first 20–30 min. Three SILP catalysts differing in the IL and Rh complex loadings showed somewhat different behavior in terms of conversion and PHIP signal enhancement as a function of time and reaction temperature. In particular, it was found that PHIP effects increase significantly with temperature in the range 25–75°C both for a fresh catalyst at the early reaction times and for the catalyst that was activated in the reaction for more than 30 min. Despite the observation of a pronounced activation period, the authors assumed that the chemical nature of the catalysts remained unchanged during the reaction. Therefore the explanation of the variation of the observed PHIP enhancements with changes in the initial composition of the catalyst and with time is based solely on the variation of the relaxation losses the product molecule experiences before it gets to the detection zone. The authors concluded that in the fresh catalyst the Rh complex concentrates near the gas/IL interface, while during the reaction the complex migrates toward the IL/solid interface and accumulates there. It is quite possible, however, that variation of conversion and PHIP signal enhancements with time on stream for these catalysts is caused by the changes in the chemical nature of the actual catalyst, e.g., by reduction of the complex and formation of metal particles in the ionic liquid supported on silica gel. HET-PHIP effects for metal Pd nanoparticles immersed in supported ionic liquids have been reported earlier [23] and are described in Sect. 3.1.3.

2.2 *Other Immobilized Noble Metal Complexes*

The instability of immobilized Rh complexes toward reduction encourages the further search for and application of alternative catalytically active species based on supported metal nanoparticles and other noble metal complexes. The advances in HET-PHIP on supported metals are considered later (Sect. 3). As for the immobilized metal complexes, a number of selected immobilized Au and Ir complexes have been examined to date.

The observation of HET-PHIP for a heterogeneous gold-based catalyst was reported by Kovtunov et al. [22] for an immobilized Au(III) Schiff base complex attached to a metal-organic framework (MOF) material. This catalyst, designated as IRMOF-3-SI-Au, proved to be stable under gas-phase hydrogenation reaction conditions [22, 53]. The catalyst was utilized in both propene and propyne hydrogenations employing the PASADENA experimental scheme. Notably, in spite of extensive heating (130°C) required for the reactions, no visual evidence of the metal complex reduction was found. On the other hand, the reaction yields were relatively low. The enhanced antiphase PASADENA polarization signals of propane and propene were observed for propene and propyne hydrogenations, respectively. In addition, the analysis of the polarization patterns revealed that propyne hydrogenation over IRMOF-3-SI-Au occurred via stereoselective *syn* addition of an H_2 molecule. The signal enhancement determined for hyperpolarized propane produced in propene hydrogenation was estimated to be $\approx$16 as compared to the thermal polarization. This value is significantly lower than what could be expected under ideal conditions. It was concluded in the original publication that most likely the polarization losses were caused by the interaction of propane molecule with the pore walls of the MOF material.

An investigation of immobilized Ir complexes in terms of HET-PHIP is the work-in-progress in our lab. For instance, it was found that the catalyst synthesized via immobilization of Vaska's complex $[Ir(CO)(PPh_3)_2Cl]$ onto phosphine-modified silica provides HET-PHIP only in the gas-phase hydrogenations, whereas in the liquid phase (in toluene-d8) the catalyst activity is extremely low. Propene and propyne were utilized as substrate gases in the PASADENA and ALTADENA experiments. The results obtained indicate a clear difference in the hydrogenations of double and triple bonds over the immobilized Vaska's catalyst. In the hydrogenation of the double bond of propene, the catalyst provided a significant reaction yield at relatively low temperatures (≈70°C) whilst the polarization was comparable to the thermal one. In the case of the triple bond hydrogenation, only very low reaction yields were observed even at higher temperatures (≈110°C) while the polarization enhancement of two orders of magnitude was detected. Moreover, it was found that the immobilized Vaska's complex does not tend to reduce under H_2-rich atmosphere even at reaction temperatures as high as 140°C. We note that the discussion of Ir complexes should be considered as a rather preliminary qualitative result, which, at the same time, valuably supplements all the data available on the behavior of immobilized complexes in HET-PHIP studies, and brings the description of the state of the art to completeness.

3 HET-PHIP with Supported Metal Catalysts

Activation of a hydrogen molecule by transition metal complexes, both dissolved in solution and immobilized on a solid support, usually takes place at a well-defined and localized active center. In many cases this appears to be sufficient to ensure the pairwise addition of H_2 to a substrate molecule upon its hydrogenation. As mentioned above, this pairwise addition is a necessary condition for PHIP to be observed. With supported metal catalysts, i.e., nanoparticles of a metal dispersed on a surface of a solid support material, the situation with hydrogen addition is entirely different.

The accepted reaction mechanism for heterogeneous hydrogenation by metals involves dissociative chemisorption of hydrogen on the metal surface, which creates a pool of surface hydrogen atoms available for the reaction [54]. These atoms can rapidly move over the metal surface, and for some catalysts can also dissolve into the metal lattice and/or spill over from the metal particles to the support. Subsequently, an unsaturated substrate physisorbs or chemisorbs on the metal surface. It is established, for instance, that the main pathway of alkene hydrogenation reaction involves a weakly π-bonded alkene, which harvests a hydrogen atom from the surface pool to form a surface alkyl moiety [55–57], e.g., $(CH_3)_2CH$–M or $CH_3CH_2CH_2$–M for propene hydrogenation. The latter harvests another H atom from the surface pool to yield the final product alkane ($CH_3CH_2CH_3$). In some studies, however, it is argued that π-bonded alkene (e.g., ethylene) is completely unreactive toward addition of surface hydrogen atoms since the latter cannot approach the π-orbitals of the substrate from the right direction [58], and that it is only (or predominantly) the H atoms dissolved in the metal lattice and emerging to the surface that are reactive enough to hydrogenate surface alkenes [59].

Pd metal in the presence of H_2 forms a hydride phase PdH_x with x changing within a narrow range from 0.7 to 0.63 when hydrogen pressure is varied from ca. 1 to 0.02 bar at room temperature [60]. The mobility of hydrogen atoms in this hydride phase is very high: on average, an atom jumps over the distance of 0.3 nm to the neighboring location once every 10^{-9} s. For surface hydrogens, the time between jumps is comparable [61] and decreases with increasing temperature. For comparison, the turnover frequency (TOF) of a very good hydrogenation catalyst is of the order of 100 s^{-1}, which means that one catalytic center, on average, produces a product molecule once every 10^{-2} s. Obviously, these time scales differ by many orders of magnitude, which leaves no doubt that a pool of completely randomized hydrogen atoms that exists on the surface and/or in the bulk of the metal is involved in the hydrogenation. This situation is clearly incompatible with the pairwise hydrogen addition required for the observation of PHIP. Furthermore, the initial correlation of the nuclear spins of H atoms starts to decay once their parent parahydrogen molecule chemisorbs on the metal surface and the equivalence of the two H atoms is lost. Therefore, even if two H atoms involved in the reaction accidentally come from the same parahydrogen molecule, by the time they take part in the reaction their original correlation should be long gone. The mechanism of hydrogenation reactions on supported metals is thus inconsistent with the pairwise hydrogen addition. Not surprisingly, such catalysts were not expected to produce any PHIP effects.

It is therefore quite remarkable that, after all, they do. This fact has been demonstrated recently [20]. It means that pairwise hydrogen addition, which quite likely is not the main reaction mechanism, is nevertheless achievable with supported metal particles. At this time it is impossible to formulate a detailed mechanism which leads to the formation of HET-PHIP in these systems. Several possible scenarios that have been considered are as follows:

- A small fraction of parahydrogen molecules is still able to add to a substrate in a pairwise manner on the surface of the supported metal catalysts – a purely statistical effect.
- Various species that exist on the metal surface during reaction, such as carbonaceous deposits, spectator species, reaction intermediates, side products, catalytic poisons, or reactants, provide some localization of the active sites and thus force the two H atoms to stay close to each other on the metal catalyst.
- There exist certain low-dimensional active sites such as corners, edges and certain facets of metal particles, particle-support interface regions, etc., that can operate as isolated active centers similar to homogeneous catalysts and immobilized metal complexes.
- Non-chemisorbed (physisorbed) hydrogen participates in the hydrogenation of a substrate, e.g., according to the Eley–Rideal mechanism of the reaction between adsorbed substrate and an incoming H_2 molecule.

Which of these scenarios is responsible for the results reported to date is still not clear, but some experimental observations that may point in the right direction are discussed below. However, it is possible that in fact more than one mechanism is responsible for HET-PHIP with supported metal catalysts. Contribution of each particular mechanism can change as the properties of these catalysts and the reaction conditions are varied. These studies are complicated by the fact that, in contrast to homogeneous catalysts, supported metal catalysts are not single-site catalysts, and in many cases it was established that multiple types of active sites are acting at the same time. Besides, metal nanoparticles can evolve (e.g., restructure, change composition, aggregate, etc.) under reactive conditions.

All issues discussed in the introductory part of Sect. 2 about the differences between immobilized and homogeneous complexes are equally applicable to supported metals. In addition, accelerated conversion of parahydrogen to normal hydrogen is known to take place on supported metals upon dissociative chemisorption of H_2 and subsequent recombination of H atoms followed by desorption of H_2 with thermally equilibrated nuclear spins [62–64]. Besides, utilization of supported metal catalysts does not automatically solve the problem of metal leaching into solution as leaching of a metal is still possible when such catalysts are employed [25]. We note once again that in all experiments described below, H_2 with the *ortho*:*para* ratio of ca. 1:1 is used, and this mixture is referred to as parahydrogen.

3.1 Gas-Phase Hydrogenations

3.1.1 Supported Pt Catalysts

Hydrogenation of Propene

In the first study where HET-PHIP was observed successfully for supported metal catalysts, heterogeneous hydrogenation of propene with parahydrogen over $Pt/\gamma\text{-}Al_2O_3$ catalysts was used. Substantial hyperpolarization of the two ^1H NMR signals of propane was successfully observed in both PASADENA and ALTADENA experiments [20, 21]. To account for the ability of supported metal catalysts to produce HET-PHIP, it was tentatively suggested that numerous surface species that are known to be present on the surface of the metal during reaction (e.g., carbonaceous deposits [59, 65, 66], π-bonded and di-σ-bonded propene, propylidyne [56, 57], etc.) can localize the active centers by statically or dynamically partitioning the surface into smaller areas and thus hinder the migration of hydrogen atoms on the metal surface. This tentative explanation was apparently in line with the observed dependence of the PHIP signal enhancement on the size of metal nanoparticles, with the smallest particles (0.6 nm) producing the largest signal enhancements, despite the fact that this catalyst was noticeably less active in the hydrogenation of propene than other $Pt/\gamma\text{-}Al_2O_3$ catalysts with larger metal particle sizes. This catalyst was shown to produce PHIP even at low flow rates of the reactant mixture, with the magnitude of the enhanced NMR signals being constant over a wide range of flow rates [20].

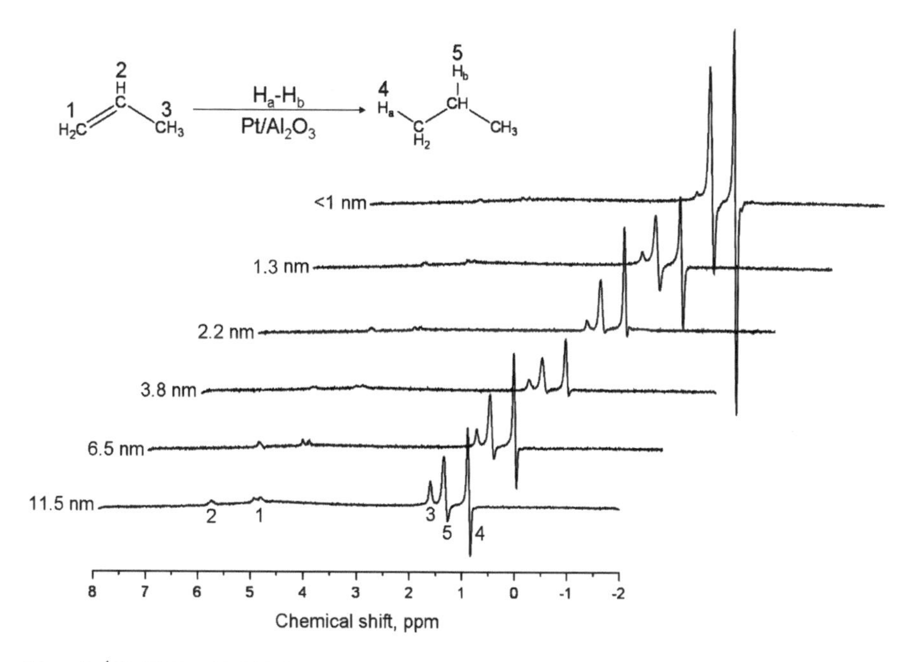

Fig. 5 ^{1}H NMR PASADENA spectra obtained during hydrogenation of propene with parahydrogen over Pt/γ-Al_2O_3 catalysts with different metal particle sizes. The reaction was performed at 345–350 K. The PASADENA polarization signals correspond to the protons of the CH_2 (5) and CH_3 (4) groups of propane. The signals from the CH_3, CH_2, and CH groups of unreacted propene are labeled as 3, 1, and 2, respectively; these signals show no hyperpolarization

A more detailed study of the structure sensitivity (particle size dependence) for hydrogenation of propene over Pt/γ-Al_2O_3 was reported later by Zhivonitko et al. [67]. For this purpose, a series of monodisperse Pt catalysts supported on γ-Al_2O_3 were prepared. The mean particle sizes and size distributions for all catalysts were evaluated using transmission electron microscopy. Heterogeneous hydrogenation reactions were carried out in a 10-mm NMR tube equipped with a screw cap. About 70 mg of a supported catalyst was loaded in the NMR tube and a mixture of gaseous reactants with the 1:4 substrate:parahydrogen ratio was passed through a 1/16" Teflon capillary extending to the bottom of the NMR tube with a flow rate of ca. 7 mL/s. During the hydrogenation experiments the sample was placed inside the NMR magnet of an NMR spectrometer (the PASADENA protocol), and the NMR spectra of the reaction mixture were detected using a single 45°-pulse without interrupting the flow [67].

In the reported experiments [67], strongly polarized PASADENA patterns were observed for the catalyst with the smallest Pt particles (<1 nm). As can be seen from Fig. 5, the dependence of signal enhancement on the Pt particle size is clearly non-monotonic (see discussion in [67] for details). To analyze the particle size dependence of the experimental observations, the following procedure was adopted. First, the values of TOF were evaluated for the catalysts studied. These values characterize the number of product molecules produced per exposed metal atom per unit time. Metal dispersion and the amount of accessible Pt atoms on the surface of nanoparticles were evaluated using irreversible H_2 chemisorption at 343 K. The TOF values were calculated separately for the overall reaction ($TOF_{overall}$) and for the pairwise H_2 addition ($TOF_{pairwise}$). The calculation of $TOF_{pairwise}$ values was based on the estimation of the contribution of the pairwise route to the hydrogenation reaction calculated from the estimated signal enhancements provided by HET-PHIP [21]. Next, the $TOF_{pairwise}$ and $TOF_{overall}$ values were related to the particle sizes using the phenomenological approach of Farin and Avnir which describes particle size effects in heterogeneous catalytic processes [68]. According to this approach, catalyst activity expressed in TOF units depends on the radius R of a metal particle as

$$\mathrm{TOF} \propto R^{D_R - 2} \tag{1}$$

where D_R is defined as reaction dimension. This parameter can be used to deduce the nature of active sites responsible for a catalytic process. When presented as a log–log plot, Eq. (1) is expected to give a linear dependence with the slope equal to $D_R - 2$.

Analysis of the results obtained with supported Pt catalysts has shown that the major (non-pairwise) reaction route was characterized by the value of $D_R \approx 2.8$ [67], which in the literature is usually associated with hydrogenations involving active sites located on the most closely packed (1 1 1) and (1 0 0) planes of platinum metals [69–71]. For pairwise H_2 addition to propene, the value obtained was $D_R \approx 0$ or 3.8 for Pt particles smaller or larger than 3 nm, respectively. This can serve as an indication that pairwise H_2 addition on highly dispersed catalysts with $2R < 3$ nm occurs mainly on the most coordinatively unsaturated corner or kink platinum atoms or other zero-dimensional defects [72], whereas on particles with $2R > 3$ nm certain active sites of multiatomic nature are responsible for pairwise H_2 addition [68]. Earlier, the contribution of the pairwise addition to the entire hydrogenation process for a Pt/γ-Al_2O_3 catalyst with 0.6-nm metal particles was estimated as ca. 3% [20, 21]. This estimate, however, assumes that no polarization is lost to nuclear spin relaxation in the reaction intermediates and products and that no equilibration of parahydrogen in the gas phase by the catalyst takes place during reaction. At the same time, the relaxation-induced losses for intermediates and products in contact with a solid material can be expected to be substantial. Thus, this estimate gives the lowest possible value of the pairwise contribution to the overall hydrogenation reaction, while the actual values may be larger.

Similar experiments were performed with Pt nanoparticles supported on ZrO_2, SiO_2, and TiO_2, and the analysis revealed similar trends for the Pt particle size dependence of the pairwise hydrogen addition [67]. For Pt catalysts supported on γ-Al_2O_3, ZrO_2, and SiO_2, the magnitudes of PHIP effects (i.e., the contribution of the pairwise H_2 addition to the double bond) were similar for similar particle sizes. However, Pt/TiO_2 supported catalysts were shown to have higher activity in pairwise H_2 addition, indicating that support can significantly affect the pairwise route of propene hydrogenation (Fig. 6). This Pt–TiO_2 synergism is presumably the

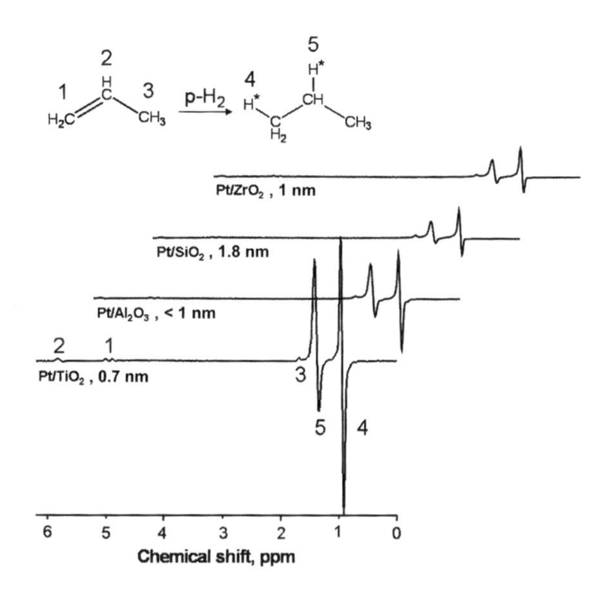

Fig. 6 ^{1}H NMR PASADENA spectra obtained during hydrogenation of propene with parahydrogen for Pt metal catalysts supported on different oxides. The reaction was performed at 345–350 K. The PASADENA polarization signals correspond to the protons of the CH_2 (5) and CH_3 (4) groups of propane. The signals from the CH_3, CH_2, and CH groups of propene are labeled as 3, 1, and 2, respectively; these signals show no hyperpolarization. The sizes of Pt nanoparticles are indicated in the figure

result of a strong metal–support interaction. These effects are well known in catalysis [73] and can induce charge transfer as well as particle morphology modifications associated with the redistribution of surface crystal faces toward less densely packed faces and defect sites and therefore could be responsible for the increase in the amount of both types of active sites that catalyze pairwise H_2 addition. Thus, both the size of metal nanoparticles and the nature of support can significantly affect the observed HET-PHIP effects.

Hydrogenation of Alkynes and Dienes

In addition to the high activity and stability, another very important property of a catalyst is its selectivity toward the required reaction product. Selective hydrogenation of alkynes and dienes in gaseous streams containing mainly C_2–C_4 alkenes is a serious challenge in the petrochemical industry [74, 75]. Before the alkenes can be used in polymerization or selective oxidation processes, trace amounts of alkynes and dienes produced in steam and catalytic cracking of hydrocarbons need to be removed selectively to avoid catalyst poisoning. Selective hydrogenation of propyne, 1,3-butadiene, and 1-butyne over supported metal catalysts is a difficult task, and solution of the problem requires a detailed understanding of the reaction mechanism [76].

As already demonstrated in preceding sections, hydrogenation of an asymmetrically substituted alkyne allows one to use PHIP to address the stereoselectivity of

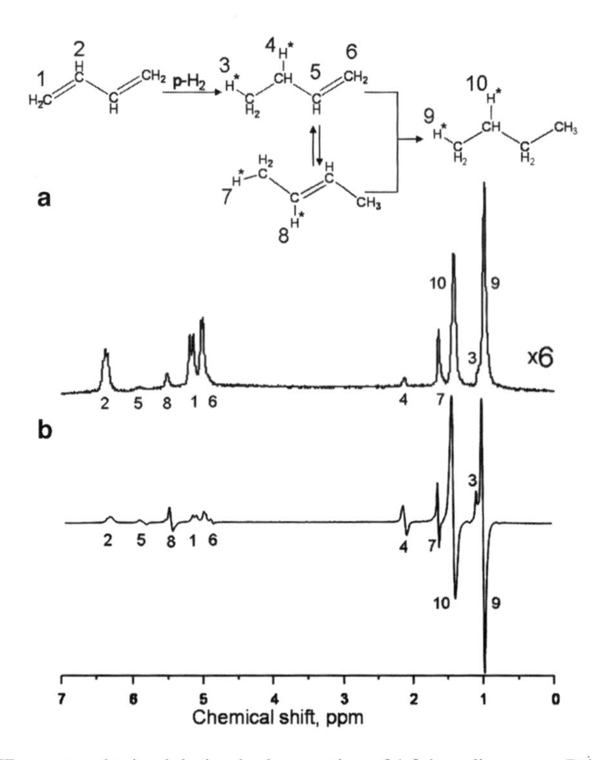

Fig. 7 ^{1}H NMR spectra obtained during hydrogenation of 1,3-butadiene over Pt/γ-Al_2O_3 catalyst with normal hydrogen (**a**) and parahydrogen (**b**). The reaction was performed at 100°C. The total hydrogenation of 1,3-butadiene (^{1}H NMR signals of CH_2 and CH groups of 1,3-butadiene are labeled as 1 and 2, respectively) provides butane (signals labeled as 9 and 10), while partial hydrogenation results in the formation of 1-butene (signals labeled as 3, 4, 5, and 6) and 2-butene (signals labeled as 7 and 8). The spectra are scaled relative to each other as indicated in the figure

hydrogenation with respect to the *syn* and *anti* addition of the two hydrogen atoms to the substrate. In particular, in propene, the product of the partial hydrogenation of propyne, the two H atoms in the CH_2 group, possess different chemical shifts. Therefore, from the PHIP NMR spectrum it is possible to obtain information regarding the *syn* and *anti* hydrogen addition by determining which of the two hydrogen atoms in the CH_2 group is polarized. For instance, in the hydrogenation of propyne with parahydrogen over the immobilized Au(III) complex discussed earlier (Sect. 2.2), only the signal of the H atom *trans* to the methyl group is polarized, which corresponds to the selective *syn* hydrogenation of propyne to propene [22]. In contrast, in the hydrogenation of propyne over supported Pt catalysts, both hydrogens of the methylene group exhibit HET-PHIP [21], presumably due to the catalytic isomerization of propene.

In 1,3-butadiene there are two conjugated double bonds, and several different products can be formed upon its hydrogenation. The total hydrogenation of 1,3-butadiene provides butane, while selective (partial) hydrogenation results in the formation of 1-butene and 2-butene (we note that *cis* and *trans* forms of 2-butene

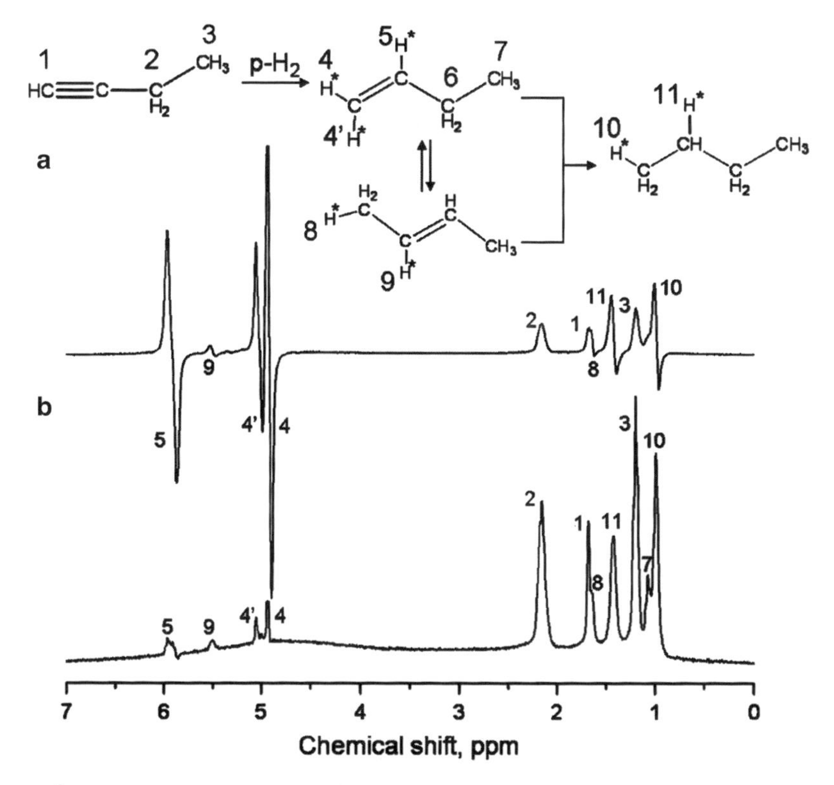

Fig. 8 ^{1}H NMR spectra obtained during hydrogenation of 1-butyne over Pt/γ-Al_2O_3 catalyst with parahydrogen (**a**) and normal hydrogen (**b**). The reaction was performed at 100°C. The total hydrogenation of 1-butyne provides butane, while selective hydrogenation results in the formation of 1-butene and 2-butene. The PASADENA polarization signals correspond to the protons of 1-butene, 2-butene and butane (**b**). The ^{1}H NMR signals of all compounds are labeled as indicated in the figure

cannot be distinguished in the spectra discussed below). The selectivity for each product and the total conversion of the substrate can be evaluated from the ^{1}H NMR spectrum of the reaction mixture (Fig. 7a). When parahydrogen was used in the hydrogenation of 1,3-butadiene, all reaction products (1-butene, 2-butene, and butane) exhibited hyperpolarization (Fig. 7b). Therefore pairwise H_2 addition contributes to the formation of each product. Based on the information provided by PHIP, it may be suggested that hydrogenation of 1,3-butadiene proceeds via semihydrogenation of 1,3-butadiene to produce butenyl intermediate which can give polarized 1-butene upon addition of the second H atom or can isomerize and yield polarized 2-butene. Hydrogenation of polarized butenes can produce polarized *n*-butane even if the addition of H_2 in this second hydrogenation is not pairwise.

Similar to propyne, in the heterogeneous hydrogenation of 1-butyne over supported platinum catalysts the products of *syn* and *anti* addition of H_2 can be distinguished by PHIP. Figure 8 shows that polarization can be successfully

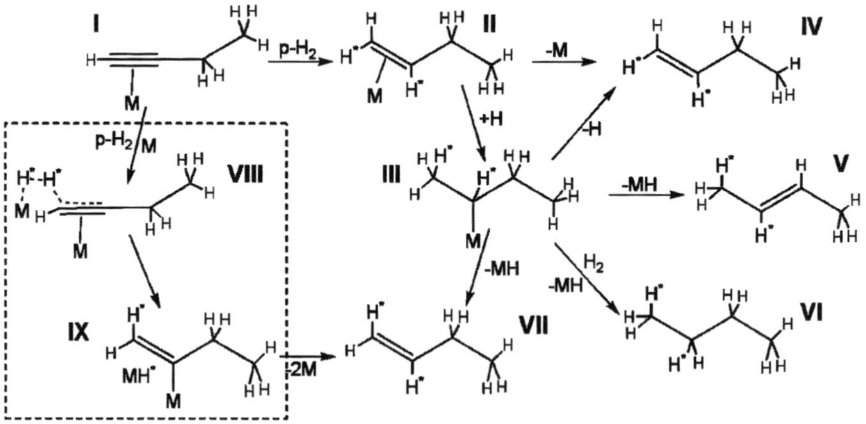

Fig. 9 The mechanistic scheme for the hydrogenation of 1-butyne with parahydrogen which includes the surface-bound 1-butyne (**I**) and a number of species produced upon its subsequent hydrogenation (**II–VII**), as well as species **VIII** and **IX** involved in the Eley–Rideal mechanism of hydrogen addition to **I**

observed for all products of 1-butyne hydrogenation. The detailed mechanism of 1-butyne hydrogenation reaction includes the surface-bound 1-butyne (**I**) and a number of species produced upon its subsequent hydrogenation (**II–VII**) (Fig. 9). Interestingly, the polarized peaks of H atoms in the CH and CH_2 groups of 1-butene that are in a *trans* position to each other (product of *anti* addition of H_2, **VII**) are stronger than the intensity of the polarized peaks of the CH_3 and CH groups of 2-butene (**V**) (Fig. 8a) even though they are expected to form from the same intermediate (**III**). Strong polarization observed for **VII** can be explained by the presence of another reaction route. While 2-butene (**V**) is formed from isobutyl intermediate (**III**), the product **VII** can also be formed in the reaction with a weakly bound H_2 molecule (species **VIII** and **IX**) by the Eley–Rideal mechanism [77, 78].

It should be noted that the product of *syn* addition of H_2 (**II**) is the first product of pairwise hydrogen addition to adsorbed 1-butyne (**I**). After that, isobutyl intermediate (**III**) is formed which yields 2-butene (**V**) in the case of elimination and *n*-butane (**VI**) in the case of isobutyl hydrogenation. Therefore, the formation of 2-butene (**V**) and 1-butene (**IV**) from the same intermediate (**III**) should give the same intensity of polarized NMR signals. However, the observation of low NMR intensity for polarized 2-butene (**V**) and high polarization for 1-butene (**VII**) confirms that *anti* addition of hydrogen to 1-butyne (**I**) prevails over the elimination of isobutyl intermediate (**III**). Therefore *anti* addition of hydrogen to 1-butyne proceeds via the stage of interaction between surface-bound 1-butyne (**I**) and weakly bound hydrogen. Our reasoning is in agreement with the literature data [79]. However, when normal hydrogen is used for hydrogenation, the concentration of 2-butene (**V**) is comparable with 1-butene (**IV** and **VII**). Therefore, the main route of 2-butene formation is 1,4-addition (not included in this scheme), leading to the loss of spin-correlation between H atoms from parahydrogen molecule. As can be seen, PHIP provides useful information about the reaction mechanisms not available with other techniques owing to the

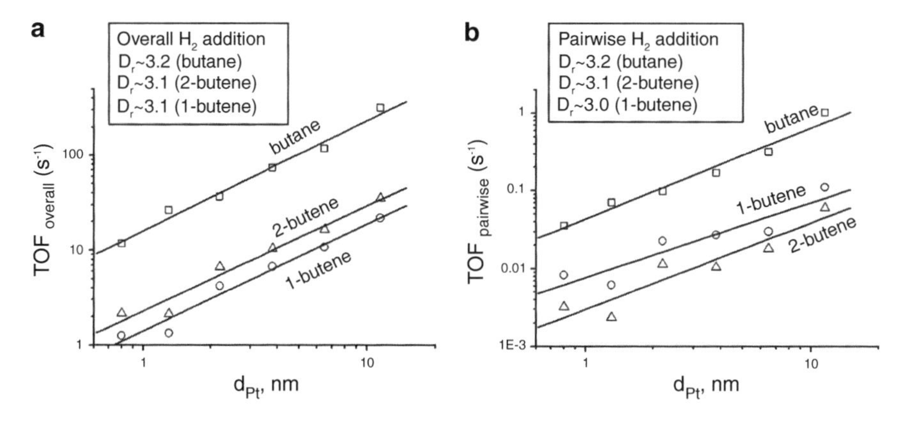

Fig. 10 The Pt particle size dependence of the specific rates of the overall 1,3-butadiene hydrogenation (**a**) and of the pairwise H_2 addition (**b**) on Pt/γ-Al_2O_3 catalysts, shown in the log–log representation. The particle sizes were determined from the TEM measurements. The TOFs were estimated from the percentage of the pairwise addition or the overall selectivity of hydrogen addition to 1,3-butadiene. The reaction dimension D_R was found to be ca. 3.1 for both overall and pairwise hydrogenation of 1,3-butadiene to 1-butene, 2-butene, and butane

ability of PHIP to track the positions of two hydrogen atoms originating from the same hydrogen molecule, in particular, when stereoselectivity of hydrogen addition and isomerization processes may be important.

As NMR spectra contain information about the concentrations of all products, the TOF values could be estimated for each product of 1,3-butadiene hydrogenation reaction. The reaction dimension [Eq. (1)] was found to be $D_R \sim 3.1$ for both overall and pairwise hydrogenation of 1,3-butadiene to 1-butene, 2-butene, and butane (Fig. 10), implying that the reaction in all cases proceeds on the flat facets of platinum, in contrast to propene hydrogenation discussed earlier where pairwise H_2 addition was observed to proceed on the active sites of different structure. Hydrogenation of 1,3-butadiene with parahydrogen on Pt/TiO_2 catalysts produced stronger polarization (Fig. 11) as compared to Pt on other oxide supports. The Pt–TiO_2 interaction effects already mentioned earlier can lead to the formation of TiO_x species on the Pt particle in the hydrogenation reaction, which can reduce the size of active centers at the (1 1 1) facets of Pt nanoparticles. In this case, H atoms have even less freedom to separate and the probability that their initial correlation is preserved may be higher.

3.1.2 Supported Pd Catalysts

Supported palladium catalysts are widely used both in industrial processes and in fundamental studies involving hydrogenation of hydrocarbons with multiple unsaturations. Specifically, selective hydrogenation of alkynes in the presence of alkenes is one of the most important industrial processes [80]. In the selective hydrogenation of propyne over a range of supported palladium catalysts, significant

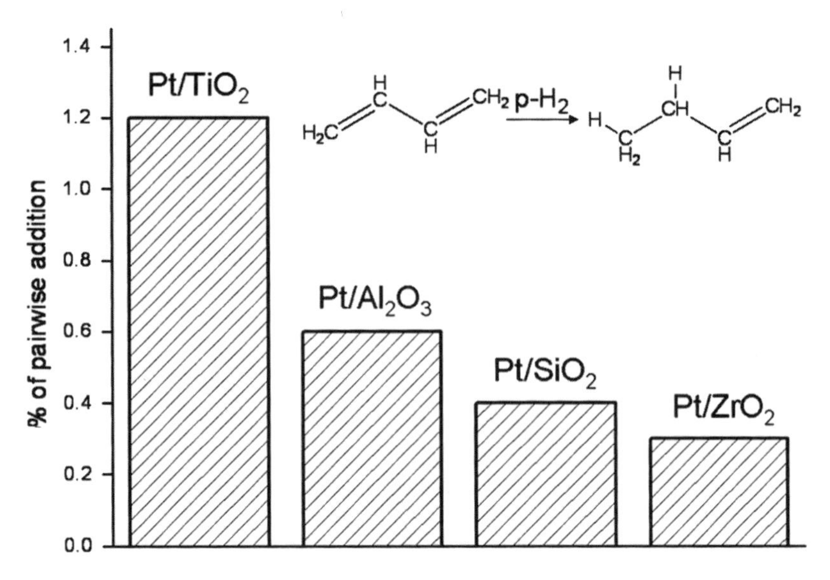

Fig. 11 The effect of the support type on the selectivity of pairwise H_2 addition to 1,3-butadiene with 1-butene formation over supported Pt catalysts. The reaction was performed at 100°C. The percentage of the pairwise hydrogen addition in the formation of 1-butene was estimated from PASADENA polarized signal of its CH_2 group

carbon deposition on the catalyst and a non-steady-state regime of the process have been observed [81]. Later studies by Kennedy et al. [82] reported the existence of two different active sites of propyne hydrogenation over supported palladium catalysts. The sites of Type I perform the total hydrogenation of propyne to propane whereas formation of carbonaceous overlayer during the early stages of the reaction defines the sites active for the selective hydrogenation of propyne to propene (Type II).

Four different series of monodisperse supported palladium catalysts (Pd/TiO_2, Pd/ZrO_2, Pd/SiO_2, Pd/γ-Al_2O_3) with different metal particle sizes were prepared and used in the heterogeneous gas-phase hydrogenation of propyne and propene. These catalysts were characterized using the same procedures as for supported Pt catalysts discussed earlier in Sect. 3.1.1. When parahydrogen was used in the hydrogenation of propene to propane, strongly polarized peaks for CH_3 and CH_2 groups of propane were observed only for Pd/TiO_2 supported catalysts, whereas for Pd/ZrO_2, Pd/SiO_2, and Pd/γ-Al_2O_3 catalysts only thermally polarized signals of propane were observed in this reaction (Fig. 12). The absence of the polarized lines in the ^{1}H NMR spectra for Pd/SiO_2, Pd/γ-Al_2O_3, and Pd/ZrO_2 catalysts in the hydrogenation of propene to propane indicates that these catalysts are unable to add molecular hydrogen to the substrate in a pairwise manner to a measurable extent. These results once again demonstrate the importance of the catalyst support for producing HET-PHIP effects. The nature of the metal is also very important, as demonstrated by the difference between these results for supported Pd catalysts and the results for Pt catalysts discussed in Sect. 3.1.1. We also note that some weak polarization of propane has been observed previously in the hydrogenation of propene over Pd/γ-Al_2O_3 catalyst [20]. This confirms the fact well-known in heterogeneous

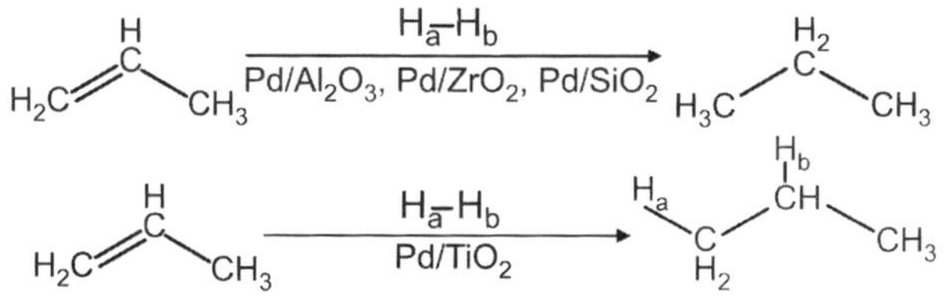

$$H_2C{=}CH{-}CH_3 \xrightarrow[Pd/Al_2O_3,\ Pd/ZrO_2,\ Pd/SiO_2]{H_a-H_b} H_3C{-}CH_2{-}CH_3$$

$$H_2C{=}CH{-}CH_3 \xrightarrow[Pd/TiO_2]{H_a-H_b} H_a{-}CH_2{-}CH(H_b){-}CH_3$$

Fig. 12 The reaction scheme that accounts for the polarized reaction products observed for propene hydrogenation at RT over supported Pd metal catalysts. Strongly polarized PASADENA peaks for the CH_3 and CH_2 groups of propane were observed only for Pd/TiO_2 supported catalysts when parahydrogen was used, whereas catalysts Pd/ZrO_2, Pd/SiO_2, and $Pd/\gamma\text{-}Al_2O_3$ were not able to produce HET-PHIP

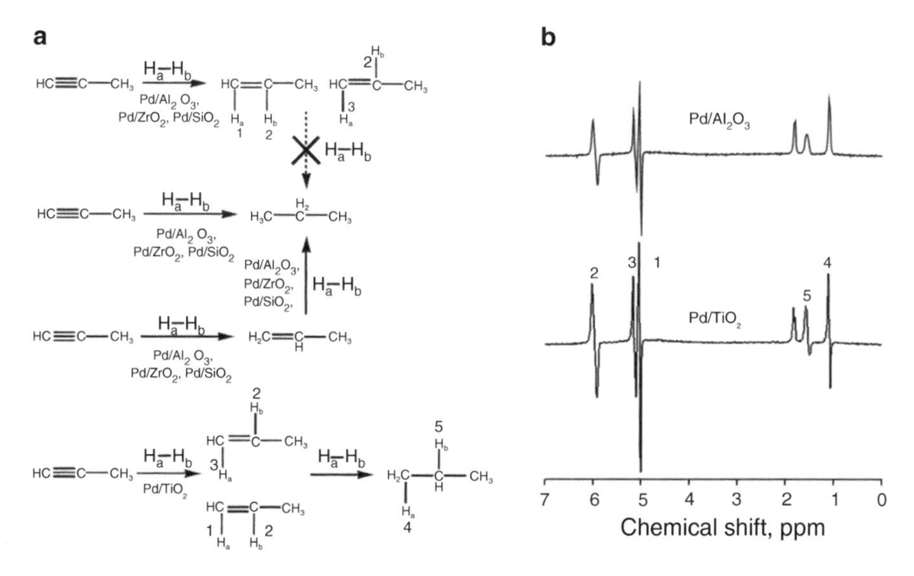

Fig. 13 ^{1}H NMR spectra detected during the hydrogenation of propyne with parahydrogen over the Pd/TiO_2 and $Pd/\gamma\text{-}Al_2O_3$ catalysts at RT (**b**) and the reaction scheme that accounts for the PASADENA polarized products observed for hydrogenation of propyne with parahydrogen over palladium supported metal catalysts at RT (**a**). The PASADENA polarized signals of propene and propane are labeled as indicated in the figure

catalysis that the details of the catalyst preparation procedure can significantly affect its catalytic behavior.

The same four series of supported palladium catalysts were used in the heterogeneous hydrogenation of propyne. When Pd/ZrO_2, Pd/SiO_2, and $Pd/\gamma\text{-}Al_2O_3$ catalysts were used in the HET-PHIP experiments, strongly polarized peaks were observed only for propene, whereas propane formed in this reaction exhibited only thermal polarization of its ^{1}H NMR signals. In contrast, when Pd/TiO_2 catalysts with various Pd particle sizes were used in this reaction, clear PASADENA polarization patterns could be observed for both propene and propane (Fig. 13b). Thus, the contribution of the pairwise H_2 addition route in the hydrogenation of propyne over supported Pd catalyst strongly depends on the nature of catalyst support

(Fig. 13a). The absence of polarization for the signals of CH_2 and CH_3 groups of propane and clear polarization patterns for the CH and CH_2 groups of propene for Pd/ZrO_2, Pd/SiO_2, and $Pd/\gamma\text{-}Al_2O_3$ catalysts allow us to conclude that the pairwise hydrogen addition route exists only for the selective hydrogenation of propyne to propene and that formation of propene and propane takes place on different active sites (Fig. 13a). These conclusions made on the basis of the observation of PHIP effects are in agreement with the data on the mechanism of propyne hydrogenation which is reported to involve two different active sites. The sites of Type I are active in the formation of propane and the sites of Type II are responsible for propene formation [82]. Our PHIP results indicate that the sites for propane and propene formation via pairwise hydrogen addition are also different (Figs. 12 and 13a).

The metal particle size dependence of hyperpolarization magnitude was found to be different for different supports. For $Pd/\gamma\text{-}Al_2O_3$ and Pd/ZrO_2 the smallest metal particle sizes gave the lowest intensities of the polarized lines of CH and CH_2 groups of propene in the hydrogenation of propyne, while for Pd/SiO_2 the polarization decreased with increasing Pd particle size. For Pd/TiO_2, the intensities of the polarized lines of CH and CH_2 groups of propene were the same for different metal particle sizes while the intensities of the polarized lines of the CH_2 and CH_3 groups of propane decreased with decreasing metal particle size.

In contrast to the hydrogenation of propyne where only the Pd/TiO_2 catalyst produced HET-PHIP for all reaction products (propene and propane), in the hydrogenation of 1-butyne all catalysts (Pd/TiO_2, Pd/SiO_2, $Pd/\gamma\text{-}Al_2O_3$, and Pd/ZrO_2) produced HET-PHIP effects for all hydrogenation products. In this case, the support influence was reflected in the different intensity of polarized signals.

In the hydrogenation of 1,3-butadiene with parahydrogen over supported palladium catalysts, only 1-butene and 2-butene were weakly polarized when Pd/SiO_2, $Pd/\gamma\text{-}Al_2O_3$, and Pd/ZrO_2 catalysts were used. However, for Pd/TiO_2 the PHIP effects could be observed for all products of 1,3-butadiene hydrogenation (Fig. 14).

3.1.3 Metal Nanoparticles Immersed in Ionic Liquid

Recently, metal nanoparticles directly formed in an ionic liquid were successfully used as catalysts for hydrogenation of unsaturated compounds both in a liquid–liquid biphasic system and in a gas–SILP (supported ionic liquid phase) system [83, 84]. In the recent HET-PHIP study [23], two different ionic liquids supported on activated carbon fiber (ACF) materials and containing Pd nanoparticles were used: Pd/[bmimOH][Tf_2N]/ACF with 5-nm Pd particles and Pd/[bmim][PF_6]/ACF with 10-nm Pd particles. It was found that these catalysts are highly active in the gas phase heterogeneous hydrogenation of propyne at 130°C and hydrogenate propyne into propene and to a lesser extent to propane, with observation of a strong PHIP effect for propene molecules and a weak one for propane (Fig. 15). The signal enhancement factor provided by PHIP was estimated by comparing the intensity of NMR lines for methyne groups of hyperpolarized and thermally polarized propene in the

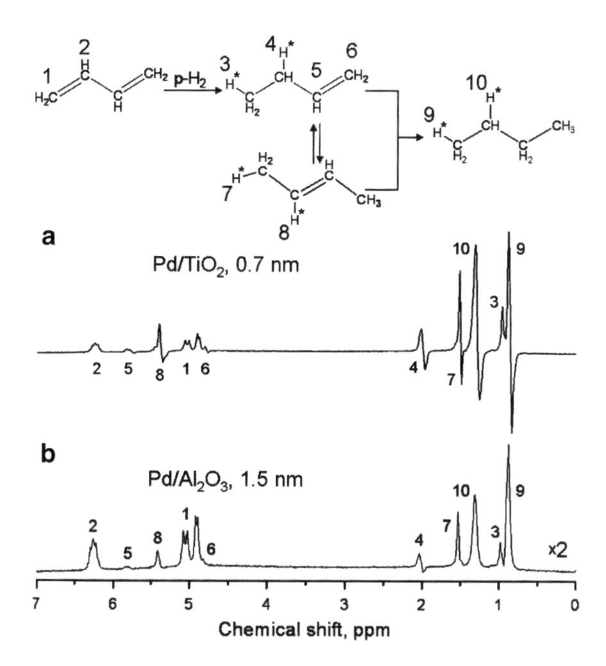

Fig. 14 ^{1}H NMR PASADENA spectra detected during hydrogenation of 1,3-butadiene with parahydrogen over Pd/TiO_2 catalyst (**a**) and $Pd/\gamma\text{-}Al_2O_3$ catalyst (**b**). The reaction was performed at 100°C. The selective pairwise hydrogen addition to 1,3-butadiene over Pd/TiO_2 supported catalyst provides formation of polarized butane, 1-butene and 2-butene (**a**), whereas over $Pd/\gamma\text{-}Al_2O_3$ supported catalyst only 1-butene is polarized (**b**). The ^{1}H NMR signals of all compounds are labeled as indicated in the figure

corresponding ^{1}H NMR spectra and was found to be ca. 8 [23]. This signal enhancement is ca. 60 times lower than the maximum possible value [21]. PHIP was observed for the methyl and the methylene groups of propane as well. If addition of H_2 to propene is non-pairwise, formation of polarized propene from propyne and subsequent hydrogenation of propene can carry over its polarization to propane. Alternatively, polarization of propane can form directly if hydrogenation of propene is partially pairwise.

3.2 *Liquid-Phase Hydrogenations*

It was successfully demonstrated that supported metal catalysts are able to produce PHIP effects in liquid-phase hydrogenations as well [21, 85, 86]. In particular, Rh/TiO_2 and Rh/AlO(OH) catalysts were used in a heterogeneous hydrogenation of dissolved propene gas. Both catalysts hydrogenated propene into propane in toluene and Rh/TiO_2 hydrogenated propene in acetone. When a mixture of parahydrogen and propene was bubbled through the NMR tube containing the solvent and the

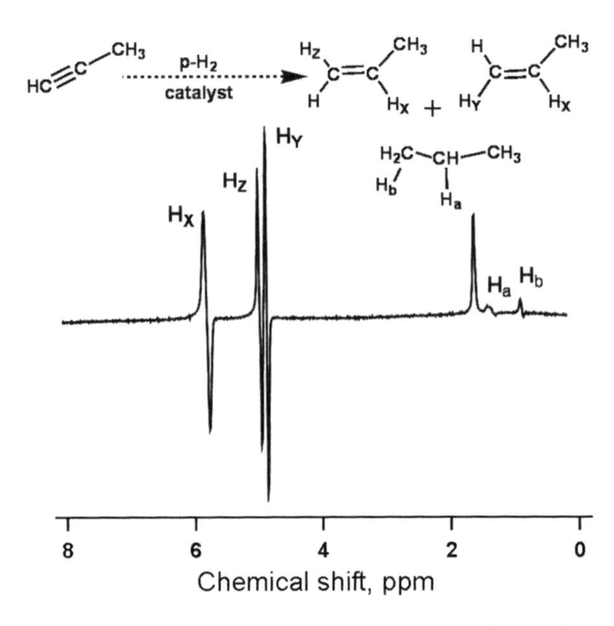

Fig. 15 ^{1}H NMR spectrum detected during the heterogeneous hydrogenation of propyne with parahydrogen over Pd/[bmimOH][Tf_2N]/ACF catalyst with 5 nm Pd particles. The polarized protons of methyne and methylene groups of propene are labeled as H_X, H_Y, and H_Z and the polarized methylene and methyl protons of propane are labeled as H_a and H_b. Adapted from [23] with permission from The Royal Society of Chemistry

solid catalyst, HET-PHIP was observed in solution for the reaction product propane. The experimental spectra detected during propene hydrogenation over Rh/TiO_2 in acetone-d6 are shown in Fig. 16, demonstrating that both ALTADENA and PASADENA experiments can produce HET-PHIP in liquid–solid heterogeneous hydrogenations. The possibility to carry out the reaction and to produce HET-PHIP outside the NMR magnet (ALTADENA) using supported metal catalysts opens up the possibility to use such catalysts for the production of continuously flowing catalyst-free polarized liquids, and in addition provides much more flexibility in the experiment design and the experimental conditions as compared to the hydrogenation performed within the restricted space of an NMR magnet inside an RF probe. For instance, much higher temperatures, pressures, and significantly larger reactor volumes can be used in an ALTADENA experiment.

Once a polarized product molecule is produced, its polarization starts to decay due to nuclear spin relaxation processes. The relaxation is expected to be much faster for molecules residing in the pores of the supported catalysts than for molecules in a bulk fluid. Compared to the gas-phase hydrogenations, in the liquid-phase hydrogenations using heterogeneous catalysts the diffusional transport of molecular species is much slower. This can lead to significant polarization losses and even to a complete destruction of polarization while the polarized product molecule diffuses out of a porous catalyst support. To verify that PHIP can be observed for slower diffusing molecules, styrene was chosen as a higher molecular weight substrate for further experiments. The same two supported metal catalysts were found to hydrogenate styrene into

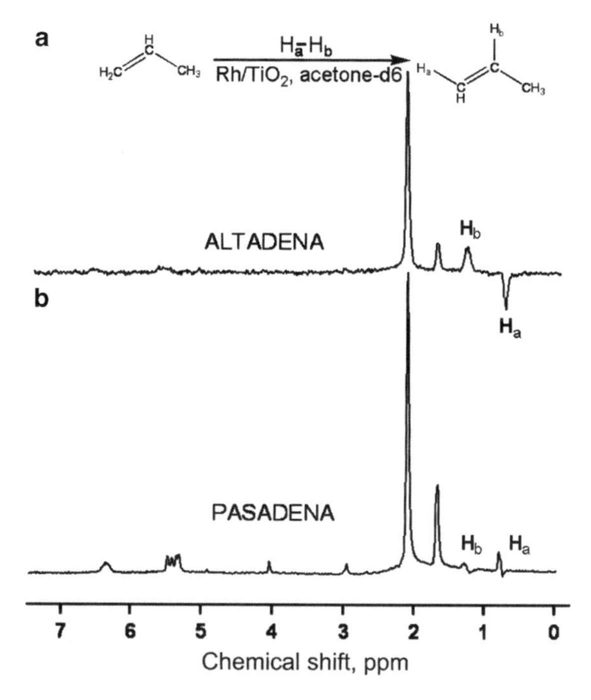

Fig. 16 ^{1}H NMR spectra detected in the heterogeneous hydrogenation of propene catalyzed by Rh/TiO_2 supported metal catalyst in acetone-d6 at room temperature in the ALTADENA (**a**) and PASADENA (**b**) experiments. The enhanced polarized lines of the CH_2 and the CH_3 groups of propane are labeled H_b and H_a, respectively

ethylbenzene in acetone-d6 and to produce PHIP effects when parahydrogen was used in the reaction [85]. The results demonstrate that observation of ALTADENA polarization patterns in liquid phase heterogeneous hydrogenations should be possible for a variety of substrates, and that the use of supported catalysts, without doubt, holds a potential for significantly broadening the range of PHIP applications.

Production of water-soluble hyperpolarized contrast agents is essential for extending PHIP MRI applications to novel in vivo studies. The Rh/TiO_2 supported catalyst was therefore further tested in the aqueous phase hydrogenation experiments. It was found that the Rh/TiO_2 catalyst hydrogenates acrylamide [85] and allyl methyl ether when the heterogeneous hydrogenation reactions are carried out in D_2O at 353 K, and that relatively weak but unambiguous PASADENA polarization patterns are observed when parahydrogen is used. The spectrum detected during the heterogeneous hydrogenation of allyl methyl ether into ethyl methyl ether in D_2O is shown in Fig. 17.

A random assortment of supported metal catalysts was tested in the liquid-phase hydrogenation of the triple bond of methyl propiolate in a PASADENA experiment at room temperature (RT) or 305 K and 2–3 bar of parahydrogen by Balu et al. [86]. For metals supported on carbon, essentially no PHIP could be observed. The activity of Au and Fe based catalysts was too low to observe any chemical conversion or hyperpolarization. A pronounced polarization of the reaction product,

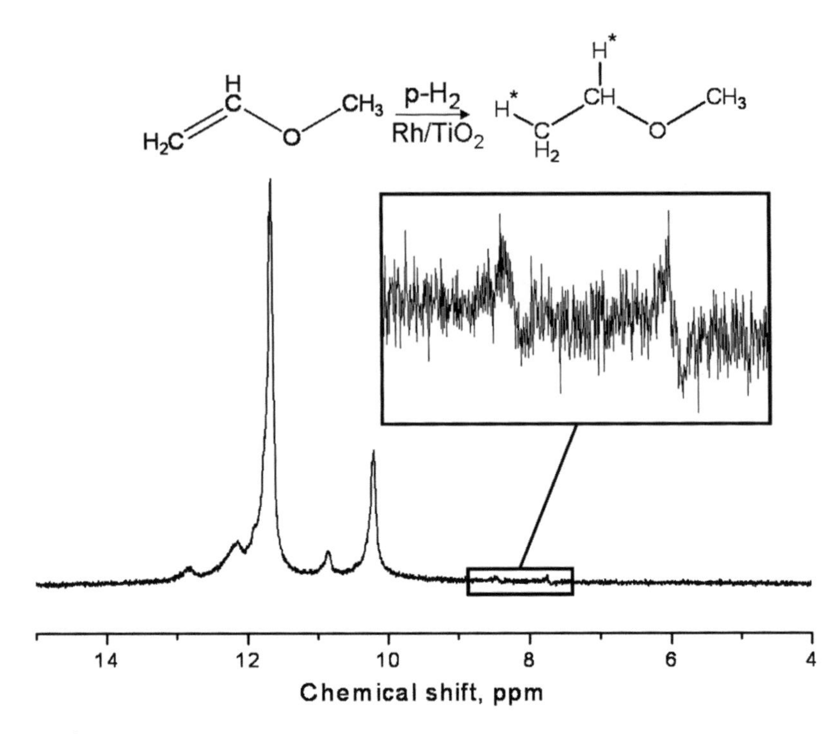

Fig. 17 ^{1}H NMR spectrum detected in the heterogeneous hydrogenation of allyl methyl ether with parahydrogen over Rh/TiO_2 supported metal catalyst in D_2O at 353 K in a PASADENA experiment

methyl acrylate, was observed in methanol-d for Pd/silica and for Pt supported on mesoporous Al-SBA-15 and Al-MCM-48 materials, but the strongest polarization was observed with the Pt/silica catalyst. The latter was further tested in the hydrogenation of a number of alkynes and alkenes. Clear PHIP signals were observed in the hydrogenation of styrene and 1-phenylpropyne, but when non-conjugated linear compounds containing double and triple bonds were employed, no PHIP could be detected despite the fact that hydrogenation products were clearly observed in the ^{1}H NMR spectra. The validity of the reaction-site localization hypothesis advanced earlier [20, 21] was questioned, but no alternative explanation of the mechanism of HET-PHIP formation was suggested.

4 HET-PHIP Applications in NMR Imaging

Conceivably, the sensitivity boosting provided by PHIP can be quite useful in all areas of magnetic resonance, but its applications are limited by the requirements associated with the practical implementation of the technique. The use of heterogeneous catalysts for producing hyperpolarization has a number of advantages compared to the homogeneous ones. At the same time, in order to apply successfully HET-PHIP in ^{1}H NMR imaging, one should carefully consider a number of important points. In particular:

- Necessary attention should be paid to the specific shape of the polarization pattern of the NMR signals of hyperpolarized molecules during an NMR pulse sequence design and image reconstruction.
- A relatively fast relaxation of the hyperpolarization requires fast fluid transfer and signal readout.
- A heterogeneous catalyst layer can significantly accelerate the relaxation of hyperpolarization compared to the catalyst-free regions.

Another general comment to add is that gas-phase HET-PHIP NMR imaging experiments are easier to implement than the analogous liquid-phase experiments. This is because molecular hydrogen is a gas under ambient conditions and can be easily premixed in any proportions with gaseous substrates such as unsaturated hydrocarbons to form a reagent mixture for the hydrogenation reaction. In contrast, the solubility of molecular hydrogen in common organic solvents is very low, and thus in the liquid-phase hydrogenations the rate of production of hyperpolarized substances is significantly reduced unless higher pressures or other sophisticated approaches [11, 87–89] are employed. This is the main reason why HET-PHIP applications in NMR imaging known to date are gas-phase experiments. Generally, these experiments were based on the production and use of hyperpolarized propane in the propene hydrogenation with parahydrogen over suitable heterogeneous catalysts. The details of these studies are briefly described and discussed below.

4.1 Gas-Phase Imaging of Relatively Large Objects

As mentioned earlier, low concentrations of molecules in gases make imaging of lungs, porous materials, and other systems with void spaces very challenging. This sensitivity problem has pushed the use of optically pumped hyperpolarized gases such as ^{129}Xe, ^{3}He, and ^{83}Kr [39–43]. Meanwhile, rather simple implementation of HET-PHIP for the production of hyperpolarized gases opens up some new attractive opportunities for gas-phase MRI.

The first instance of HET-PHIP application for gas-phase imaging [44] utilized heterogeneous hydrogenation of propene with parahydrogen catalyzed by an immobilized Wilkinson's catalyst ($[RhCl(PPh_3)_2PPh_2(CH_2)_2]–SiO_2$) to produce a continuous flow of hyperpolarized propane gas. The flow diagram of the experimental procedure is shown in Fig. 18.

At the initial stage of these experiments, parahydrogen (a hydrogen mixture enriched to 50% *para*-H_2) was produced by passing pure hydrogen through a tube packed with *ortho–para* conversion catalyst (FeO(OH)) at the liquid nitrogen temperature (Fig. 18a). After that, the mixture of parahydrogen and propene in a 4:1 ratio was prepared in a gas cylinder (Fig. 18b). Hydrogenation of propene with

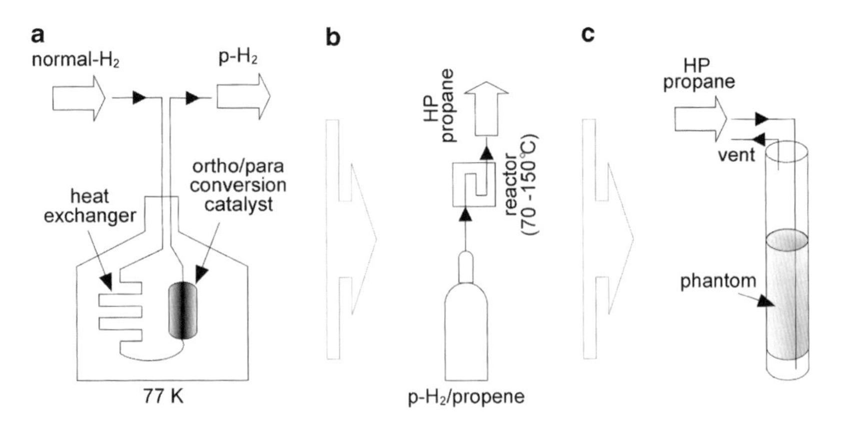

Fig. 18 Flow diagram for the gas-phase experiment. (**a**) The parahydrogen enrichment was performed at 77 K by passing the normal hydrogen through the *ortho–para* conversion catalyst, FeO(OH), immersed in liquid nitrogen. (**b**) The mixture of parahydrogen and propene (in a 4:1 ratio, respectively) prepared in a gas cylinder was passed through the reactor comprising an S-shaped catalytic cell packed with immobilized Wilkinson's catalyst that was held at 70–150°C. This procedure produced hyperpolarized propane at the Earth's magnetic field. (**c**) The hyperpolarized propane was transferred to the 7 T field of an NMR magnet where it passed through the model phantoms for the NMR imaging experiments. The transfer produced an ALTADENA polarization pattern for propane

parahydrogen was then carried out by supplying the mixture into an S-shaped catalytic reactor made of copper tubing packed with a certain amount of the immobilized Wilkinson's catalyst. The reactor (polarizer) was kept at ca. 150°C and served as the source of hyperpolarized propane gas. The estimated reaction yield was about 5%; therefore the resulting reactant–product mixture contained only about 1 vol.% of hyperpolarized propane. After that, the reaction mixture was adiabatically transferred to the high field of an NMR magnet resulting in two strongly enhanced ALTADENA multiplets of opposite sign (emissive and absorptive) in the ^{1}H NMR spectrum, which was detected using a single-pulse excitation (Fig. 19).

These enhanced signals were utilized to image simple phantoms made of Teflon that were placed inside a 10-mm NMR tube. For this, hyperpolarized propane was allowed to flow through the phantoms from the bottom of the tube to the top (see Fig. 18c), and images of the gas flowing in the void spaces of the phantoms were acquired using pure phase-encoding protocol for the spatial domain [90] so that the frequency information in the spectroscopic domain (i.e., the NMR spectrum) was preserved in the experiment. One of the enhanced ALTADENA peaks was integrated to generate the images shown in Fig. 20c, e. For comparison, conventional ^{1}H MR image of one of the phantoms with the void spaces filled with water is shown in Fig. 20a.

The SNR was estimated to be on the order of 150–200 for the images corresponding to hyperpolarized propane (Fig. 20c, e), while no or very little signal could be seen in the images generated using the CH_2 peak of unreacted thermally

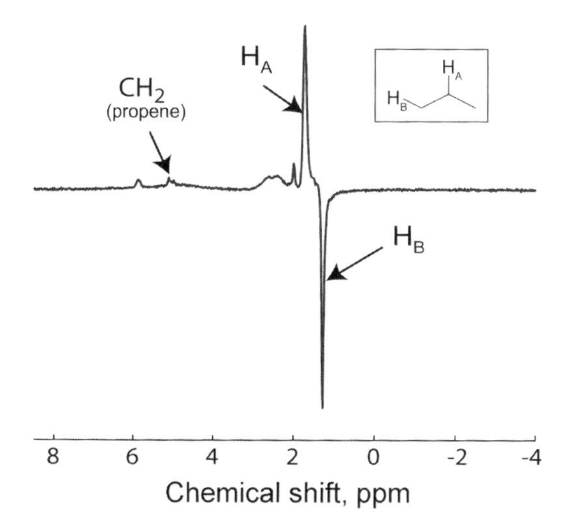

Fig. 19 ^{1}H NMR spectrum of the mixture of propene and polarized propane gas produced upon hydrogenation in the catalytic cell under flowing conditions. The peaks from the parahydrogen-derived propane protons and the CH_2 peak in the unreacted propene are marked in the spectrum [44]. Copyright Wiley-VCH Verlag GmbH & Co. KGaA, Weinheim. Reproduced with permission

polarized propene (Fig. 20b, d) which was present in the gas mixture in much larger quantities as compared to propane.

The low signal of thermally polarized propene made it impossible to estimate accurately the polarization enhancement directly from these images. Instead, the enhancement was determined from the ^{1}H NMR spectra similar to that shown in Fig. 19. First, the reaction yield (conversion of propene to propane) was estimated to be 5% (1/20). Second, it was found that under flowing conditions the ^{1}H NMR signal intensity of propene was only about one half of its thermal equilibrium value. Third, the ratio of peak areas for the ALTADENA-enhanced propane protons to the thermally polarized propene protons was determined to be 30:1. Altogether this gave the value of the enhancement factor of $30 \times 20 \times 1/2 = 300$. To be more accurate, the enhancement should be determined by comparing the signals of propane in two separate experiments with normal hydrogen and with parahydrogen. Nevertheless, one can conclude that an enhancement of at least two orders of magnitude relative to the thermally polarized gas is a reliable estimate for these gas-phase imaging experiments.

4.2 *Combining HET-PHIP and Remote Detection MRI for Microfluidic Gas-Flow Imaging*

Another example of MRI applications of hyperpolarized gases produced using HET-PHIP concerns microfluidic gas-flow imaging [46]. In principle, NMR is

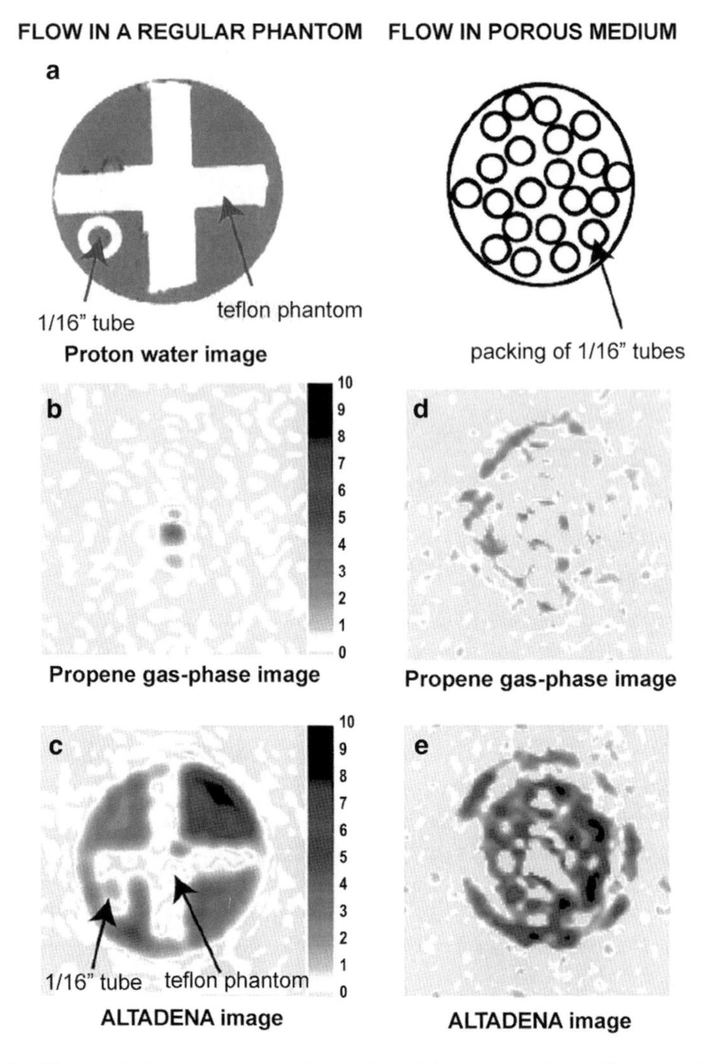

Fig. 20 (**a**) High-resolution water proton image for a thin cross-sectional slice through a cross-shaped Teflon phantom. The *circle* depicts the Teflon tube (0.16 cm or 1/16 in.) which delivers the gas. (**b**) Propene gas phase image based on the signal intensity from the CH_2 peak. (**c**) ALTADENA image from the propane CH_3 peak. Parts (**b**) and (**c**) were reconstructed from the same experiment and are plotted with respect to the same color scale. Images (**d**) and (**e**) correspond to a phantom consisting of a random packing of capillaries to demonstrate gas-phase MRI in a porous medium [44]. Copyright Wiley-VCH Verlag GmbH & Co. KGaA, Weinheim. Reproduced with permission

one of only a few techniques suitable for a noninvasive and traceless monitoring of microfluidic flow phenomena. At the same time, the problem of its low sensitivity is especially severe when small amounts of fluids in microchannels are imaged using relatively large conventional RF detectors. This issue is complicated further when gases, whose molecular density number is several orders of magnitude lower than in liquids, are under investigation.

The sensitivity issue can be resolved to a good extent by using an alternative signal detection scheme named "remote-detection" (RD) [91–95]. In this technique the encoding of spatial information and the detection of the signal are performed using two different RF coils in different spatial locations. Typically, a microfluidic device is placed inside an encoding RF coil which is large enough to house the entire device. After the spatial encoding step, the fluid flows out of the device through a thin capillary and subsequently arrives at the detection coil for signal read-out. The size and circuit characteristics of the detection coil are adjusted to have ultrasensitive detection properties. As the fluid molecules have to travel from the encoding coil to the detection coil, the remote-detection technique inherently also provides the time-of-flight information.

Depending on the sample, RD can provide the sensitivity enhancement of up to several orders of magnitude in microfluidic flow studies, but even under these circumstances an additional sensitivity boost is desired for microfluidic gas-flow imaging. In the study reported by Telkki et al. [46], this additional sensitivity boost was achieved by hyperpolarizing propane gas using HET-PHIP. As compared to the phantom experiments described above [44], the important difference of these experiments is that the total volume of the channels in the microfluidic systems studied is several orders of magnitude lower than the volume of the voids in the phantoms studied using conventional MRI. Therefore it is essential to achieve the maximum possible sensitivity in the RD experiments. As the relaxation time of propane gas is relatively short (on the order of 900 ms), it is essential to implement a reasonably fast transfer of the hyperpolarized propane from the reactor where it is produced to the NMR system where the microfluidic device and the RF coils are located (in the experiments discussed here the two RF coils are located close to each other within the same NMR magnet [46]). At the same time, the gas flow in the microfluidic device has to be much slower in order to have reasonable residence times of the gas in the device and in the detection coil. In practice, it was accomplished by connecting an additional outlet gas line branch close to the inlet part of the microfluidic chip. The needle valve 1 at the end of this branch (Fig. 21a) allowed one to maintain the required fast transfer of the gas but efficiently decreased the rate of gas flow through the microfluidic system under study (Fig. 21b). The needle valve 2 was connected to the outlet part of the microfluidic system for a more precise regulation of gas flow rate.

Supported Wilkinson's catalyst or [Rh(COD)(sulfos)]/SiO_2 (sulfos = $^{-}O_3S(C_6H_4)CH_2C(CH_2PPh_2)_3$) were utilized as the hydrogenation catalyst precursors for the propene hydrogenation process to produce the hyperpolarized propane in a similar way to that described in the previous section. Both catalysts provided signal enhancements of about 80 compared to thermal polarization in the 7 T magnetic field, while the estimated reaction yield was 60%. Because of different reaction conditions, the product yield was higher than in the experiments described previously (ca. 5%) [44]. Notably, this much better yield resulted in about three times stronger signal of the hyperpolarized propane, even though the polarization

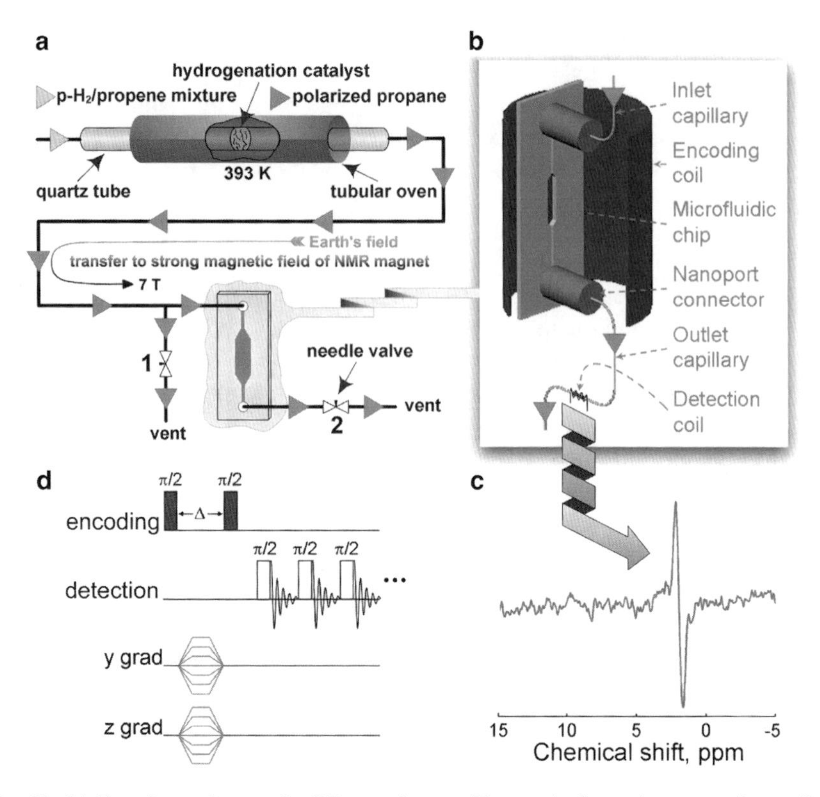

Fig. 21 (**a**) Experimental setup for RD experiments. The parahydrogen/propene mixture flows through the hydrogenation catalyst layer packed between plugs of glass wool inside a heated quartz tube. After the hydrogenation reaction, the polarized propane gas flows from the tube into the microfluidic chip positioned inside an NMR magnet. (**b**) Remote-detection MRI experimental setup. (**c**) ^{1}H NMR spectrum of hyperpolarized propane gas measured by the detection coil. (**d**) Remote-detection MRI pulse sequence, in which phase encoding of spatial coordinates is carried out in the *y* and *z* directions. Time-of-flight information is obtained using a train of $\pi/2$ pulses applied in the detection coil [46]. Copyright Wiley-VCH Verlag GmbH & Co. KGaA, Weinheim. Reproduced with permission

enhancement in this study was somewhat lower. The sensitivity of the detection microcoil and the PHIP signal enhancement were sufficient to observe an ALTADENA spectrum of continuously flowing polarized propane at 1 atm pressure in a 150-μm capillary after accumulation of eight scans (Fig. 21c). The gas volume inside the detection coil was only 53 nL. The SNR was estimated to be about 8.8 per scan and atmosphere, which is 88 times higher compared to that for hyperpolarized xenon in a $Xe/He/N_2$ gas mixture in a continuous flow experiment performed using an almost identical microcoil [94].

Three different microfluidic systems were examined in the imaging experiments (Fig. 22). First, a 150-μm capillary tubing was set to lead through the encoding and the detection coils, and two-dimensional RD time-of-flight images were acquired using the pulse sequence shown in Fig. 21d. The images corresponding to the

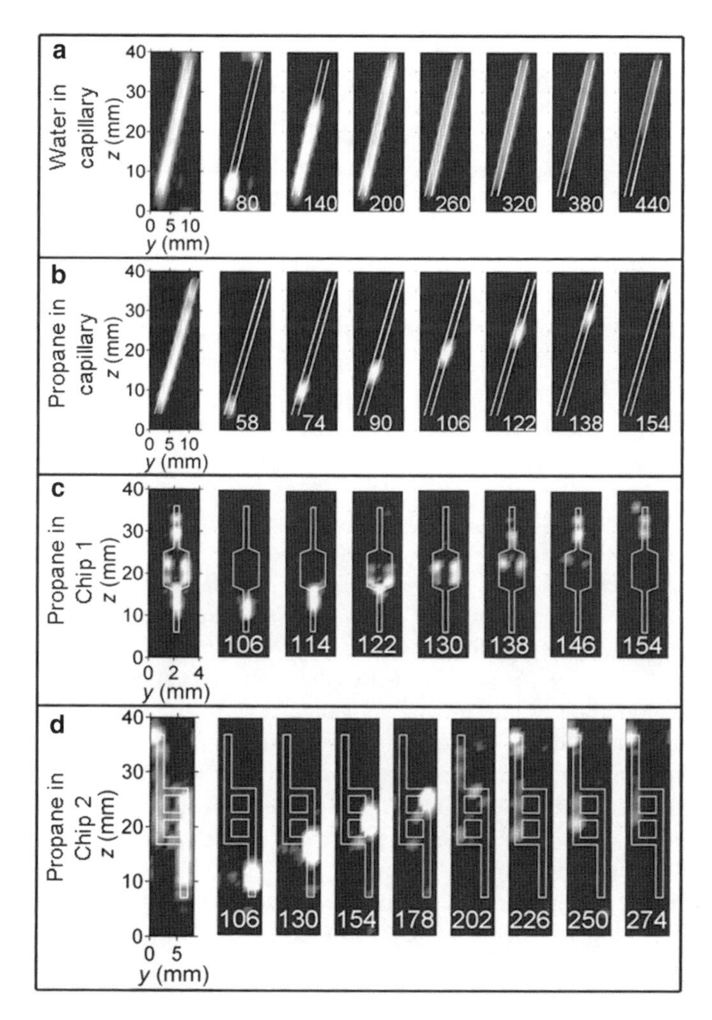

Fig. 22 Remote-detection time-of-flight images measured from microfluidic systems. Water (**a**) and hyperpolarized propane (**b–d**) flowing in the capillary tubing leading through the encoding coil (**a**, **b**), and in microfluidic chips [Chip 1 (**c**) and Chip 2 (**d**)]. The flow channels are outlined with *white lines*. The first images on the *left* are the result of summation of subsequent images measured at different travel-time instants (time projection). The travel time instants are indicated in milliseconds. The spatial resolution in the y and z directions is 0.5–1.6 mm and 1.5–2.5 mm, respectively, depending on the experiment [46]. Copyright Wiley-VCH Verlag GmbH & Co. KGaA, Weinheim. Reproduced with permission

hyperpolarized propane are shown in Fig. 22b. The signal patches in the panels acquired at different travel time instants are relatively short, indicating that gas dispersion along the flow direction is small and therefore the diffusional mixing in the capillary transverse to the direction of flow is very efficient. Comparison with a similar experiment performed using water, which has three orders of magnitude slower self-diffusion, instead of hyperpolarized propane revealed the presence of significant flow dispersion (Fig. 22a). For example, at $t = 200$ ms one can observe water in the entire encoding coil region.

The second microfluidic system examined was Chip 1, which had a 2 mm wide enlargement section and a depth of channels of 50 μm (Fig. 22c). Although the channel thickness was expected to be constant at different channel cross-sections, the RD time-of-flight images indicated that the gas mostly flowed close to the edges of the enlarged section, while almost no flow streams passed through the central part. This observation led to the manufacturing imperfections being exposed, since the central part was found to be much thinner than the expected thickness of 50 μm. This fact was eventually confirmed by visual inspection and conventional ^{1}H MRI imaging of water in the channels of Chip 1. It is worth noting that the channel geometry of Chip 1 was also used in the experiments with hyperpolarized ^{129}Xe under comparable conditions [94]; however, the acquisition time in the xenon experiment (9 h) was much longer than in the experiment with hyperpolarized propane (10 min), likely because of a much lower sensitivity in the xenon experiment.

The imperfections of the geometry were also revealed in Chip 2 with the ladder-like channel structure (Fig. 22d). In this case, the channel depth was expected to be constant and equal to 100 μm, but the signal amplitudes in the time projection (Fig. 22d, leftmost image) indicate, for instance, that the cross-sectional area of the left vertical channel is smaller than that of the right vertical channel. It should be noted that, in addition to information on geometry of the objects, the RD time-of-flight experiments allowed one to measure flow velocities in different parts of the examined microfluidic systems, providing comprehensive information about mass transport in these systems. For further details we refer the reader to the original publication [46], and another describing the travel time distribution analysis in RD-HET-PHIP experiments [96].

In addition to the imaging of microfluidic chips, the combined use of RD and PHIP has a promising potential for boosting the sensitivity in the studies of operating chemical microfluidic reactors [97]. The discussion of this subject is included in the next section.

4.3 NMR Imaging of Catalytic Hydrogenation

In all the experiments described above, parahydrogen naturally takes part in heterogeneous catalytic hydrogenations as a reactant. This provides a unique opportunity for the in situ MRI studies of catalytic processes with reasonably high sensitivity even in gas-phase hydrogenations. The goal of such investigations is a deeper understanding of physico-chemical processes influencing the performance of a catalytic reactor, which is commonly considered as a "black box" that converts reactants into products. Extremely versatile information which can be obtained using MRI can provide means for better optimization of important catalytic processes. In addition, the studies of heterogeneous processes involving parahydrogen are required in order to understand better the factors which control the formation of highly hyperpolarized substances in HET-PHIP experiments, since both the build-up and the decay of hyperpolarization should depend strongly on the transport processes taking place in catalytic layers.

In the first reported implementation of HET-PHIP for studying model catalytic microreactors [45], hydrogenation of propene over Wilkinson's catalyst (most likely, reduced) supported on modified silica gel was used as a model reaction process. One of the challenges of that study was the demonstration of the significant sensitivity improvement in microreactor imaging due to the use of HET-PHIP. The sensitivity enhancement of 300 compared to thermally polarized propane was measured previously in an ALTADENA experiment using immobilized Wilkinson's catalyst [44]. The degree of sensitivity enhancement was confirmed again in the imaging study [45]. At the same time, one significant difference faced was that the reactor under study was located inside a 7 T NMR magnet. The maximum signal amplitude in this PASADENA experiment is observed with a $\pi/4$-pulse, and the ^{1}H NMR signals of hyperpolarized propane have an antiphase pattern. This is much less convenient for imaging compared to the in-phase ALTADENA patterns, since an attempt to generate an image by integrating the signal is bound to cancel the enhancement. Thus a spin-echo readout ($\pi/4$–τ–π–τ–acquisition) was employed along with the pure phase encoding protocol. It served several purposes: to refocus the dephasing of magnetization due to magnetic susceptibility gradients, to remove the broad ^{1}H NMR signals of orthohydrogen and solid materials, and to allow a certain amount of evolution under homonuclear J-coupling which converts some of antiphase magnetization into the in-phase magnetization.

Two microreactors made of 1/8″ nylon tubing ($\approx$2 mm ID) were used in the experiments. The propene–parahydrogen mixture flowed through the microreactors containing hydrogenation catalyst to produce hyperpolarized propane. The first reactor was tightly packed with about 0.1 g of immobilized Wilkinson's catalyst to form a 5-mm catalyst bed surrounded by two layers of glass beads. This reactor was used for the demonstration of the in situ imaging of the production of hyperpolarized propane, the visualization of active regions of the catalyst bed, flow mapping, and the demonstration of a controlled transport of the polarization out of the catalyst bed. As a particular example, the ^{1}H NMR image acquired from the hyperpolarized propane in the catalyst bed is shown in Fig. 23a. The SNR was good enough to generate a velocity map of the gas flowing through the reactor (Fig. 23b) using a flow-encoded MRI pulse sequence [45]. The second reactor contained a powder of loosely packed pure Wilkinson's complex mixed with glass beads, and served as a model of a more heterogeneous packing as well as for demonstration of the polarized outstream control.

Amongst other things, the study demonstrated an approach for reducing the influence of the nuclear spin relaxation in HET-PHIP NMR imaging using long-lived spin states [98, 99]. This approach allowed the observation of delivery of the hyperpolarized propane beyond the catalyst beds in both microreactors.

Recently, Zhivonitko et al. [97] have shown that combination of RD and HET-PHIP described in Sect. 4.2 provides a means for a versatile characterization of microfluidic reactors of much smaller sizes compared to the reactors discussed

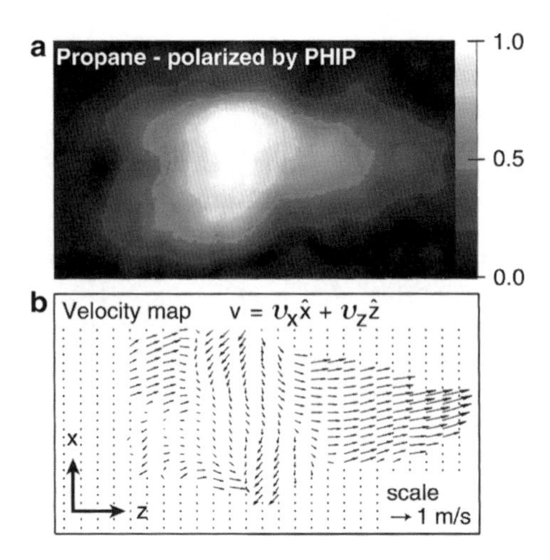

Fig. 23 ^{1}H NMR image (field of view 2.3 mm (*x*) × 7.0 mm (*z*); spatial resolution 20 μm × 60 μm) acquired from a tightly packed catalyst bed (catalyst layer thickness is ~5 mm) and the corresponding flow map of the gas. (**a**) Parahydrogen-polarized propane image. (**b**) Flow map in the *xz* plane generated with the use of the hyperpolarized propane. The orientation and length of the *arrows* show the direction and magnitude of the local flow velocity, respectively. The resolution of the flow map is intentionally decreased by retaining only 1 of every 16 *arrows* to avoid excessive overlap of the *arrows*. From [45]. Adapted with permission from AAAS

above. The microfluidic reactors they studied had single cylindrical channels of various diameters, from 800 to 150 μm, packed with heterogeneous catalysts [97]. Propene hydrogenation was employed as a model reaction process. The study demonstrated that RD-HET-PHIP provides the sensitivity enhancement of almost five orders of magnitude compared to a fictitious direct MRI experiment performed without RD and PHIP. Such a significant improvement in sensitivity allowed the authors to study mass transport and to obtain quantitative estimates of gas flow velocities, to visualize the reaction product distribution, and to address adsorption processes that take place in the microfluidic packed bed reactors. These data also provided an insight into the processes influencing the build-up of hyperpolarization in the packed layers of heterogeneous catalysts.

As an example of the results obtained, the two-dimensional time-of-flight images demonstrating the build-up of the hyperpolarized product propane are shown in Fig. 24. The leftmost images are the time projections generated by summing all other panels corresponding to different travel time instants from the encoding to the detection coil. It is clearly seen that for longer travel times, the signal intensities are lower compared to those for shorter travel times, which indicates that the amount of hyperpolarized propane increases from the inlet toward the outlet of the reactors (from the upper to the lower parts of the panels). The shortest travel times (86, 62, and 133 ms for Fig. 24a–c, respectively) correspond to the gas which was encoded when it already flowed out the catalyst bed region and entered the outlet capillary leading to the detection coil. A theoretical analysis

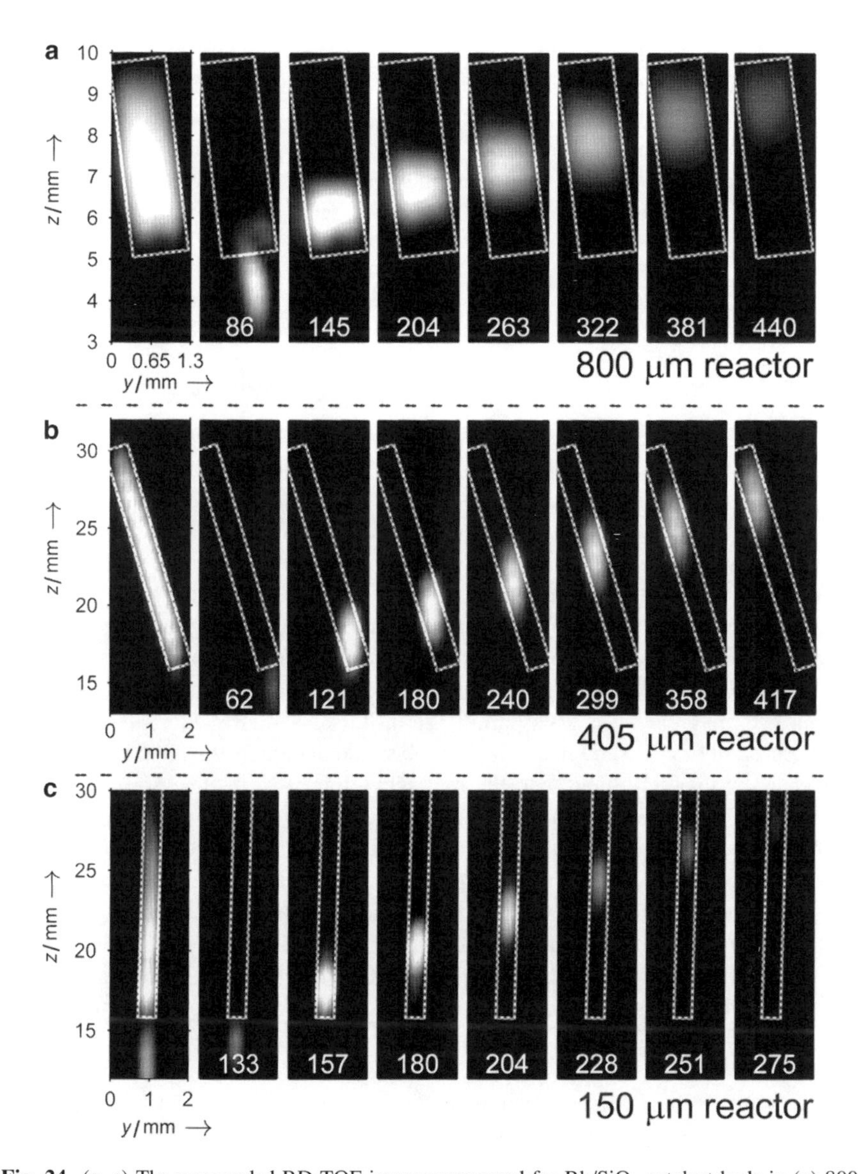

Fig. 24 (**a**–**c**) The *yz*-encoded RD TOF images measured for Rh/SiO_2 catalyst beds in (**a**) 800, (**b**) 405, and (**c**) 150 μm reactors at 60°C. The *leftmost* images are the time projections. Travel time instants in milliseconds are indicated in the panels. The catalyst bed regions are outlined with *white dashed lines*. The complete data set for each reactor was acquired in 13 min with a time resolution of 12 ms and a spatial resolution of 160–250 μm in the *y* direction and 0.62–2.2 mm in the *z* direction [97]. Copyright Wiley-VCH Verlag GmbH & Co. KGaA, Weinheim. Reproduced with permission

performed in the study brought authors to the conclusion that the decay of hyperpolarization due to nuclear spin relaxation processes within the catalyst beds of the reactors is insignificant under given experimental conditions. Moreover, the examined devices can be considered as the first prototypes of microfluidic PHIP polarizers that provide signal enhancements on the order of two orders of magnitude, bringing together the HET-PHIP research with the new rapidly emerging capabilities provided by microfluidics.

5 Evaluation of Signal Enhancement in Continuous-Flow PHIP Experiments

When signal enhancements in PHIP and other hyperpolarization NMR experiments are evaluated, the hyperpolarized NMR signals are usually compared to the signals of the same samples in thermal equilibrium. It is therefore important to ensure that the observed intensities of non-hyperpolarized signals do indeed have thermal equilibrium values at the time of their detection. This is not necessarily the case if such quantitative NMR experiments are performed in a continuous flow regime.

Indeed, before the flowing fluid enters an NMR magnet from the outside, its nuclear spin magnetization is essentially zero. It takes (3–5) $\times$ T_1 to attain thermal equilibrium, but as the fluid flows toward the signal detection zone it experiences an ever growing magnetic field, and the magnetization of its nuclear spins is permanently trying to catch up with the gradually increasing thermal equilibrium value. Therefore, even if the temperatures of the fluid and the RF probe are the same, for high enough flow rates the magnetization of the fluid in the RF coil may be measurably smaller than the thermal equilibrium value at the nominal magnetic field of the NMR spectrometer. As a result, the NMR signal of the "thermally polarized" fluid may be significantly reduced. This effect will be particularly pronounced for fluids with very long relaxation times, but even for liquids with the proton T_1 times of a few seconds such effects may be easily observed. This is also a problem with continuously flowing gases as flow rates used are usually higher, whereas the T_1 times of some of the gases used are not much shorter than those of liquids. For instance, for 1 bar of propane at 7 T the proton T_1 time is ca. 0.9 s, and for other gaseous hydrocarbons discussed above the values of T_1 are also on the order of 1 s. Differences in the flow regimes of gases (plug flow with little flow dispersion) and liquids (laminar flow with high dispersion) may also affect the degree of the NMR signal suppression of the flowing "thermally polarized" fluids.

To explore and quantify these effects, our experimental setup was modified to accommodate a cell with activated charcoal in the NMR tube just above the sensitive zone of the RF probe so that the inflowing gas passes through the bed of paramagnetic charcoal before it reaches the catalyst. As a representative example, Fig. 25 presents the results obtained with 1,3-butadiene using this setup. When the ^{1}H NMR spectrum of 1,3-butadiene flowing at ca. 7 mL/s is detected without the

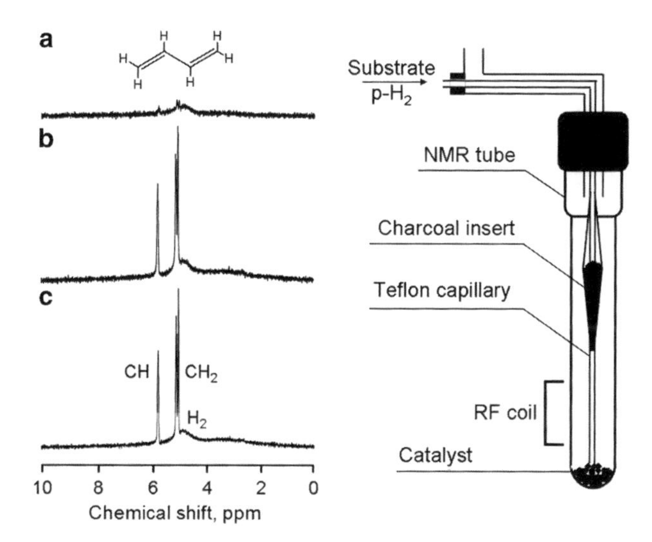

Fig. 25 ^{1}H NMR spectra of the mixture of 50%-enriched parahydrogen and 1,3-butadiene (4:1 by volume) (*left*) and the schematic drawing of the experimental geometry (*right*). The spectra were detected for (**a**) flowing gas without the paramagnetic charcoal insert; (**b**) static gas under stopped flow conditions; (**c**) flowing gas with the insert in place. The spectra (a–c) were acquired without the catalyst at the bottom of the tube

charcoal insert (Fig. 25a), its signals have low intensity, but become much larger when the insert is in place and the nuclear spin relaxation of the gas is accelerated before its signals are detected (Fig. 25c). A simple way to see if the incomplete relaxation may be a potential problem is to compare the spectra detected for the flowing (Fig. 25a, c) and the static gas (Fig. 25b). This comparison shows that the paramagnetic insert allows one to quantify properly the thermal equilibrium intensities of the NMR signals of the flowing gas, whereas in the experiment without the insert the "thermal equilibrium" signals are in fact severely diminished in intensity. We note here that the signal attenuation observed in Fig. 25a is not caused by the fact that an FID is being detected while the gas is flowing out of the RF coil. This outflow could shorten the FID and thus broaden the observed NMR lines, but the integrated signal intensity should be largely unaffected. This is not what is observed in the spectra of Fig. 25: the linewidths in the spectra detected under static and flowing conditions are similar, whereas the signals of flowing 1,3-butadiene in the absence of the paramagnetic insert are dramatically suppressed. In addition, comparison of Fig. 25b, c shows that the intensities of the signals are not affected by flow when the insert is in place. Thus, signal suppression is clearly caused by a rapid inflow of the gas into the magnet from a low magnetic field which makes the magnetization of the gas much smaller than the thermal equilibrium value in the high magnetic field of the NMR instrument at the time of NMR signal detection.

In the work of Sharma and Bouchard [100], hydrogenation of propene with parahydrogen was studied using supported Pt catalysts, with Pt nanoparticles stabilized using *p*-mercaptobenzoic acid. The catalyst modification procedure, which the authors refer to as "rational design", is well known in heterogeneous catalysis for a long time as catalyst poisoning. Catalyst poisoning can be very useful in modifying the selectivity of a catalyst to get higher yields of the desired product by selectively deactivating (poisoning) those active centers that exhibit low selectivity toward the target product. The conclusions the authors make about the mechanism of HET-PHIP generation in this system and the contribution of the pairwise hydrogen addition to the overall hydrogenation reaction on this catalyst are based on their estimate of the signal enhancement factor. However, a careful inspection of their spectra indicates that, quite likely, they have overlooked the effects described above, and that the actual enhancements were largely overestimated.

Figure 26 compares the ^{1}H NMR spectrum from the work of Sharma and Bouchard [100] and our ^{1}H NMR spectrum of a static gaseous sample. Both spectra were detected under very similar conditions except that the gas in the experiment of Fig. 26a was static. Both samples contained ca. 1 bar of the 4:1 mixture of 50%-enriched parahydrogen and propene in a 10-mm NMR tube and were detected using four signal averages. It can be seen that the signal of propene is dramatically stronger for the static sample as evidenced by its SNR which is larger by a factor of ~30 even though the spectrum in Fig. 26b was acquired at a higher magnetic field (9.4 T vs. 7.05 T for Fig. 26a). This major difference in SNR points to the fact that propene signals in the spectra reported by Sharma and Bouchard [100] (Fig. 26b) are suppressed well below the thermal equilibrium value. This is further confirmed by the fact that the ratio of the signals of propene and *ortho*-H_2 in Fig. 26b is obviously much lower than in Fig. 26a (see also Fig. 1b of [100] for the spectrum acquired using 4,000 accumulations), even for the spectrum of the static gas with the propene signals intentionally broadened to emulate the outflow effects (Fig. 26a, inset). The signal of flowing *ortho*-H_2 is not suppressed because its spin–lattice relaxation is much faster than that of propene. An additional confirmation of the suppression of the non-hyperpolarized signals is the statement made by the authors that polarization efficiency decreased significantly when the gas flow rate was reduced [100]. This is not surprising since lower gas flow velocities result in a larger "thermal equilibrium" magnetization before the gas enters the RF coil.

Fortunately, the situation for propane is somewhat different from that for propene. For propane produced as the product of hydrogenation reaction within the NMR tube residing in the NMR magnet, it is almost impossible to suppress the thermal equilibrium signal completely. The very short T_1 time of *ortho*-H_2 somewhat increases when H_2 is mixed with heavier gases such as hydrocarbons, but in any case does not exceed a few milliseconds. Therefore, when normal hydrogen is used in the reaction, at any practically feasible gas flow rate the nuclear spins of H_2 hydrogen atoms are expected to be in thermal equilibrium by the time H_2 enters the reaction. As a result, one of the six methyl protons and one of the two methylene protons of propane that come from H_2 will be in thermal equilibrium even if other hydrogens of propane are not. Thus, in a catalytic experiment with normal hydrogen the signal intensities of propane could be suppressed at the high gas flow rates by the factors of up to two and six for the methylene and methyl resonances,

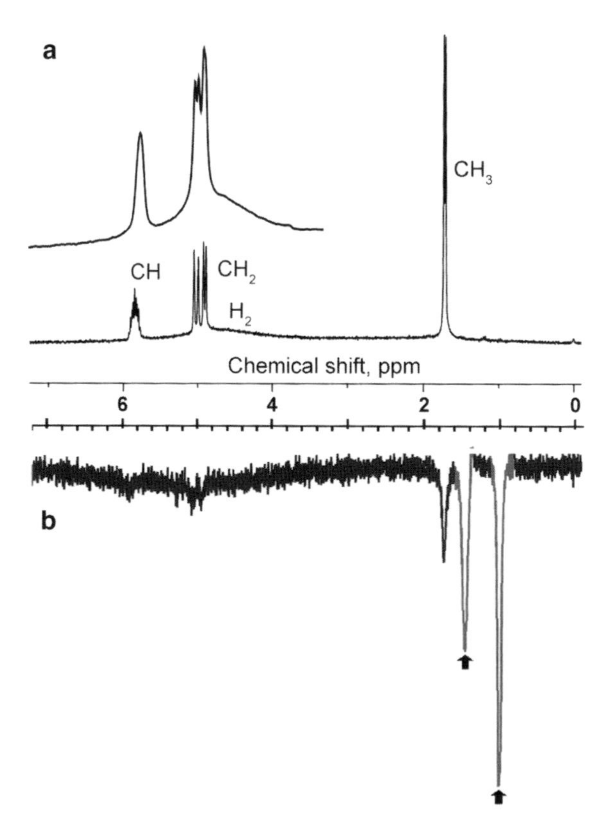

Fig. 26 ^1H NMR spectra detected for the 4:1 mixture of 50%-enriched parahydrogen and propene at 1 bar in a 10 mm NMR tube using a 45°-pulse for excitation and four signal averages. (**a**) The spectrum detected at 7.05 T for a static gas in the absence of any catalyst. No apodization was used in processing of the FID. NMR signals of H_2 and the three groups of propene are labeled. The *inset* shows part of the same spectrum with the lines artificially broadened using exponential apodization of the FID to mimic the flow-induced line broadening effect. (**b**) The spectrum detected by Sharma and Bouchard [100] at 9.4 T for the gas flowing through the tube containing supported Pt catalyst. The spectrum in (**b**) is presented in the absolute value (magnitude) mode, and the entire spectrum is inverted in the figure. The *lines* marked with *arrows* belong to hyperpolarized propane. See text for further details. (**b**) was adapted with permission from [100]

respectively. As the signal enhancement in [100] was estimated from the resonance of the methyl group of propane, it can be concluded that the degree of polarization and the degree of pairwise hydrogen addition were overestimated by at least a factor of 6. Combined with the fact that catalyst poisoning resulted in an almost negligible conversion of propene to propane, the results of Sharma and Bouchard can hardly be considered as a significant advance with respect to the results reported earlier.

The authors also draw the conclusion that the ligands present at the Pt particle surface can confine the catalytically active sites, apparently confirming the possible mechanism of HET-PHIP formation on the supported metals advanced earlier [20, 21]. At the same time, the results can equally well be explained in a different

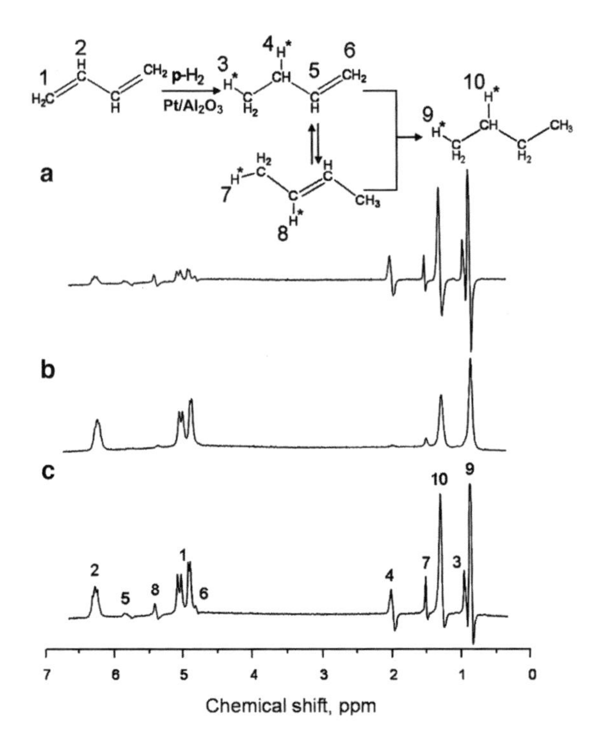

Fig. 27 ^{1}H NMR spectra detected during the hydrogenation of 1,3-butadiene with parahydrogen over Pt/γ-Al_2O_3. The reaction was performed at 100°C. The spectra were detected for (**a**) flowing gas without the paramagnetic charcoal insert; (**b**) static gas under stopped flow conditions; (**c**) flowing gas with the insert in place

way. Poisoning of the catalyst with *p*-mercaptobenzoic acid significantly reduces its activity as evidenced by the dramatic reduction in the product yield. This poisoning of catalytically active sites could be selective, e.g., it could preferentially suppress the active centers that produce no PHIP, which in fact could potentially increase the degree of pairwise hydrogen addition at the expense of a significantly decreased conversion. At the same time, since the enhancement factor and the degree of pairwise H_2 addition [100] were most likely significantly overestimated, neither conclusion seems justified without further careful experimental verification.

Finally, it is worth noting that this effect of suppression of thermally polarized signals by fluid flow can be used to advantage in the continuous-flow PHIP experiments to reveal weak polarizations potentially masked by the much stronger thermally polarized signals of the reactants and especially products. As an example, Fig. 27 presents the results of the PASADENA experiments performed with 1,3-butadiene as a substrate. Figure 27c shows the HET-PHIP spectrum detected with the charcoal insert present. The polarization contributions to the NMR signals of the products are smaller than the equilibrium signals of the same products, as revealed by the small size of the emissive components of the lines compared to the absorptive components. For even smaller levels of polarization, these negative parts may be overwhelmed by the much larger absorptive contributions and would not be observed. At the same time, when no paramagnetic insert is present, the lines of the non-polarized products are significantly suppressed and the polarization patterns are much closer to the symmetric absorption-emission multiplets expected in a PASADENA experiment (Fig. 27a).

6 Conclusions and Outlook

The concept of HET-PHIP has clearly proven its viability. While this sub-field of hyperpolarization in magnetic resonance is still in its infancy, at this point the demonstration that many types of heterogeneous catalysts have an intrinsic ability to produce PHIP is quite an important and encouraging achievement. Heterogeneous processes may have certain limitations, but they also have many advantages over their homogeneous counterparts. In particular, while it is possible in principle to remove a dissolved catalyst from solution of a hyperpolarized molecule, practical attempts to do this lead to a dramatic reduction in the levels of signal enhancement by several orders of magnitude [101]. In this respect, HET-PHIP appears to be a more direct way to catalyst-free hyperpolarized liquids and solutes, and quite likely the only way to utilize parahydrogen to produce hyperpolarized gases. Another advantage of HET-PHIP is that the duration of the hydrogenation event can easily be controlled by selecting the desired contact time of the reaction medium with the heterogeneous catalyst, something which is not easily achievable with homogeneous hydrogenations as dissolved parahydrogen can ensure that reaction continues for an undefined period of time even after bubbling of parahydrogen through solution is terminated. This feature may be useful in the studies of the dynamics of PHIP formation and transfer to other nuclei in a coupled system of nuclear spins. It is also expected that HET-PHIP could become a useful tool in the mechanistic studies of heterogeneous catalytic processes, as demonstrated by some of the examples discussed above. In addition, HET-PHIP could potentially address the low-sensitivity issue in the NMR studies of surfaces, as demonstrated by the observation of PHIP-enhanced ^{1}H NMR spectrum of surface-bound H atoms upon heterolytic activation of parahydrogen on the surface of ZnO [102].

Further development of HET-PHIP will continue in several different directions. These include the contribution of HET-PHIP to the theory and practice of PHIP in general, the efforts to gain a much better understanding of the mechanisms responsible for HET-PHIP, the design of better catalysts and catalytic processes to produce large quantities of strongly polarized liquids and gases, the development of HET-PHIP as a tool to study the mechanisms of heterogeneous catalytic processes and the processes in operating catalytic reactors, and the application of HET-PHIP in hypersensitive spectroscopic and imaging studies including in vivo magnetic resonance. It is quite likely that significant progress can be achieved in this novel field of research, but this will require very careful experimentation and knowledgeable reasoning.

Significant progress has been achieved lately in the development and application of other nuclear spin hyperpolarization techniques. Furthermore, a number of other major developments in modern magnetic resonance that took place in recent years are also very important in the context of hyperpolarization studies. Those developments that already revealed their utility in combination with PHIP include the demonstration and use of the unusual properties of the long-lived (singlet) spin states [99, 103–105], the growing number of the types of molecules that can be hyperpolarized and used as, e.g., advanced contrast agents for MRI [106–112], the development of the new approaches and instrumentation to generate hyperpolarization [11, 15, 87–89], the techniques for the optimized conversion of the initial spin order of the product molecule into the observable polarization [113–115] and for magnetization transfer to other nuclei [34, 87, 116–124], and more. The tendency to overcome the isolation between different research directions and to develop the group of hyperpolarization techniques and their applications as one big and rapidly growing family, in combination with the joint collaborative efforts of the magnetic resonance community, will definitely go a long way in accelerating progress in this important field of modern magnetic resonance.

Acknowledgments This work was partially supported by the grants from RFBR (## 11-03-93995-CSIC-a, RFBR 11-03-00248-a, RFBR 12-03-00403-a), RAS (# 5.1.1), SB RAS (## 60, 61, 57, 122), the program of support of leading scientific schools (# NSh-2429.2012.3), and the program of the Russian Government to support leading scientists (# 11.G34.31.0045).

References

1. Seidler PF, Bryndza HE, Frommer JE, Stuhl LS, Bergman RG (1983) Synthesis of trinuclear alkylidyne complexes from dinuclear alkyne complexes and metal hydrides. CIDNP evidence for vinyl radical intermediates in the hydrogenolysis of these clusters. Organometallics 2:1701–1705
2. Bowers CR, Weitekamp DP (1986) Transformation of symmetrization order to nuclear spin magnetization by chemical reaction and nuclear magnetic resonance. Phys Rev Lett 57:2645–2648
3. Bowers CR, Weitekamp DP (1987) Parahydrogen and synthesis allow dramatically enhanced nuclear alignment. J Am Chem Soc 109:5541–5542
4. Duckett SB, Wood NJ (2008) Parahydrogen-based NMR methods as a mechanistic probe in inorganic chemistry. Coord Chem Rev 252:2278–2291
5. Blazina D, Duckett SB, Dunne JP, Godard C (2004) Applications of the parahydrogen phenomenon in inorganic chemistry. Dalton Trans 2601–2609
6. Duckett SB, Sleigh CJ (1999) Applications of the para hydrogen phenomenon: a chemical perspective. Prog Nucl Magn Reson Spectrosc 34:71–92
7. Natterer J, Bargon J (1997) Parahydrogen induced polarization. Prog Nucl Magn Reson Spectrosc 31:293–315
8. Eisenberg R (1991) Parahydrogen-induced polarization: a new spin on reactions with H_2. Acc Chem Res 24:110–116
9. Canet D, Aroulanda C, Mutzenhardt P, Aime S, Gobetto R, Reineri F (2006) Para-hydrogen enrichment and hyperpolarization. Concepts Magn Reson 28A:321–330

10. Green RA, Adams RW, Duckett SB, Mewis RE, Williamson DC, Green GGR (2012) The theory and practice of hyperpolarization in magnetic resonance using parahydrogen. Prog Nucl Magn Reson Spectrosc. http://dx.doi.org/10.1016/j.pnmrs.2012.03.001
11. Golman K, Axelsson O, Johannesson H, Mansson S, Olofsson C, Petersson JS (2001) Parahydrogen-induced polarization in imaging: subsecond ^{13}C angiography. Magn Reson Med 46:1–5
12. Bhattacharya P, Chekmenev EY, Perman WH, Harris KC, Lin AP, Norton VA, Tan CT, Ross BD, Weitekamp DP (2007) Towards hyperpolarized ^{13}C-succinate imaging of brain cancer. J Magn Reson 186:150–155
13. Bhattacharya P, Harris K, Lin AP, Mansson M, Norton VA, Perman WH, Weitekamp DP, Ross BD (2005) Ultra-fast three dimensional imaging of hyperpolarized ^{13}C in vivo. MAGMA 18:245–256
14. Mansson S, Johansson E, Magnusson P, Chai C-M, Hansson G, Petersson JS, Stahlberg F, Golman K (2006) ^{13}C imaging – a new diagnostic platform. Eur Radiol 16:57–67
15. Adams RW, Aguilar JA, Atkinson KD, Cowley MJ, Elliott PIP, Duckett SB, Green GGR, Khazal IG, Lopez-Serrano J, Williamson DC (2009) Reversible interactions with parahydrogen enhance NMR sensitivity by polarization transfer. Science 323:1708–1711
16. Dumesic JA, Huber GW, Boudart M (2008) Principles of heterogeneous catalysis. Wiley-VCH Verlag GmbH & Co. KGaA, Weinheim
17. Permin AB, Eisenberg R (2002) One-hydrogen polarization in hydroformylation promoted by platinum-tin and iridium carbonyl complexes: a new type of parahydrogen-induced effect. J Am Chem Soc 124:12406–12407
18. Fox DJ, Duckett SB, Flaschenriem C, Brennessel WW, Schneider J, Gunay A, Eisenberg R (2006) A model iridium hydroformylation system with the large bite angle ligand xantphos: reactivity with parahydrogen and implications for hydroformylation catalysis. Inorg Chem 45:7197–7209
19. Koptyug IV, Kovtunov KV, Burt SR, Anwar MS, Hilty C, Han S, Pines A, Sagdeev RZ (2007) para-Hydrogen-induced polarization in heterogeneous hydrogenation reactions. J Am Chem Soc 129:5580–5586
20. Kovtunov KV, Beck IE, Bukhtiyarov VI, Koptyug IV (2008) Observation of parahydrogen-induced polarization in heterogeneous hydrogenation on supported metal catalysts. Angew Chem Int Ed 47:1492–1495
21. Kovtunov KV, Koptyug IV (2008) Parahydrogen-induced polarization in heterogeneous catalytic hydrogenations. In: Codd S, Seymour JD (eds) Magnetic resonance microscopy. Spatially resolved NMR techniques and applications. Wiley-VCH, Weinheim
22. Kovtunov KV, Zhivonitko VV, Corma A, Koptyug IV (2010) Parahydrogen-induced polarization in heterogeneous hydrogenations catalyzed by an immobilized Au(III) complex. J Phys Chem Lett 1:1705–1708
23. Kovtunov KV, Zhivonitko VV, Kiwi-Minsker L, Koptyug IV (2010) Parahydrogen-induced polarization in alkyne hydrogenation catalyzed by Pd nanoparticles embedded in a supported ionic liquid phase. Chem Commun 46:5764–5766
24. Eichhorn A, Koch A, Bargon J (2001) In situ PHIP NMR – a new tool to investigate hydrogenation mediated by colloidal catalysts. J Mol Catal A Chem 174:293–295
25. Crabtree RH (2012) Resolving heterogeneity problems and impurity artifacts in operationally homogeneous transition metal catalysts. Chem Rev 112:1536
26. Barbaro P, Bianchini C (2002) Recent aspects of asymmetric catalysis by immobilized transition metal complexes. Top Catal 19:17–32
27. Cole-Hamilton DJ, Tooze RP (eds) (2006) Catalyst separation, recovery and recycling: chemistry and process design, vol 30, Catalysis by metal complexes. Springer, Dordrecht
28. End N, Schoning K-U (2004) Immobilized catalysts in industrial research and application. Top Curr Chem 242:241–271
29. Merckle C, Blumel J (2005) Improved rhodium hydrogenation catalysts immobilized on silica. Top Catal 34:5–15

30. Wegener SL, Marks TJ, Stair PC (2012) Design strategies for the molecular level synthesis of supported catalysts. Acc Chem Res 45:206–214
31. Skovpin IV, Zhivonitko VV, Koptyug IV (2011) Parahydrogen-induced polarization in heterogeneous hydrogenations over silica-immobilized Rh complexes. Appl Magn Reson 41:393–410
32. Gutmann T, Ratajczyk T, Xu Y, Breitzke H, Grunberg A, Dillenberger S, Bommerich U, Trantzschel T, Bernarding J, Buntkowsky G (2010) Understanding the leaching properties of heterogenized catalysts: a combined solid-state and PHIP NMR study. Solid State Nucl Magn Reson Spectrosc 38:90–96
33. Gordon B, Butler JS, Harrison IR (1987) Rhodium(I) catalyst supported on polymer crystal surfaces: further hydrogenation studies. J Polym Sci Part A: Polym Chem 25:2139–2142
34. Kuhn LT, Bommerich U, Bargon J (2006) Transfer of parahydrogen-induced hyperpolarization to ^{19}F. J Phys Chem A 110:3521–3526
35. Kirss RU, Eisenberg R (1989) Di(phosphine)-bridged complexes of palladium. Parahydrogen-induced polarization in hydrogenation reactions and structure determination of tris(μ-bis(diphenylphosphino)methane)dipalladium, $Pd_2(dppm)_3$. Inorg Chem 28:3372
36. Schleyer D, Niessen HG, Bargon J (2001) In situ ^{1}H-PHIP-NMR studies of the stereoselective hydrogenation of alkynes to (E)-alkenes catalyzed by a homogeneous $[Cp^*Ru]^+$ catalyst. New J Chem 25:423–426
37. Lopez-Serrano J, Duckett SB, Aiken S, Lenro KQA, Drent E, Dunne JP, Konya D, Whitwood NC (2007) A para-hydrogen investigation of palladium-catalyzed alkyne hydrogenation. J Am Chem Soc 129:6513–6527
38. Osborn JA, Jardine FH, Young JF, Wilkinson G (1966) The preparation and properties of tris (triphenylphosphine)halogenorhodium(I) and some reactions thereof including catalytic homogeneous hydrogenation of olefins and acetylenes and their derivatives. J Chem Soc A 12:1711–1732
39. Goodson BM (2002) Nuclear magnetic resonance of laser-polarized noble gases in molecules, materials, and organisms. J Magn Reson 155:157–216
40. Brunner E, Haake M, Kaiser L, Pines A, Reimer JA (1999) Gas flow MRI using circulating laser-polarized ^{129}Xe. J Magn Reson 138:155–159
41. Ruppert K, Mata JF, Brookeman JR, Hagspiel KD, Mugler JP (2004) Exploring lung function with hyperpolarized ^{129}Xe nuclear magnetic resonance. Magn Reson Med 51:676–687
42. Mair RW, Hrovat MI, Patz S, Rosen MS, Ruset IC, Topulos GP, Tsai LL, Butler JP, Hersman FW, Walsworth RL (2005) ^{3}He lung imaging in an open access, very-low-field human magnetic resonance imaging system. Magn Reson Med 53:745–749
43. Pavlovskaya GE, Cleveland ZI, Stupic KF, Basaraba RJ, Meersmann T (2005) Hyperpolarized krypton-83 as a contrast agent for magnetic resonance imaging. Proc Nat Acad Sci USA 102:18275–18279
44. Bouchard L-S, Kovtunov KV, Burt SR, Anwar MS, Koptyug IV, Sagdeev RZ, Pines A (2007) Parahydrogen-enhanced hyperpolarized gas-phase magnetic resonance imaging. Angew Chem Int Ed 46:4064–4068
45. Bouchard L-S, Burt SR, Anwar MS, Kovtunov KV, Koptyug IV, Pines A (2008) NMR imaging of catalytic hydrogenation in microreactors with the use of para-hydrogen. Science 319:442–445
46. Telkki V-V, Zhivonitko VV, Ahola S, Kovtunov KV, Jokisaari J, Koptyug IV (2010) Microfluidic gas-flow imaging utilizing parahydrogen-induced polarization and remote-detection NMR. Angew Chem Int Ed 49:8363–8366
47. Harthun A, Giernoth R, Elsevier CJ, Bargon J (1996) Rhodium- and palladium-catalysed proton exchange in styrene detected in situ by para-hydrogen induced polarization. Chem Commun 2483–2484
48. Niessen HG, Schleyer D, Wiemann S, Bargon J, Steiner S, Driessen-Holscher B (2000) In situ PHIP-NMR studies during the stereoselective hydrogenation of sorbic acid with a $[Cp^*Ru]^+$ catalyst. Magn Reson Chem 38:747–750

49. Gutmann T, Sellin M, Breitzke H, Stark A, Buntkowsky G (2009) Para-hydrogen induced polarization in homogeneous phase – an example of how ionic liquids affect homogenization and thus activation of catalysts. Phys Chem Chem Phys 11:9170–9175
50. Mehnert CP, Mozeleski EJ, Cook RA (2002) Supported ionic liquid catalysis investigated for hydrogenation reactions. Chem Commun 3010–3011
51. Virtanen P, Salmi T, Mikkola J-P (2009) Kinetics of cinnamaldehyde hydrogenation by supported ionic liquid catalysts (SILCA). Ind Eng Chem Res 48:10335–10342
52. Gong Q, Klankermayer J, Blumich B (2011) Organometallic complexes in supported ionic-liquid phase (SILP) catalysts: a PHIP NMR spectroscopy study. Chem Eur J 17:13795–13799
53. Zhang X, Llabres i Xamena FX, Corma A (2009) Gold(III) – metal organic framework bridges the gap between homogeneous and heterogeneous gold catalysts. J Catal 265:155–160
54. Cremer PS, Su X, Shen YR, Somorjai GA (1996) Hydrogenation and dehydrogenation of propylene on Pt(111) studied by sum frequency generation from UHV to atmospheric pressure. J Phys Chem 100:16302–16309
55. Wasylenko W, Frei H (2007) Dynamics of propane in silica mesopores formed upon propylene hydrogenation over Pt nanoparticles by time-resolved FT-IR spectroscopy. J Phys Chem C 111:9884–9890
56. Wasylenko W, Frei H (2005) Direct observation of surface ethyl to ethane interconversion upon C_2H_4 hydrogenation over Pt/Al_2O_3 catalyst by time-resolved FT-IR spectroscopy. J Phys Chem B 109:16873–16878
57. Cremer PS, Su X, Shen YR, Somorjai GA (1996) Ethylene hydrogenation on Pt(111) monitored in situ at high pressures using sum frequency generation. J Am Chem Soc 118:2942–2949
58. Haug KL, Burgi T, Trautman TR, Ceyer ST (1998) Distinctive reactivities of surface-bound H and bulk H for the catalytic hydrogenation of acetylene. J Am Chem Soc 120:8885–8886
59. Teschner D, Vass E, Havecker M, Zafeiratos S, Schnorch P, Sauer H, Knop-Gericke A, Schlogl R, Chamam M, Wootsch A, Canning AS, Gamman JJ, Jackson SD, McGregor J, Gladden LF (2006) Alkyne hydrogenation over Pd catalysts: a new paradigm. J Catal 242:26–37
60. Rayhel LH, Corey RL, Shane DT, Cowgill DF, Conradi MS (2011) Hydrogen NMR of palladium hydride: measuring the hydride-gas exchange rate. J Phys Chem C 115:4966–4970
61. Renouprez A, Fouilloux P, Stockmeyer R, Conrad HM, Goeltz G (1977) Diffusion of chemisorbed hydrogen on a nickel catalyst. Ber Bunsenges Phys Chem 81:429–432
62. Farkas A, Farkas L (1938) The catalytic interaction of ethylene and heavy hydrogen on platinum. J Am Chem Soc 60:22–28
63. Farkas A, Farkas L (1942) The mechanism of the catalytic conversion of para-hydrogen on nickel, platinum and palladium. J Am Chem Soc 64:1594–1599
64. Scholten JJF, Konvalinka JA (1966) Hydrogen-deuterium equilibration and parahydrogen and orthodeuterium conversion over palladium: kinetics and mechanism. J Catal 5:1–17
65. Somorjai GA, Zaera F (1982) Heterogeneous catalysis on the molecular scale. J Phys Chem 86:3070–3078
66. Bond GC (1997) The role of carbon deposits in metal-catalysed reactions of hydrocarbons. Appl Catal A 149:3–25
67. Zhivonitko VV, Kovtunov KV, Beck IE, Ayupov AB, Bukhtiyarov VI, Koptyug IV (2011) Role of different active sites in heterogeneous alkene hydrogenation on platinum catalysts revealed by means of parahydrogen-induced polarization. J Phys Chem C 115:13386–13391
68. Farin D, Avnir D (1988) The reaction dimension in catalysis on dispersed metals. J Am Chem Soc 110:2039–2045
69. McLeod AS (2004) The influence of catalyst geometry and topology on the kinetics of hydrocarbon reactions. Chem Eng Res Des 82:945–951
70. Borodzinski A (2001) The effect of palladium particle size on the kinetics of hydrogenation of acetylene-ethylene mixtures over Pd/SiO_2 catalysts. Catal Lett 71:169–175

71. Rioux RM, Hsu BB, Grass ME, Song H, Somorjai GA (2008) Influence of particle size on reaction selectivity in cyclohexene hydrogenation and dehydrogenation over silica-supported monodisperse Pt particles. Catal Lett 126:10–19
72. Norskov JK, Bligaard T, Hvolbaek B, Abild-Pedersen F, Christensen CH (2008) The nature of the active site in heterogeneous metal catalysis. Chem Soc Rev 37:2163–2171
73. Tauster SJ, Fung SC, Garten RL (1978) Strong metal-support interactions. Group 8 noble metals supported on titanium dioxide. J Am Chem Soc 100:170–175
74. Arnold H, Döbert F, Gaube J (2008) Selective hydrogenation of hydrocarbons. In: Ertl G, Knözinger H, Weitkamp J (eds) Handbook of heterogeneous catalysis. Wiley, Weinheim
75. Bond GC (2005) Metal-catalysed reactions of hydrocarbons (fundamental and applied catalysis). Springer, New York
76. Olah GA, Molnar A (2003) Hydrocarbon chemistry. Wiley, New York
77. Rideal EK (1939) A note on a simple molecular mechanism for heterogeneous catalytic reactions. Proc Cambridge Philos Soc 35:130–132
78. Eley DD, Rideal EK (1940) Parahydrogen conversion on tungsten. Nature 146:401–402
79. Maetz P, Touroude R (1994) Mechanism of but-1-yne hydrogenation on platinum catalysts: deuterium tracer study. J Mol Catal 91:259–275
80. Bos ANR, Westerterp KR (1993) Mechanism and kinetics of the selective hydrogenation of ethyne and ethane. Chem Eng Process 32:1–7
81. Jackson SD, Casey NJ (1995) Hydrogenation of propyne over palladium catalysts. J Chem Soc Faraday Trans 91:3269–3274
82. Kennedy DR, Webb G, Jackson SD, Lennon D (2004) Propyne hydrogenation over alumina-supported palladium and platinum catalysts. Appl Catal A 259:109–120
83. Dupont J, Fonseca GS, Umpierre AP, Fichtner PFP, Teixeira SR (2002) Transition-metal nanoparticles in imidazolium ionic liquids: recycable catalysts for biphasic hydrogenation reactions. J Am Chem Soc 124:4228–4229
84. Ruta M, Laurenczy G, Dyson PJ, Kiwi-Minsker L (2008) Pd nanoparticles in a supported ionic liquid phase: highly stable catalysts for selective acetylene hydrogenation under continuous-flow conditions. J Phys Chem C 112:17814–17819
85. Koptyug IV, Zhivonitko VV, Kovtunov KV (2010) New perspectives for parahydrogen-induced polarization in liquid phase heterogeneous hydrogenation: an aqueous phase and ALTADENA study. Chemphyschem 11:3086–3088
86. Balu AM, Duckett SB, Luque R (2009) Para-hydrogen induced polarisation effects in liquid phase hydrogenations catalysed by supported metal nanoparticles. Dalton Trans 5074–5076
87. Hovener J-B, Chekmenev EY, Harris KC, Perman WH, Robertson LW, Ross BD, Bhattacharya P (2009) PASADENA hyperpolarization of ^{13}C biomolecules: equipment design and installation. MAGMA 22:111–121
88. Waddell KW, Coffey AM, Chekmenev EY (2011) In situ detection of PHIP at 48 mT: demonstration using a centrally controlled polarizer. J Am Chem Soc 133:97–101
89. Roth M, Kindervater P, Raich H-P, Bargon J, Spiess HW, Munnemann K (2010) Continuous ^{1}H and ^{13}C signal enhancement in NMR and MRI using parahydrogen and hollow-fiber membranes. Angew Chem Int Ed 49:8358–8362
90. Callaghan PT (1991) Principles of nuclear magnetic resonance microscopy. Clarendon, Oxford
91. Moule AJ, Spence MM, Han S, Seeley JA, Pierce KL, Saxena S, Pines A (2003) Amplification of xenon NMR and MRI by remote detection. Proc Natl Acad Sci USA 100:9122–9127
92. Seeley JA, Han S, Pines A (2004) Remotely detected high-field MRI of porous samples. J Magn Reson 167:282–290
93. Granwehr J, Harel E, Han S, Garcia S, Pines A (2005) Time-of-flight flow imaging using NMR remote detection. Phys Rev Lett 95:075503
94. Hilty C, McDonnell EE, Granwehr J, Pierce KL, Han S, Pines A (2005) Microfluidic gas-flow profiling using remote-detection NMR. Proc Natl Acad Sci USA 102:14960–14963

95. McDonnell EE, Han S-I, Hilty C, Pierce KL, Pines A (2005) NMR analysis on microfluidic devices by remote detection. Anal Chem 77:8109–8114
96. Telkki V-V, Zhivonitko VV (2011) Analysis of remote detection travel time curves measured from microfluidic channels. J Magn Reson 210:238–245
97. Zhivonitko V, Telkki V-V, Koptyug IV (2012) Characterization of microfluidic gas reactors using remote-detection MRI and parahydrogen-induced polarization. Angew Chem Int Ed 51:8054–8058
98. Carravetta M, Johannessen OG, Levitt MH (2004) Beyond the T_1 limit: singlet nuclear spin states in low magnetic field. Phys Rev Lett 92:153003
99. Carravetta M, Levitt MH (2004) Long-lived nuclear spin states in high-field solution NMR. J Am Chem Soc 126:6228–6229
100. Sharma R, Bouchard LS (2012) Strongly hyperpolarized gas from parahydrogen by rational design of ligand-capped nanoparticles. Sci Rep 2:277
101. Reineri F, Viale A, Dastru W, Gobetto R, Aime S (2011) How to design ^{13}C para-hydrogen-induced polarization experiments for MRI applications. Contrast Media Mol Imaging 6:77–84
102. Carson PJ, Bowers CR, Weitekamp DP (2001) The PASADENA effect at a solid surface: high-sensitivity nuclear magnetic resonance of hydrogen chemisorption. J Am Chem Soc 123:11821–11822
103. Tayler MCD, Levitt MH (2011) Singlet nuclear magnetic resonance of nearly-equivalent spins. Phys Chem Chem Phys 13:5556–5560
104. Tayler MCD, Levitt MH (2011) Paramagnetic relaxation of nuclear singlet states. Phys Chem Chem Phys 13:9128–9130
105. Warren WS, Jenista E, Branca RT, Chen X (2009) Increasing hyperpolarized spin lifetimes through true singlet eigenstates. Science 323:1711–1714
106. Gloggler S, Muller R, Colell J, Emondts M, Dabrowski M, Blumich B, Appelt S (2011) Parahydrogen induced polarization of amino acids, peptides and deuterium-hydrogen gas. Phys Chem Chem Phys 13:13759–13764
107. Reineri F, Santelia D, Viale A, Cerutti E, Poggi L, Tichy T, Premkumar SSD, Gobetto R, Aime S (2010) Para-hydrogenated glucose derivatives as potential ^{13}C-hyperpolarized probes for magnetic resonance imaging. J Am Chem Soc 132:7186–7193
108. Chekmenev EY, Hovener J, Norton VA, Harris K, Batchelder LS, Bhattacharya P, Ross BD, Weitekamp DP (2008) PASADENA hyperpolarization of succinic acid for MRI and NMR spectroscopy. J Am Chem Soc 130:4212–4213
109. Shchepin RV, Coffey AM, Waddell KW, Chekmenev EY (2012) PASADENA hyperpolarized ^{13}C phospholactate. J Am Chem Soc 134:3957–3960
110. Roth M, Koch A, Kindervater P, Bargon J, Spiess HW, Munnemann K (2010) ^{13}C hyperpolarization of a barbituric acid derivative via parahydrogen induced polarization. J Magn Reson 204:50–55
111. Aime S, Dastru W, Gobetto R, Viale A (2005) para-Hydrogenation of unsaturated moieties on poly(lysine) derived substrates for the development of novel hyperpolarized MRI contrast agents. Org Biomol Chem 3:3948–3954
112. Trantzschel T, Bernarding J, Plaumann M, Lego D, Gutmann T, Ratajczyk T, Dillenberger S, Buntkowsky G, Bargon J, Bommerich U (2012) Parahydrogen induced polarization in face of keto-enol tautomerism: proof of concept with hyperpolarized ethanol. Phys Chem Chem Phys 14:5601–5604
113. Kadlecek S, Emami K, Ishii M, Rizi R (2010) Optimal transfer of spin-order between a singlet nuclear pair and a heteronucleus. J Magn Reson 205:9–13
114. Reineri F, Bouguet-Bonnet S, Canet D (2011) Creation and evolution of net proton hyperpolarization arising from para-hydrogenation. J Magn Reson 210:107–112
115. Bretschneider C, Karabanov A, Nielsen NC, Kockenberger W (2012) Conversion of parahydrogen induced longitudinal two-spin order to evenly distributed single spin polarisation by optimal control pulse sequences. J Chem Phys 136:094201

116. Johannesson H, Axelsson O, Karlsson M (2004) Transfer of para-hydrogen spin order into polarization by diabatic field cycling. CR Phys 5:315–324
117. Chekmenev EY, Norton VA, Weitekamp DP, Bhattacharya P (2009) Hyperpolarized ^{1}H NMR employing low gamma nucleus for spin polarization storage. J Am Chem Soc 131:3164–3165
118. Goldman M, Johannesson H (2005) Conversion of a proton pair para order into ^{13}C polarization by RF irradiation, for use in MRI. CR Phys 6:575–581
119. Duckett SB, Newell CL, Eisenberg R (1993) More than INEPT: parahydrogen and INEPT+ give unprecedented resonance enhancement to ^{13}C by direct ^{1}H polarisation transfer. J Am Chem Soc 115:1156–1157
120. Haake M, Natterer J, Bargon J (1996) Efficient NMR pulse sequences to transfer the parahydrogen-induced polarization to hetero nuclei. J Am Chem Soc 118:8688–8691
121. Aime S, Gobetto R, Reineri F, Canet D (2003) Hyperpolarization transfer from parahydrogen to deuterium via carbon-13. J Chem Phys 119:8890–8896
122. Aime S, Gobetto R, Reineri F, Canet D (2006) Polarization transfer from para-hydrogen to heteronuclei: effect of H/D substitution. The case of $AA'X$ and $A_2A'_2X$ spin systems. J Magn Reson 178:184–192
123. Kuhn LT, Bargon J (2007) Transfer of parahydrogen-induced hyperpolarization to heteronuclei. Top Curr Chem 276:25–68
124. Bommerich U, Trantzschel T, Mulla-Osman S, Buntkowsky G, Bargond J, Bernarding J (2010) Hyperpolarized ^{19}F-MRI: parahydrogen-induced polarization and field variation enable ^{19}F-MRI at low spin density. Phys Chem Chem Phys 12:10309–10312

Top Curr Chem (2013) 338: 181–228
DOI: 10.1007/128_2013_436

Published online: 7 July 2013

Dynamic Nuclear Polarization Enhanced NMR in the Solid-State

Ümit Akbey, W. Trent Franks, Arne Linden, Marcella Orwick-Rydmark, Sascha Lange, and Hartmut Oschkinat

Abstract Nuclear magnetic resonance (NMR) spectroscopy is one of the most commonly used spectroscopic techniques to obtain information on the structure and dynamics of biological and chemical materials. A variety of samples can be studied including solutions, crystalline solids, powders and hydrated protein extracts. However, biological NMR spectroscopy is limited to concentrated samples, typically in the millimolar range, due to its intrinsic low sensitivity compared to other techniques such as fluorescence or electron paramagnetic resonance (EPR) spectroscopy.

Dynamic nuclear polarization (DNP) is a method that increases the sensitivity of NMR by several orders of magnitude. It exploits a polarization transfer from unpaired electrons to neighboring nuclei which leads to an absolute increase of the signal-to-noise ratio (S/N). Consequently, biological samples with much lower concentrations can now be studied in hours or days compared to several weeks.

This chapter will explain the different types of DNP enhanced NMR experiments, focusing primarily on solid-state magic angle spinning (MAS) DNP, its applications, and possible means of improvement.

Keywords Dynamic Nuclear Polarization (DNP) · Hyperpolarization · Sensitivity · Solid-state MAS NMR · Structural Biology

Contents

Ü. Akbey (✉), W.T. Franks, A. Linden, M. Orwick-Rydmark, S. Lange, and H. Oschkinat
NMR Supported Structural Biology, Leibniz Institute für Molekulare Pharmakologie, Robert Roessle Str. 10, 13125 Berlin, Germany
e-mail: akbey@fmp-berlin.de; oschkinat@fmp-berlin.de

Abbreviations

APG	Alanyl-prolyl-glycine
B_0	External magnetic field
B_1	RF field strength
B_{1e}	MW field strength
bCTbK	bis-cyclohexyl-TEMPO-bisketal
BDPA	1,3-Bisdiphenylene-2-phenyl allyl
BT2E	bis-TEMPO-2-ethylene oxide
bTbk	bis-TEMPO-bisketal
bTbtk-py	bis-TEMPO-bis-thioketal-tetra-tetrahydropyran
BTOX	bis-TEMPO tethered by oxalate
BTOXA	bis-TEMPO tethered by oxalyl amide
BTUrea	bis-TEMPO tethered by urea
BWOs	Backward wave oscillators
c	Concentration
CE	Cross effect
CIDNP	Chemically induced dynamic nuclear polarization
CP	Cross polarization
DMSO	Dimethyl sulfoxide
DNP	Dynamic Nuclear Polarization
DOTAPO-TEMPO	4-[*N*,*N*-Di-(2-hydroxy-3-(TEMPO-40-oxy)-propyl)]-amino-TEMPO
DQ	Double quantum
EIKs	Extended interaction klystrons
EIOs	Extended interaction oscillators
EPR	Electron paramagnetic resonance
FT	Fourier transform

IMPATT	Impact ionization avalanche transit-time
INEPT	Insensitive nuclei enhanced by polarization transfer
K	Kelvin
MAS	Magic angle spinning
MRI	Magnetic resonance imaging
MW	Microwave
nAChR	Nicotinic acetylcholine receptor
N_e	Number of electrons
N-f-MLF-OH	*N*-Formyl-Met-Leu-Phe-OH
NMR	Nuclear magnetic resonance
NT-II	Neurotoxin II
OX063 (Trityl)	Tris{8-carboxyl-2,2,6,6-benzo(1,2-d:4,5-d)-bis(1,3)dithiole-4-yl}methyl sodium salt
PE	Paramagnetic effects
PHIP	Para-hydrogen induced polarization
PRE	Paramagnetic relaxation enhancement
Q-factor	Quality factor
S/N	Signal-to-noise ratio
SD	Spin diffusion
SE	Solid effect
SQ	Single quantum
T	Tesla
T_{1DQ}	Double quantum relaxation time
T_{1e}	Electron spin-lattice relaxation time
T_{1n}	Nuclear spin-lattice relaxation time
$T_{1\rho}$	Nuclear relaxation time in the rotating frame
T_{1ZQ}	Zero quantum relaxation time
T_{2e}	Electron spin–spin relaxation time
T_{2n}	Nuclear spin–spin relaxation time
TEMPO	2,2,6,6-Tetramethylpiperidinoxyl
TJ-DNP	Temperature-jump dynamic nuclear polarization
TM	Thermal mixing
TOTAPOL	1-(TEMPO-4-oxyl)-3-(TEMPO-4-amino)-propan-2-ol
W	Watt
ZQ	Zero quantum
Δ	Inhomogeneous breadth of the EPR spectrum
δ	Homogeneous EPR linewidth
ε	Enhancement
γ_e	Gyromagnetic ratio of electron
γ_n	Gyromagnetic ratio of nucleus
κ	Sensitivity
τ_B	Polarization buildup time constant
τ_R	Rotor period
ω_{0e}	Electron Larmor frequency
ω_{0I}	Nuclear Larmor frequency
ω_R	Spinning frequency

1 Introduction and Motivation

Nuclear magnetic resonance (NMR) spectroscopy is a powerful tool for studying structure and dynamics of biological macromolecules at atomic resolution. Magic-angle spinning (MAS) NMR can be used to acquire high-resolution data on solid samples, and has been applied to many biological systems [1, 2] including amyloid [3–5], nanocrystalline [6, 7], and membrane proteins [8–12].

However, NMR is limited to the study of samples with high concentrations of magnetically active nuclei, typically requiring labeling strategies for biomolecular applications. Furthermore, the NMR transition energy is very low when compared to the transition energy for infrared or ultraviolet/visible spectroscopy. Therefore, only an exceedingly small excess Boltzmann polarization can be induced (~1 in 1,000,000) by an external magnetic field.

The sensitivity of NMR spectroscopy has dramatically increased throughout the lifetime of the technique. Instrumental innovations such as superconducting magnets and cryo-probes have proved to be invaluable to the NMR spectroscopist. Methodological developments such as Fourier transform (FT) spectroscopy, improved polarization transfer schemes [i.e., cross-polarization (CP) and insensitive nuclei enhanced by polarization transfer (INEPT)], and optimized detection techniques have made NMR a versatile tool in chemistry, biology, and materials science [13].

An additional approach for increasing NMR sensitivity is to use hyperpolarization to produce spin population differences far greater than that obtained at the thermal Boltzmann distribution. Several hyperpolarization techniques have been demonstrated such as chemically induced dynamic nuclear polarization (CIDNP) [14–19], para-hydrogen induced polarization (PHIP) [20–22], optical pumping [20, 23–25], and dynamic nuclear polarization (DNP) [26]. Each of these methods increases the sensitivity by several orders of magnitude. The first two approaches generate hyperpolarized states by special chemical reactions, while optical pumping is limited to noble gases. DNP, however, can be applied whenever the sample contains an unpaired electron source, provided that the electron relaxation times and concentrations are favorable. The primary focus of this chapter will be DNP enhanced solid-state MAS NMR spectroscopy.

DNP was first proposed by Overhauser in 1953 [26] and was soon experimentally verified in conducting solid metals by the work of Carver and Slichter [27], and later on in solutions [28–30]. DNP is currently achieved using continuous high-power microwave (MW) irradiation of unpaired electron spins from polarizing agents such as the stable nitroxide radicals, TEMPO and TOTAPOL [31]. The initial hyperpolarization is usually achieved at cryogenic temperatures (from ~100 K to as low as 1–2 K) to improve the relaxation processes that help determine the transfer efficiency. Solution state DNP experiments either polarize the sample at room temperature [32] or at liquid nitrogen and helium temperatures, then followed by a fast temperature jump [33, 34]. MAS DNP [35–38] is usually performed at ~100 K with a high-power, high-frequency gyrotron MW source [39, 40]. Commercialization of high-field MAS DNP instrumentation has made the technique widely available [41]. In routine MAS

DNP NMR applications to biological and macromolecular molecules, DNP-mediated increases in the NMR signal between ~10 and 100 (DNP enhancement, ε) have been obtained. Such signal enhancements decrease the signal averaging time by 100- to 10,000-fold (~ ε^2). The extent of this signal enhancement by DNP becomes more pronounced when the additional polarization increase due to cryogenic temperatures is considered. Thus, an additional factor of ~3 is possible for the absolute sensitivity (DNP + cryogenic enhancement), e.g., for experiments performed at ~100 K compared to ~300 K.

In principle, the DNP enhancement is proportional to the gyromagnetic ratio of the electron (γ_e) and nuclear spins (γ_I). Therefore, a signal enhancement of $\varepsilon = (\gamma_e/\gamma_I) \approx 660$ can theoretically be obtained for ^{1}H, but in practice $\varepsilon = 100$–200 is more commonly observed when using model compounds [42]. In biological systems such as lysozyme [40] and bacteriorhodopsin [43–45] the enhancements are smaller, $\varepsilon = 40$–50. In systems with longer relaxation times due to crystallinity and/or deuteration, considerably larger enhancements were observed ($\varepsilon = \sim 100$ in the amyloidogenic model peptide $GNNQQNY_{7-13}$ [46] and ~120 for a deuterated protein [99]).

DNP-enhanced NMR spectroscopy has been used to study biological and functional materials at low concentrations in reasonable experimental times that were inaccessible without DNP [44, 47–50]. Moreover, large biological molecules, reaction dynamics, and high-throughput screening may now be studied with improved sensitivity using DNP techniques. Hyperpolarized molecules have also been used to monitor reaction intermediates in magnetic resonance imaging (MRI).

In this chapter we present several approaches that have been taken to utilize DNP. A brief summary is first given regarding solution state DNP and dissolution DNP. The chapter's main emphasis is on solid-state MAS DNP. The theoretical aspects of solid-state MAS DNP are briefly summarized followed by a third section on specialized DNP instrumentation. Fourth, electron polarization sources are discussed. Practical aspects of MAS DNP will be summarized in a fifth section.

2 Experimental Approaches for DNP

Different instrumental methods to exploit DNP include solution-state, shuttling, dissolution and solid-state methods. In the following text, "solution-DNP" will refer to hyperpolarization and detection at ambient temperatures in the solution phase of a sample. "Shuttle DNP" will refer to hyperpolarization of a solution at low magnetic field followed by the physical shuttling of the sample into a higher magnetic field for NMR detection. "Dissolution-DNP" will refer to hyperpolarization of a frozen solution at ~1–2 K followed by fast melting of the sample to ambient temperatures for detection of a solution phase sample. "Solid-state DNP" will refer to the hyperpolarization of a frozen solid sample at temperatures below 200 K and detection at this temperature.

2.1 Solution-State and Shuttle DNP

Solution DNP hyperpolarizes nuclear spins at ambient temperatures. It has several advantages over dissolution and solid state DNP in regards to instrumentation, sample repeatability, and broad applicability, though the signal enhancements are smaller compared to other DNP methods. Solution DNP can provide detailed local structural information since the technique relies on the nuclear–electron interaction in a site specific manner [51].

Spin polarization is mediated by the Overhauser effect in solution DNP. Hence solution state DNP is also referred to as Overhauser DNP [26, 30]. Spin polarization is transferred between a hyperfine coupled electron and nuclear spin via an allowed single quantum (SQ) electron paramagnetic resonance (EPR) transition excited by the MW irradiation. MW irradiation saturates the electrons, and as a result increases the polarization of the hyperfine coupled nucleus. The polarization transfer is successful due to the difference in the zero-quantum (ZQ) and double-quantum (DQ) relaxation rates, which favor hyperpolarization of the nucleus. A detailed description of solution state DNP theory and the nature of the polarization transfer at several different magnetic fields can be found in the literature [30, 32, 51, 52, 53].

Polarization transfer at high magnetic fields is instrumentally demanding due to the difficulty in saturating the broad EPR signal and the intrinsic relaxation of the sample [54, 55]. Solution state DNP was initially demonstrated at 1.4 T (60 MHz ^{1}H frequency) where an enhancement of -9 was reported for the water signal using the trityl radical [56]. The enhancement was later increased to -17 by using a high-pressure DNP setup, as a result of better nuclear–electron interactions at liquid/solid-radical interfaces [57]. At 3.4 T (145 MHz ^{1}H frequency), enhancements of up to -65 were reported for water containing the 4-hydroxy-TEMPO radical using a milliwatt (mW) microwave power source [58]. At 5 T (214 MHz ^{1}H frequency), enhancements of -181 and -41 were reported for ^{31}P and ^{13}C, respectively, using BDPA as the polarizing agent [59].

The highest field for which a solution state DNP enhancement has been reported so far is 9.4 T (400 MHz ^{1}H frequency) [60, 61]. An enhancement of approximately -9 was achieved using a 45-mW microwave source and ^{15}N-Fremy's Salt as the radical (Fig. 1). DNP enhancement of -29 was achieved using ~20 W of MW power produced by a gyrotron. This enhancement is larger than the theoretically predicted value [62] indicating that the current understanding of the polarization mechanism is not complete, especially with regards to the effects of molecular dynamics [63, 64]. As promising examples of solution state DNP–NMR experiments, several interesting systems have been investigated, including site specific hydration dynamics measurements on membranes and lipid bilayers, as well as NMR flow imaging [65–69].

Shuttle DNP is an alternative method to overcome the poor saturation of the EPR resonance and B_0-dependence observed in solution DNP. Shuttle DNP hyperpolarizes the solution sample at low magnetic fields, followed by rapid sample transfer (shuttle) into a high magnetic field to observe a high resolution NMR spectrum [70–72].

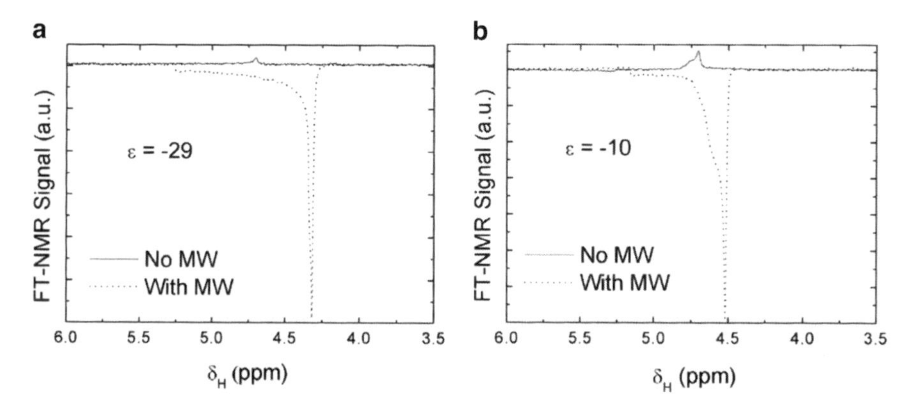

Fig. 1 Solution state DNP enhanced NMR spectroscopy at 400 MHz ^{1}H Larmor frequency on 40 mM aqueous Fremy's Salt solution. (**a**) With a high power gyrotron MW source. (**b**) With a low power solid-state MW source. Reproduced from Denysenkov et al. with permission of the PCCP Owner Societies [62]

The first demonstration of shuttle DNP was performed with an initial polarization step at 0.34 T followed by detection at 14 T. DNP enhancements of −2.6 and 15 were observed for water and ^{13}C labeled chloroform, respectively, using TEMPO-based radicals [70, 71]. An analogous, continuous-flow approach was used for MRI and spectroscopy with hyperpolarization solutions having enhancements of around −15 [73–75].

2.2 *Dissolution DNP and MRI Applications*

Dissolution DNP utilizes low temperatures (~1−2 K) and magnetic fields (>1 T) to obtain hyperpolarization with very high efficiencies (Fig. 2) [33]. For the first experimental demonstration by Ardenkjaer-Larsen et al., a sample containing trityl radical was first cooled to liquid helium temperature in a ~3.35-T magnetic field. MW irradiation at 94 GHz was used to saturate the EPR transitions, resulting in the hyperpolarization of the surrounding nuclei in the frozen solution. The sample was then rapidly melted (dissolution step) and transferred into an NMR spectrometer or MRI scanner for detection [33, 76]. The total dissolution and acquisition period only takes ~10 s, although the polarization period may require several hours to complete. With this method, remarkable enhancements of ~44,400 (37% efficiency; where efficiency is ~$\varepsilon_{\text{observed}}/\varepsilon_{\text{theoretical}}$) and 23,500 (7.8% efficiency) were obtained for carbon and nitrogen nuclei, respectively. The large enhancements result from a combination of the hyperpolarization and the rapid temperature-jump of the hyperpolarized sample from low to high temperatures.

The losses from nuclear relaxation due to the paramagnetism of the radicals, however, limit efficiency of dissolution DNP. Nevertheless, fast sample transfer

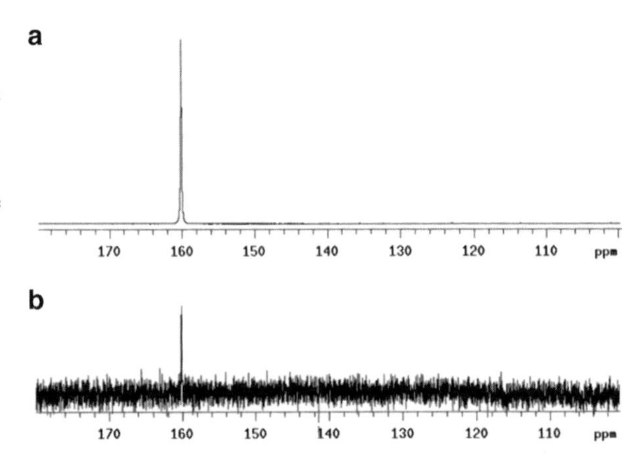

Fig. 2 (**a**) A single scan dissolution state DNP enhanced ^{13}C NMR spectrum of unlabeled urea, with a hyperpolarization of ~20 %. (**b**) Solution state NMR spectrum of the same sample at 9.4 T and at room temperature, which is acquired in 65 h with many signal-averaging repetitions. Reproduced from Larsen et al. with permission [33]. Copyright (2003) National Academy of Sciences, USA

times of ~6 s result in only a minor loss of polarization during sample transfer due to relaxation [33]. The dissolution DNP polarization transfer mechanism (with the trityl radical) is dominated by thermal mixing, with a minor contribution of the solid effect (see theory section). It is possible to remove up to 99% of the radical from the hyperpolarized solvent by filtration prior to transfer into an NMR or MRI instrument for in vivo applications [33].

Dissolution DNP may dramatically change the way conventional NMR and MRI are currently done, and open many new research perspectives. However, the hyperpolarized sample can be measured only once and then needs to be polarized again and the exact characteristics of the sample can differ from scan to scan. Nevertheless, it was shown that small flip angle experiments or single-scan approaches can make it possible to record multidimensional NMR spectra [77–79].

In addition to NMR applications, DNP can also help in the field of MRI, which is a non-invasive imaging technique. Most MRI applications create an image using proton detection due to sensitivity considerations. However, MRI based on ^{13}C spectroscopy might better fit to the diagnostic purposes, though the sensitivity is poor due to its low gyromagnetic ratio and low natural abundance. A huge increase in ^{13}C sensitivity would be needed for such an image. The solution is through the combination of MRI with DNP [80–84]. The DNP enhancement in MRI experiments has been reported in several instances. These studies use isotopically labeled hyperpolarized small metabolites, e.g., pyruvate. The metabolite can then be used for a cell-cycle-targeted imaging experiment by utilizing the increased signals ~10^4–10^5-fold, Fig. 3 [33, 85–95].

2.3 *Solid-State DNP*

The first DNP experiment was carried out in the solid state in the early 1950s under static conditions (Fig. 4, left) [27]. Incorporating MAS into DNP experiments to improve the spectroscopic resolution was a milestone for solid state DNP NMR

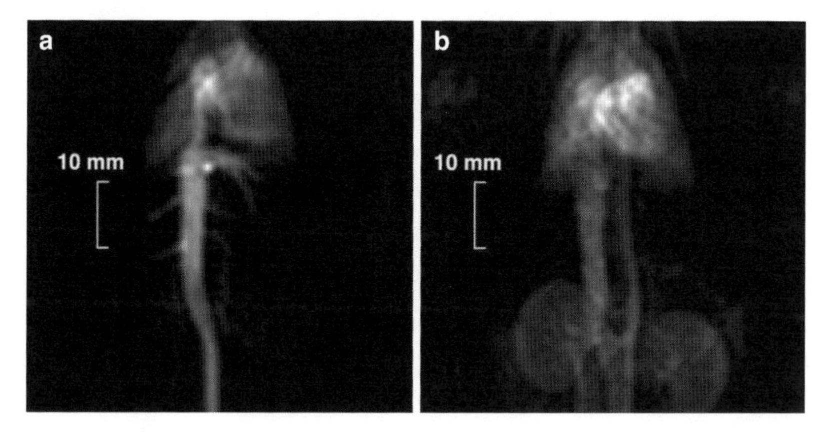

Fig. 3 The images obtained by ^{13}C MRI (**a**) immediately and (**b**) 2 s after the injection of the hyperpolarized contrast agent into a rat. Reproduced from Golman et al. with permission [76]. Copyright (2003) National Academy of Sciences, USA

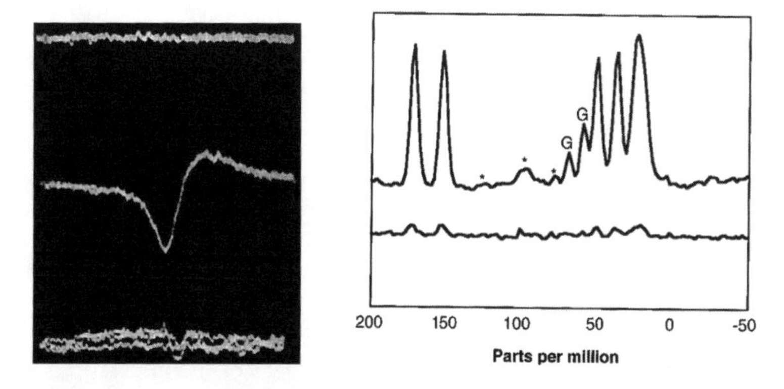

Fig. 4 *Left*: The first experimental demonstration of DNP enhancement observed on Li metal. The *middle line* represents the enhanced signal by electron irradiation, in comparison to the *top line* representing thermal equilibrium signal. Reprinted with permission from Slichter et al. [27] Copyright (1953) by the American Physical Society. *Right*: The representation of the DNP enhanced ^{13}C CPMAS NMR spectrum utilizing MAS on a uniformly labeled arginine sample containing 40 mM 4-amino-TEMPO radical. The MW on and off spectra were recorded at ~5 T (200 MHz ^{1}H frequency). Adapted from Hall et al. [40]. Reprinted with permission from AAAS

which was introduced in the mid-1980s (Fig. 4, right) [35, 39, 40]. Before explaining the practical aspects of DNP enhanced NMR in the solid state, the mechanisms utilized in the DNP enhanced MAS NMR for hyperpolarization of solid samples will be discussed briefly below. Moreover, the effect of the spin diffusion to the polarization transfer will be mentioned.

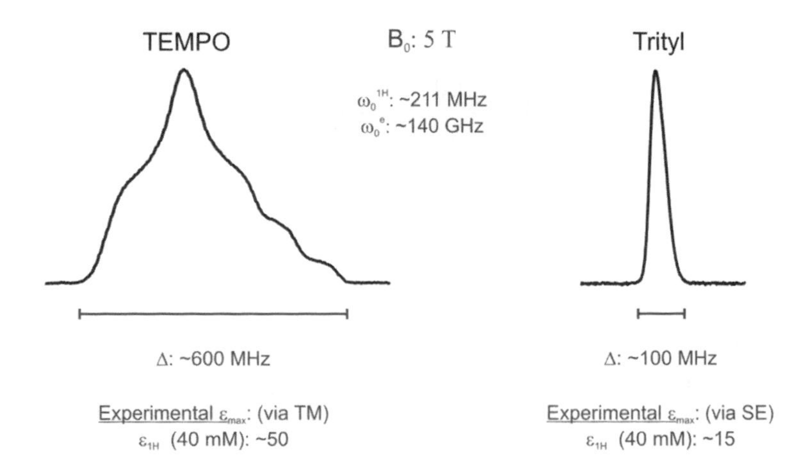

Fig. 5 Representation of the EPR spectra of TEMPO and trityl radicals, along with the spectral widths and obtained DNP enhancement values. The spectra were kindly provided by Thorsten Maly

2.3.1 Basics of Polarization Transfer in Solid-State DNP

There are three mechanisms for polarization transfer from an electron to its surrounding nuclei in dielectric solids: the solid effect (SE), the cross effect (CE), and thermal mixing (TM). Parameters determining the DNP mechanism in solids are the inhomogeneous breadth of the EPR spectrum (Δ, up to several hundreds of MHz for TEMPO at 5 T), the homogeneous EPR linewidth (δ), and the nuclear Larmor frequency (ω_{0I}, i.e., 400 MHz ^{1}H frequency at 9.4 T). Δ denotes the full breadth of the entire EPR spectrum, which can be directly obtained from an EPR spectrum. δ denotes the linewidth of a single electron component (orientation) that could be determined from, for example, a hole-burning experiment. The EPR spectra of TEMPO and trityl radicals, along with the magnitudes of Δ and ω_{0I}, are shown in Fig. 5.

The efficiency of DNP is affected by experimental parameters such as γ_e and γ_n, B_{1e} (the microwave power), B_0 (the static magnetic field), N_e (the electron concentration), δ, and T_{1e} and T_{1n} (the electron and nuclear spin-lattice relaxation times), respectively [96]. Moreover, the extent to which polarization is transferred to the surrounding nuclei via spin diffusion affects the overall DNP efficiency. Factors which influence the efficiency of spin polarization for the SE and CE mechanisms are discussed below, and are valid in the limits of fast spin diffusion, low MW power, and low radical concentration. Homogeneous electron lineshapes are assumed, and sample rotation is not taken into account [35, 97]. B_{1e} and N_e may have a linear effect on DNP enhancement due to saturation at high radical concentration and high MW powers [41, 98]. The concentration of protons affects DNP enhancement through T_{1n} and should be carefully optimized [99], which will be discussed in the following sections.

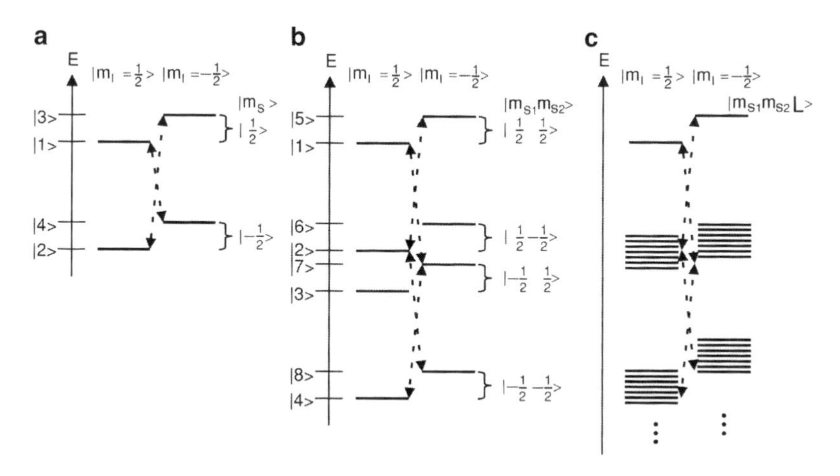

Fig. 6 The representation of the energy diagrams for (**a**) SE, (**b**) CE and (**c**) TM types of DNP mechanisms. Reprinted with permission from Hu et al. [100]. Copyright (2011), American Institute of Physics

The Solid Effect

The SE utilizes the hyperfine coupling between electron and nuclear spins to transfer polarization. A four-level energy diagram describes the system (Fig. 6). MW irradiation is applied at either the ZQ or DQ transitions (at the so-called forbidden-transitions) [101–103]. If the DQ transition is saturated, the population of the irradiated states will equalize, leading to hyperpolarization of the nuclear states.

The ZQ and DQ transitions result in negative (at frequency $\omega_{0e} - \omega_{0I}$) and positive (at frequency $\omega_{0e} + \omega_{0I}$) DNP enhancements, respectively, separated by twice the nuclear Larmor frequency (ω_{0I}). The polarization transfer is more rigorously described via rate equations that depend on the spin-lattice relaxation times of the nucleus, electron, and the cross relaxation terms T_{1DQ} and T_{1ZQ}. A detailed mathematical treatment is beyond the scope of this work, and can be found for the SE in the literature [89, 90, 100, 104–111] including calculations on moderate and large spin systems [112–115].

The SE is more favorable at low B_0 fields because the enhancement depends on the static magnetic field by the inverse square:

$$\varepsilon_{SE} \propto \frac{\gamma_e}{\gamma_n}\left(\frac{B_{1e}}{B_0}\right)^2 \frac{N_e}{\delta} T_{1n} \tag{1}$$

The SE occurs when a polarizing agent has a homogeneous EPR linewidth and an inhomogeneous EPR spectral breadth that are both smaller than the nuclear Larmor frequency (δ, $\Delta < \omega_{0I}$) [42, 107]. This requirement is met, for instance, when hyperpolarizing protons with a Larmor frequency of ~210 MHz at 5 T using trityl radicals which have a spectral breadth of ~40–50 MHz. The SE DNP scales quadratically with the inverse of the magnetic field (B_0^{-2}) due to the inefficiency of electron excitation, and as a result it is much less efficient at high magnetic fields. Nevertheless, in recent work by Corzilius et al. a remarkably high ^{1}H DNP enhancement of ~90 was achieved using 40 mM trityl at 5 T [110], and higher MW powers (>10 W) demonstrating the possibility of SE-induced DNP at high magnetic fields. A strong MW power dependence was observed in this work and high enhancements were only observed at high MW power levels. The enhancement decreased dramatically at lower MW power (~27 at 3 W). This experimental result demonstrates the possibility of increasing SE type DNP efficiency at high magnetic fields by scaling the MW power levels, which will allow a similar extend of excitation efficiency and the DNP enhancements will not necessarily be compromised.

The Cross Effect

The CE is a three spin process that requires a strongly dipole–dipole coupled electron pair, and a nucleus hyperfine-coupled to these electrons. It was discovered by Kessenikh et al. [116] in polystyrene samples that were doped with radiation defects as electron sources, and further investigated by Hwang and Hill [104]. In these studies, a typical SE DNP enhancement was observed for samples containing low radical concentrations. However, it was observed that for the samples with higher radical concentrations, the enhancement profile became broader due to increased electron–electron dipolar couplings. More generally, the cross effect is most effective when biradicals are used [31]. A complex eight-level energy diagram then describes the system (Fig. 6). MW radiation is used to excite one electron, which then excites the dipolar coupled electron via a DQ or ZQ transition. If the Larmor frequencies of the electrons differ by the nuclear Larmor frequency (i.e., $|\omega_{0e1} - \omega_{0e2}| = \omega_{0I}$), a nuclear spin is also excited, which is the basis of CE DNP. A more detailed mathematical treatment of the CE DNP can be found in several excellent articles and reviews, and will not be covered here [42, 100, 105, 117]. The effect of MAS on CE DNP will be discussed in Sect. 5.5.

The CE occurs when the homogeneous EPR linewidth is smaller than the nuclear Larmor frequency (ω_{0I}), and the inhomogeneous breadth of the EPR spectrum is larger than ω_{0I} ($\delta < \omega_{0I} < \Delta$) [42]. The efficiency of the polarization transfer strongly depends on the frequency difference between the two coupled electrons. The CE is more efficient when the inter-electron distance and the relative orientation of the electron g-tensor is optimized, as is the case for the biradical, TOTAPOL [31]. The CE mechanism for DNP is mediated by the homogeneous linewidth, as described by (2):

$$\varepsilon_{CE} \propto \frac{\gamma_e}{\gamma_n}\left(\frac{B_{1e}^2}{B_0}\right)\frac{N_e^2}{\delta^2}T_{1n} \qquad (2)$$

The inverse linear B_0 dependence suggests that the CE mechanism is more efficient at high fields relative to the SE [see (1)]. As a result, the CE is the most commonly utilized mechanism to obtain DNP enhancements at high B_0.

Thermal Mixing

TM is a multi-electron mediated polarization transfer mechanism that occurs when the homogeneously broadened EPR linewidth is larger than the nuclear Larmor frequency ($\delta > \omega_{0I}$). This situation is typically observed at high radical concentrations that increase the overall dipolar couplings of electrons and broaden the δ [42]. A multi-level energy diagram describes the system (Fig. 6). A detailed description of the effect of homogeneous and inhomogeneously broadened EPR lineshape on the efficiency of the TM mechanism is described in recent work by Hovav et al. [118]. In this work it was proposed that both TM and SE result in DNP enhancements in a homogeneously broadened line, while in the inhomogeneous case, the combined effect of SE and CE results in DNP enhancements. Generally, the TM mechanism is more efficient at high magnetic fields compared to the SE, but smaller compared to the CE (see Sect. 4.1 and Hu et al. [119]).

To describe the mechanism for TM, advanced models that account for both homogeneous broadening and partial inhomogeneous effects are required. The details of TM and its relation to the so-called "spin temperature" phenomenon are not discussed further here [35, 106, 107, 118, 120–127].

Influence of Spin Diffusion

Spin diffusion (SD) is a process by which the dipolar coupled spins exchange magnetization among each other [128]. It is used to explain the T_1 relaxation times in dielectric solids via rate equations describing the evolution of local spin polarizations [129]. The SE and CE transfer polarization from electrons to the core nuclei (nuclei which are hyperfine coupled with electrons) in close proximity and as a result modify the spatial polarization distribution. The hyperpolarized NMR signal must originate from the bulk nuclei (nuclei which are not hyperfine coupled to electrons due to larger distance, and which are surrounding the core nuclei) because the strong hyperfine couplings broaden the signals of core nuclei below detection [113]. The transfer of polarization from core nuclei to the bulk nuclei determines the overall DNP efficiency, and observed enhancement. Moreover, in the case of a solid protein sample dissolved in a solvent, an additional transfer of polarization from the bulk nuclei to the protein sample is required via SD.

To quantify the effect of spin diffusion on DNP processes via SE, simulations were performed to understand polarization transfer progressing from core to bulk in a chain of nuclei, some of which are not directly hyperfine coupled to the electron [113]. Results demonstrate that the DNP polarization of the bulk nuclei propagates via nuclear–dipolar interactions, and as a result shows the sizeable effects of SD.

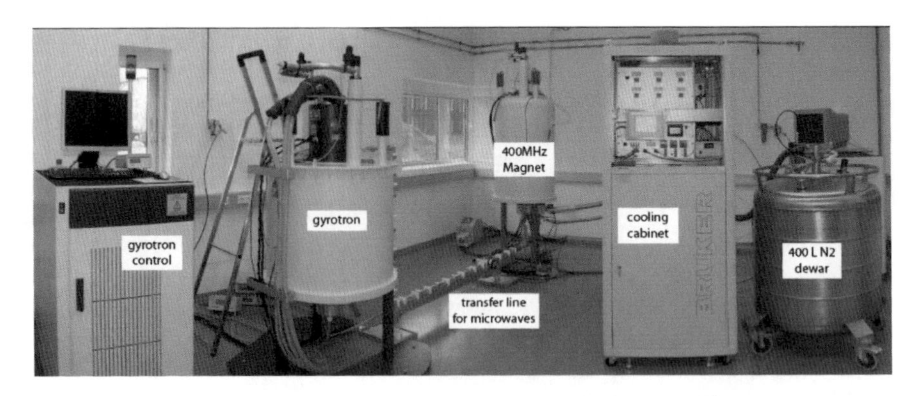

Fig. 7 The commercial Bruker DNP solid-state NMR system with a commercial gyrotron, corrugated wave guide, cooling cabinet and 9.4 T magnet at the Leibniz Institute für Molakulare Pharmakologie, Campus Buch, Berlin, Germany. MWs are delivered to the sample through the probe base. The picture was kindly provided by Arne Linden

3 Instrumentation for Solid-State DNP

While the first successful DNP experiments were carried out at low magnetic fields (<1 T) [13, 35, 130], advancements in hardware technology and the availability of gyrotrons as high power and high frequency MW sources enabled solution and MAS DNP NMR at higher magnetic fields (>5 T) [35, 39, 40, 96]. In the following sections we will summarize the instrumentation in three parts: MW sources to achieve polarization transfer, waveguides to deliver MWs, and the cryo-probes to perform the NMR experiment (Fig. 7).

3.1 Microwave Sources

High power (~ several watts) and high frequency (100–500 GHz) MWs are necessary for solid state DNP NMR, where the DNP enhancement is proportional to the MW power up to saturation (by B_{1e} or B_{1e}^{2}). Gunn diodes, IMPATT (IMPact ionization Avalanche Transit-Time) diodes and pumped far-infrared CO_2 lasers generate MWs in the respective GHz range needed, though with lower power output at higher frequencies. A detailed description of this topic can be found in a review by Maly et al. [42]. Nevertheless, a ~30 mW diode microwave source was used to produce enhancements of up to 80 at ~20 K in a 9.4-T field [131]. This experimental setup was then used for high resolution imaging, which resulted in an ~0.03 μm^3 voxel size, nearly three orders of magnitude smaller than usually obtained in contemporary experiments [131].

Vacuum electronic devices are separated into slow-wave and fast-wave devices [42]. Slow-wave devices such as extended interaction oscillators (EIOs), extended interaction klystrons (EIKs), backward wave oscillators (BWOs), and orotrons produce high power MW up to ~140 GHz. At higher frequencies their output power drops dramatically (Fig. 8). Fast-wave devices such as gyrotrons have been used successfully in DNP experiments to generate high MW power at frequencies up to 460 GHz (suitable for experiments at 700 MHz ^{1}H Larmor frequency and 16.4 T) [39, 41, 42, 62, 97, 133–144].

3.2 Waveguides

The intensity of the MWs delivered to the sample site is a major factor in the overall DNP efficiency. Low-loss transmission lines are used to guide MWs to the sample in the NMR probe including fundamental, oversized or corrugated waveguides [96, 145]. Corrugated waveguides are rectangular or circular tubes with periodic wall perturbations (corrugations, with depth around a quarter of the MW wavelength), to suppress the wall currents for reducing losses. Corrugated waveguides are more efficient and thus currently preferred [41, 42].

3.3 Cryo-Probes

Special probes are required for solid-state DNP NMR experiments that can operate at cryogenic temperatures, deliver MW irradiation, and detect NMR signals. The probe can be either for static or for rotating samples (Fig. 9) [42]. For static samples, a MW resonant cavity design around the samples is advantageous because the cavity provides a high quality factor (Q-factor) at EPR frequencies, which circumvents the need for high-power MW sources. However, cavity designs are not easily incorporated into the MAS apparatus: the size of high-frequency resonators may cause problems when fitting conventional MAS rotors inside. This makes it necessary to use high-power MW sources for high resolution MAS experiments.

In MAS probes, MWs are spread to the coil and sample in a direction either parallel to [36, 37, 96] or perpendicular to the rotor axis [148, 149]. With the parallel-to-axis irradiation, the MWs penetrate the bottom of the rotor, leading to inefficient polarization of the entire sample. The perpendicular-to-axis design more efficiently penetrates the rotor with MWs through the NMR coil turns [150]. While both orientations may lose penetration efficiency at higher magnetic fields due to the shorter MW wavelength, MW penetration into an MAS rotor can be adequately achieved at temperatures as low as 20 K with a careful design of the MAS stator [41, 131].

For the static case, the EPR and NMR requirements are matched by combining an NMR coil with an MW waveguide in a cavity [146]. With such a design MW field strengths (B_{1e}) of about 2.5 MHz could be achieved for a double resonance circuit using only ~20 mW power, and a ^{1}H DNP enhancement of ~400 was obtained at 12 K

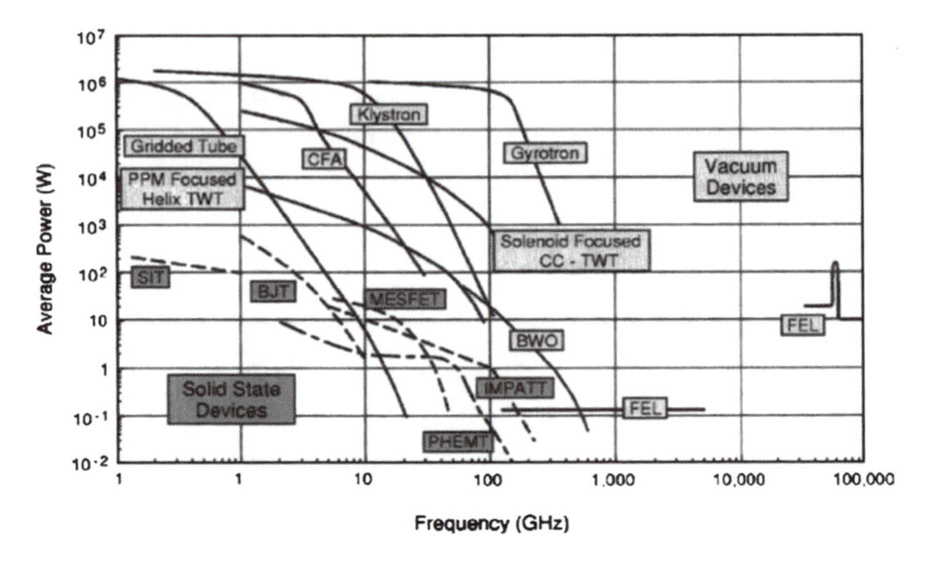

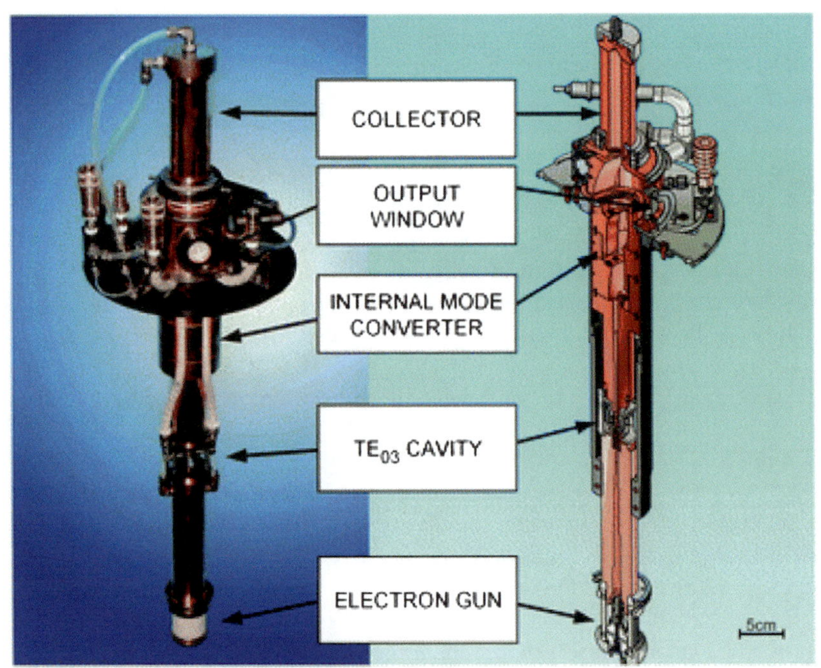

Fig. 8 *Top*: Power output profile of MW sources vs. frequency. Reprinted with permission from Granatstein et al. [132]. Copyright (1999) IEEE. *Bottom*: A gyrotron resonator and cross section schematic. Reproduced from Rosay et al. with permission of the PCCP Owner Societies [41]

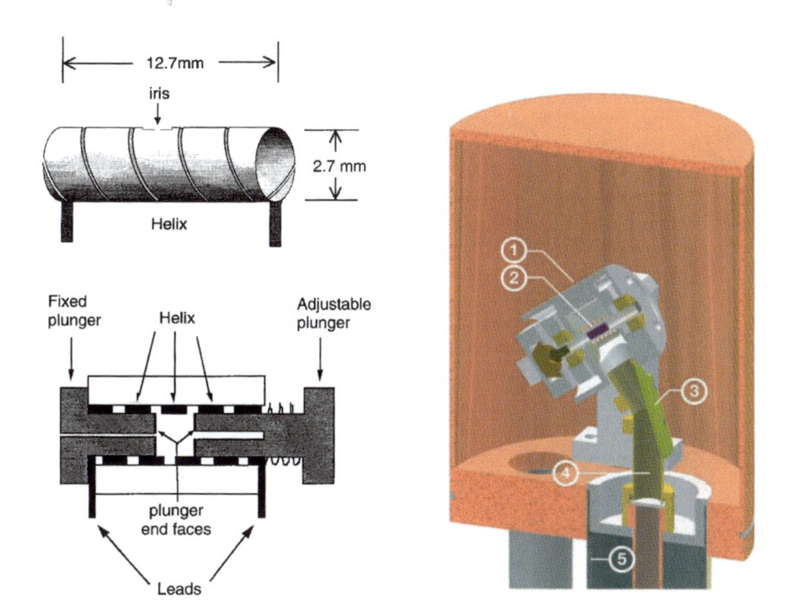

Fig. 9 *Left*: A cavity probe design for static applications. Reprinted from Weis et al. with permission from Elsevier [146]. Copyright (1999). *Right*: An MAS probe for high-resolution applications: (*1*) stator; (*2*) sample/coil; (*3*) mirror miter; (*4*) corrugated waveguide; (*5*) thermal insulation. Reprinted from Barnes et al. with permission from Springer [147]

compared to the spectrum recorded under same conditions without MW. Tycko and coworkers presented another example of static DNP at low temperatures by using low MW powers. ^{1}H DNP enhancements of ~20 and ~40 were obtained with TOTAPOL and DOTAPO-TEMPO radicals, respectively, at ~10 K, 9.4 T, and ~80 mM nitroxide concentration [151].

4 Polarizing Agents

The spin polarization mechanism in a DNP NMR experiment strongly depends on the type and concentration of radical used. The radical's solubility, relaxation properties, and temperature dependence all affect the efficiency of DNP polarization. Stable organic radicals (Fig. 10) are the preferred electron polarization source due to their favorable spectroscopic properties and somewhat tunable solubility [32, 100]. They are typically incorporated into the sample as a homogeneous mixture, usually by dilution. An endogenously doped sample is one where the radical is inseparable from the analyte, such as samples that contain intrinsic free electrons, materials with surface electronic defects, paramagnetic proteins, or chemically bound radicals [152]. We will focus here on exogenous radicals (TEMPO-based and others) used in solid state DNP NMR.

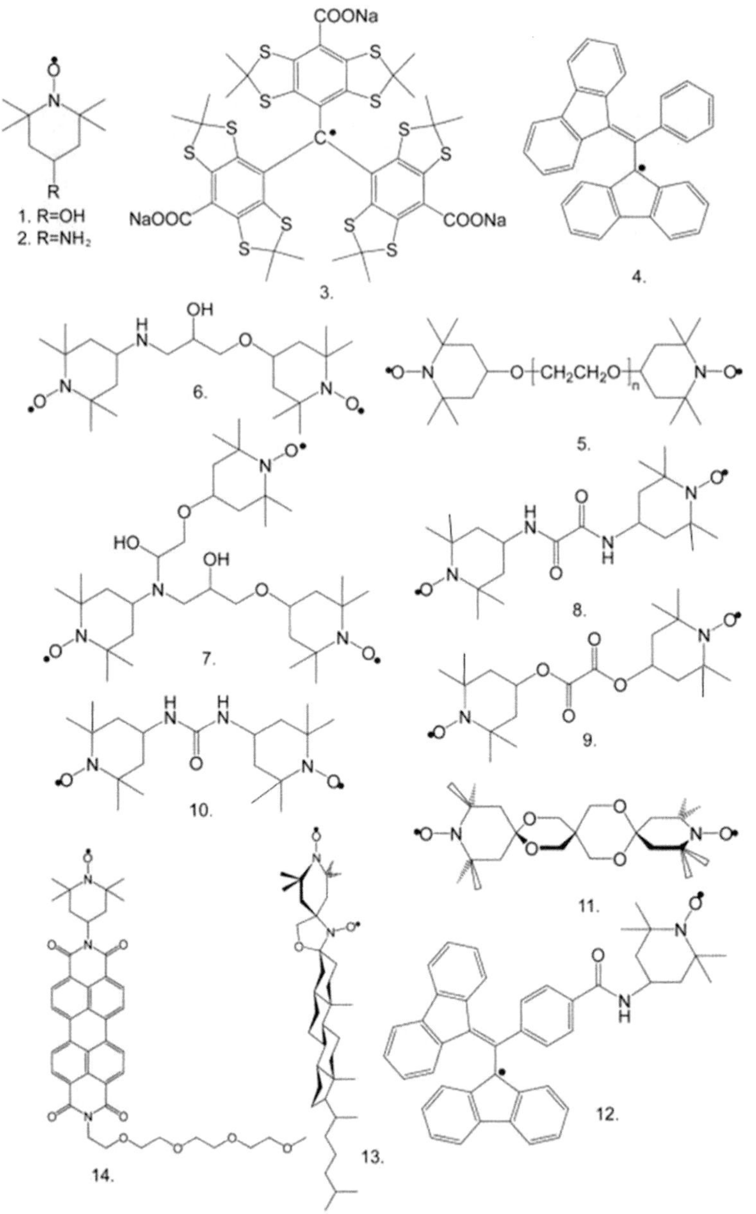

Fig. 10 Radicals that have been used for DNP enhanced NMR applications. (*1*) 4-Hydroxy-TEMPO, (*2*) 4-amino-TEMPO, (*3*) commonly referred to as "Trityl", (*4*) BDPA, (*5*) BTnE, (*6*) TOTAPOL, (*7*) DOTOPA-TEMPO, (*8*) BTOXA, (*9*) BTOX, (*10*) BTurea, (*11*) bTbk, (*12*) BDPA-TEMPO, (*13*) BTcholesterol, (*14*) pyrelene-TEMPO. Reprinted from Hu et al. with permission from Elsevier [100]. Copyright (2011)

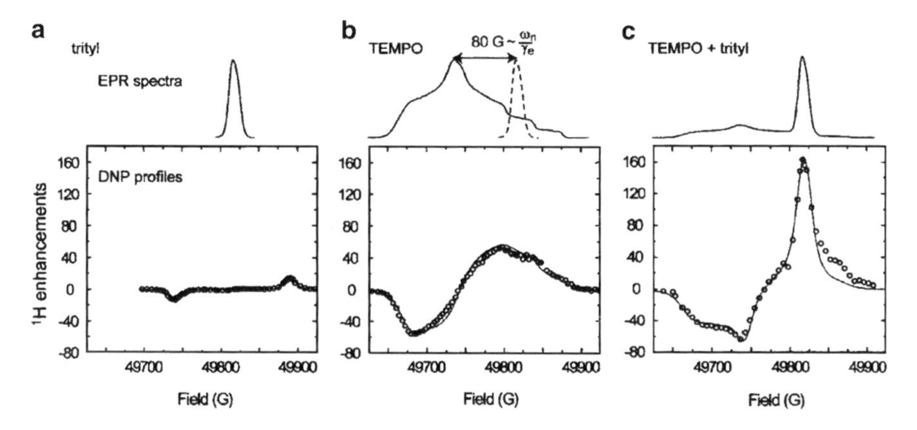

Fig. 11 The EPR spectra and magnetic field dependent ^{1}H DNP enhancement profiles of water/DMSO mixtures with (**a**) trityl, (**b**) TEMPO and (**c**) trityl/TEMPO mixture (50:50). The data were recorded at 5 T. The *circles* are the experimental data and the *solid lines* are the simulations. EPR data were obtained by using 1 mM radicals at 20 K, whereas, the DNP data were recorded by using 40 mM radicals at 90 K. Reprinted with permission from Hu et al. [119]. Copyright (2007), American Institute of Physics

4.1 Monoradical Polarizing Agents

Monoradicals are commonly used for SE and TM polarization transfers. Trityl (OX063: tris{8-carboxyl-2,2,6,6-benzo(1,2-*d*:4,5-*d*)-bis(1,3)dithiole-4-yl}methyl sodium salt) [33, 153], and BDPA (bis-α,γ-diphenylene-β-phenyl-allyl) [126, 127, 154] are popular radicals for the direct electron to nucleus transfer with SE mechanism. Due to their threefold molecular symmetry, the EPR spectra of BDPA and trityl radicals are very narrow (~20 and ~40–50 MHz at ~5 T for BDPA and trityl, respectively). BDPA has been used as a polarization source for hydrophobic polymers [36–38, 40, 155]. Trityl has been used to study biological materials in aqueous solutions. A proton enhancement of ~15 was obtained on a ^{13}C-urea sample using 40 mM trityl radical at ~5 T; see Fig. 11 [119].

A commonly used nitroxide based radical is TEMPO (4-oxo-2,2,6,6-tetramethyl-piperidine-1-oxyl). The EPR spectral breadth is approximately 600 MHz in a 5-T magnetic field (Fig. 11). Inhomogeneous broadening occurs due to the radical's lack of symmetry and amorphous nature, causing different orientations with respect to the external magnetic field to resonate at different frequencies (Fig. 12). Experiments performed with 40 mM TEMPO resulted in a TM-mediated ^{1}H DNP enhancement of $\varepsilon = {\sim}50$ at 5 T and 90 K [119]. At ~9 T, this ^{1}H DNP enhancement decreased to $\varepsilon = {\sim}17$ at ~100 K and 60 mM TEMPO concentration [97, 119].

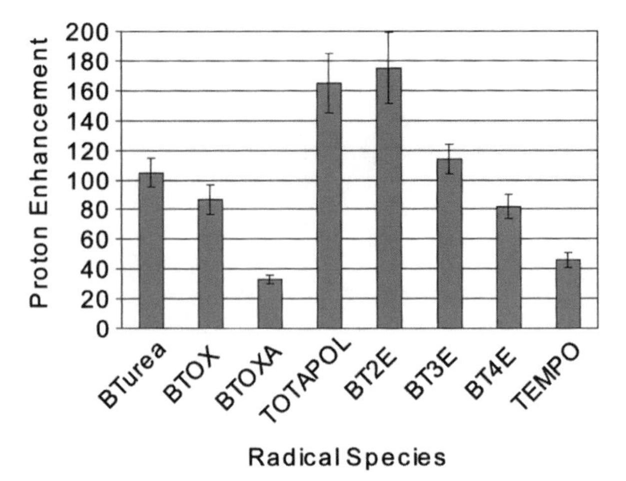

Fig. 12 Comparison of the DNP enhancements using different radicals at 5 T and 90 K by using 10 mM radical concentration on 2 M ^{13}C labeled urea samples. Reprinted with permission from Hu et al. [98, 100]. Copyright (2008), American Institute of Physics

4.2 Biradical Polarizing Agents

At high magnetic fields the efficiency of SE DNP with monoradicals is very low and only small enhancements are observed [42, 100]. TM- or CE-mediated DNP with nitroxide biradicals are more efficient at high magnetic fields compared to SE mediated DNP. Employing biradicals is promising because larger DNP enhancements are obtained as compared to the use of monoradicals (Fig. 12) [31].

BT2E (bis-TEMPO-2-ethylene oxide) consists of two TEMPO molecules separated by two ethylene glycol groups [156]. When the molecule was designed, the electron–electron dipolar coupling was chemically varied between ~22 and 11 MHz by changing the spacer length (2–4 glycol moieties; the shorter the radical, the stronger the dipolar coupling between them). The dipolar coupling between electrons for a biradical solution is significantly larger compared to solutions containing monoradicals with similar concentrations. For example, TEMPO has an intermolecular electron–electron dipolar coupling of ~0.3 MHz for a 10 mM solution [98]. Approximately four times larger ^{1}H DNP enhancements of ~175 were obtained via CE DNP compared to monoradicals under comparable concentrations. BT2E is poorly water-soluble and its use in biological applications is limited, though in materials science this radical should have an impact. Structurally similar biradicals, such as BTOX (bis-TEMPO tethered by oxalate), BTOXA (bis-TEMPO tethered by oxalyl amide), and BTUrea (bis-TEMPO tethered by urea), did not perform as well as BT2E, possibly due to the overall larger electron dipole coupling and more rigid nature of those radicals compared to BT2E, leading to a poorer CE matching condition [98]. The BTOXA radical was the worst among these polarizing agents (performing worse than TEMPO), due to the very similar EPR frequencies of the two TEMPO units in this configuration, which deteriorates the CE matching condition.

The water-soluble radical TOTAPOL (1-(TEMPO-4-oxy)-3-(TEMPO-4-amino) propan-2-ol) contains two tempo molecules separated by a flexible linker [31].

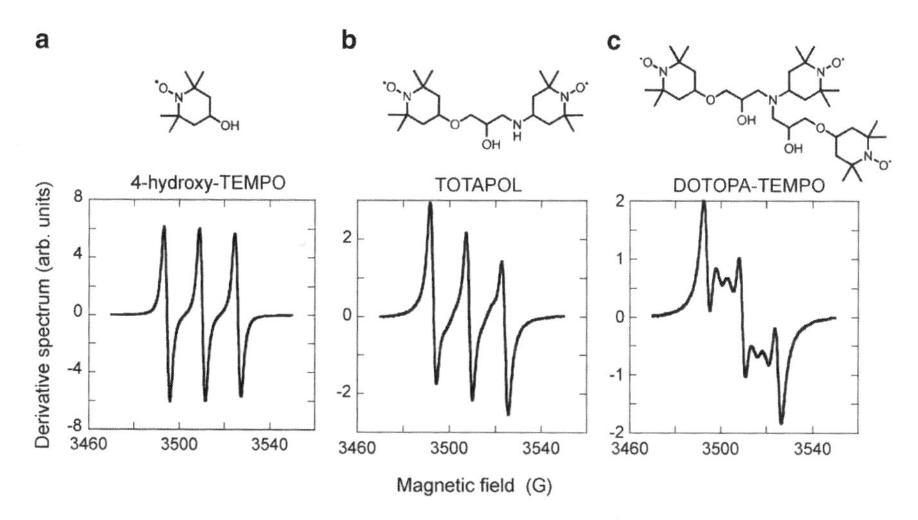

Fig. 13 Comparison of the X-band EPR spectra of monoradical (**a**: TEMPO), biradical (**b**: TOTAPOL) and triradical (**c**: DOTOPA-TEMPO) measured on water/ethanol solutions containing 0.5 mM radicals at ambient temperature. Reprinted from Thurber et al., Copyright (2010), with permission from Elsevier [131]

TOTAPOL is more soluble in water than the BTnE class of biradicals, and still produces similar enhancements. The flexible linker allows different conformations, which helps to achieve the CE matching condition. The electron–electron dipolar coupling of TOTAPOL is approximately 22 MHz. Large ^{1}H DNP enhancements of ~165 are possible with lower radical concentration (10–20 mM). The TOTAPOL EPR spectrum is shown in Fig. 13.

bTbk (bis-TEMPO-bisketal) is a rigid biradical that was designed to improve the relative orientation of the tethered nitroxide radicals compared to BT2E [157]. bTbK leads to larger DNP enhancements (~250 at 5 T) than BT2E. Though bTbk is not very water soluble, it can be used in the presence of biological membranes [158]. Bis-TEMPO-bis-thioketal-tetra-tetrahydropyran (bTbtk-py), a TOTAPOL-based biradical, yields slightly better DNP enhancements with a proton DNP enhancement of ~230 at 5 T (~20% more than TOTAPOL) [159, 160].

4.3 *Outlook and Optimization*

The design principles for multiradical compounds have been investigated via computer simulations. The effect of the relative subunit orientation has been calculated for multiple biradicals in the static case [161]. To obtain the largest DNP enhancements, rigid structures should be produced with the proper relative N–O bond orientation and distance. Additionally, longer T_{1e} favors higher enhancements [162].

Mixed radical solutions containing radicals with narrow and broad EPR line shapes, such as trityl mixed with TEMPO, may be an attractive alternative to biradical optimization [119]. This approach yielded significantly better DNP efficiency under otherwise similar conditions (~160 vs ~50) compared to solutions containing TEMPO as the sole polarizing source. The radical mixture mimics a theoretical CE polarizing agent but with well-resolved and narrow EPR lines. Moreover, the difference between the TEMPO's g_{yy} and trityl's g_{iso} is ~214 MHz, which is very close to the Larmor frequency of proton of 211 MHz at 5 T. An important point to note is that the frequency at which the maximum ^{1}H DNP enhancement observed in this study is different for the radical mixture (trityl + TEMPO) compared to the experiments with only trityl or TEMPO. As a result, to utilize the larger DNP enhancements, the frequency of MW irradiation or the B_0 should be adjusted accordingly.

The DOTOPA-TEMPO triradical was introduced to satisfy better the CE condition due to many correct orientation possibilities [131]. Larger ^{1}H DNP enhancements were observed using this triradical, ~75 at 16 K and 20 mM in comparison to ~26 obtained by using TOTAPOL by their experimental setup. Transition metals, such as Mn^{2+} and Gd^{3+}, can be used as polarizing agents in the solid-state [163], with SE DNP enhancements of ~2 and ~13, respectively, at 5 T and ~84 K. Moreover, a naturally occurring radical in a flavodoxin semiquinone protein produced DNP enhancement of up to ~15 [152].

5 Practical Aspects of DNP in Biological MAS NMR

In this section, important practical considerations of MAS DNP applied to solid biological samples are discussed, covering effects of sample preparation, radical concentration, temperature and resolution, magnetic field, MW power, MAS, relaxation times, and deuteration. The different factors influencing the observed DNP enhancement are summarized in the theory part (Sect. 2); here their relevance for protein DNP NMR is discussed.

In a DNP enhanced MAS NMR experiment, hyperpolarization of protons is most commonly exploited. The polarization is then transferred to heteronuclei (^{13}C or ^{15}N) by a cross-polarization (CP) step for detection (referred to as "indirect DNP"). Examples of 1D and 2D NMR pulse sequences are shown in Fig. 14. Enhancements of ~30–40 are routine on biological systems using TOTAPOL at 9.4 T, and enhancements of ~100 and ~120 are possible in special cases [46, 99].

In addition to indirect-DNP experiments, direct hyperpolarization of low-γ nuclei is also possible (called "direct DNP") and can be utilized when the hyperpolarization of protons followed by polarization transfer to a heteronucleus is not desired [99, 164]. Direct ^{13}C DNP on polystryrene samples doped with BDPA resulted in an ε_{13C} = ~40 at room temperature by using a gyrotron [39].

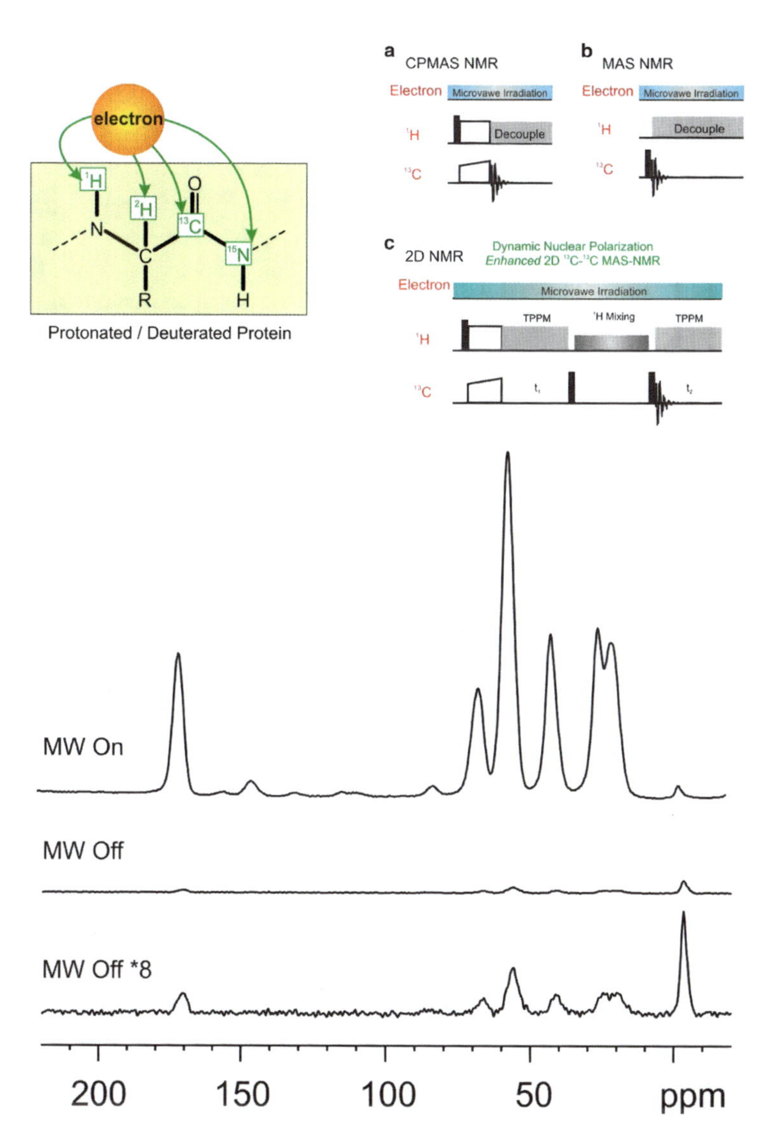

Fig. 14 (*Top*) The electron to nucleus polarization transfer paths in a biological samples, and direct and indirect DNP NMR pulse schemes. (*Bottom*) The 1D conventional and DNP enhanced ^{13}C CPMAS NMR spectra of a proline sample in a sapphire rotor at ~100 K and 9.4 T containing 20 mM TOTAPOL radical. A DNP enhancement of ~60–70 is observed

5.1 Sample Preparation and Radical Concentration

Sample preparation is a crucial step for achieving sufficient DNP enhancement, ensuring resolution, and maintaining sample integrity at cryogenic temperatures. A glass-forming solvent is necessary to prevent the formation of ice crystals, cryoprotects the sample at cryogenic temperatures, and ensures a uniform mixture of radical and analyte, leading to more favorable relaxation times and larger enhancements [40]. A glycerol–water mixture (60 % d_8-glycerol, 30 % D_2O, 10 % H_2O) is usually used for protein samples [166]. However, this mixture may affect proteins by changing their hydration shells [167, 168].

The proton concentration of the solvent (as well as of the solute) is adjusted to achieve maximum DNP polarization. On the one hand protons are needed to distribute the hyperpolarization throughout the entire sample, on the other hand they facilitate unfavorable relaxation [99]. Changing the proton and radical concentrations, as well as temperature, alters the relaxation properties of the sample and solvent, which results in different DNP enhancements.

The radical concentration determines the overall DNP efficiency in several ways. Relaxation times (T_1, spin–spin relaxation time T_2*, and nuclear relaxation time in the rotating frame $T_{1\rho}$) depend upon the radical (TOTAPOL) concentration (c_T). The radical introduces paramagnetic effects (PE) such as paramagnetic relaxation enhancement (PRE) and paramagnetic shifts, which effectively 'remove' a portion of the NMR signals. As a positive effect, the decrease in T_1 relaxation times allows for shorter repetition delays. The homogeneous linewidth, ~$1/T_2$*, affects both sensitivity and resolution. T_2* becomes worse with higher c_T, but if the radical can be spatially separated from the analyte, T_2* may be less affected [46, 49, 50, 156]. The efficiency of heteronuclear polarization transfer is mediated by $T_{1\rho}$ which is also shortened at large radical concentrations. Balancing these effects is necessary to maximize sensitivity [169]. The sensitivity (κ) is defined by Lange et al. as

$$\kappa = \frac{\mathrm{S/N}}{n_\mathrm{A} \cdot \sqrt{1.3 \cdot T_1 \cdot n_\mathrm{s}}} \tag{3}$$

where $1.3 \cdot T_1$ is the optimum recycle delay, n_s is the number of scans, n_A is the number of analyte molecules, and S/N is the signal-to-noise ratio [169].

The S/N of an NMR spectrum recorded without MW on a proline solution decreases with increasing c_T and a constant number of scans. If the experiment time is held constant, by changing the number of scans and setting the relaxation delay to the 1.3 * T_1, the S/N reaches a maximum at a c_T of 26 mM and then decreases at higher c_T for the same proline solution (Fig. 15). At higher radical concentrations, DNP enhancements decrease more slowly than the sensitivity.

To analyze the effect of radical concentration on the sensitivity and enhancement further, the paramagnetic signal 'bleaching effect' was modeled. The radical and the bleached area around which no NMR signal is observed as a result of hyperfine couplings are represented by a rod-shaped volume. The bleaching barrier was estimated to be ~9–10 Å. The calculation showed that only ~50% of the nuclei are observable at a radical concentration of 50 mM in a proline solution [169].

For DNP applications on insoluble samples, the incorporation of a sufficient amount of radical is sometimes difficult. A method called "incipient wetness impregnation" was introduced by Lesage et al. This method is simply physically mixing the

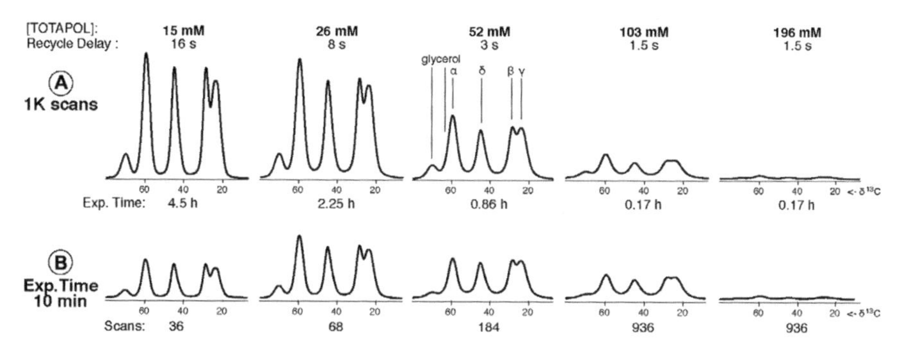

Fig. 15 1D DNP enhanced ^{13}C CPMAS NMR spectra of proline samples as a function of TOTAPOL concentration. The spectra were recorded (**a**) either with the same number of scans or (**b**) within the same experimental time (10 min). Reproduced from Lange et al., Copyright (2012), with permission from Elsevier [169]

powder sample and radical-containing solvents, and as a result wetting powder samples to incorporate a sufficient amount of radical [170]. This method results in large DNP enhancements ($\varepsilon_{1H} = 20$–100), e.g., for silica-based mesoporous materials. The hydrophobic radical BDPA can be suitable for such applications. More recently, the radical bCTbK, which can be effectively solved in organic solvents, was utilized to achieve a ^{1}H DNP enhancement of ~100 for a powder material [162].

Another factor that determines the amount of enhancement concerns the MW absorption by the MAS rotor walls. Sapphire rotors yield larger enhancements (~ 20–25%) compared to zirconia rotors, since sapphire is more transparent to MW radiation [99]. Moreover, they prevent large temperature gradients across the sample compared to zirconia rotors. The rotor diameter and wall thickness affect the DNP enhancements for similar sample preparations [31, 42]. A 2.5-mm sapphire rotor yielded a ~290 ^{1}H indirect DNP enhancement, while DNP enhancement of ~170 was observed using a 4-mm rotor, most probably due to different MW penetration efficiency into the sample.

5.2 *Cryogenic Temperatures and Resolution*

DNP-enhanced NMR of solids requires cryogenic temperatures for slowing the nuclear and electron relaxation processes. From temperatures above 100 K up to room temperature, electron and nuclear T_1 and T_2 decrease dramatically.

However, a sufficient understanding of spectra from biological samples at low temperatures is lacking. The structure and dynamics of proteins can vary significantly depending on the experimental temperature [49]. Dynamics is the most affected feature that changes upon cooling, due to effects such as glass or ice formation by the solvent, the solvent-accessible side chains interacting with

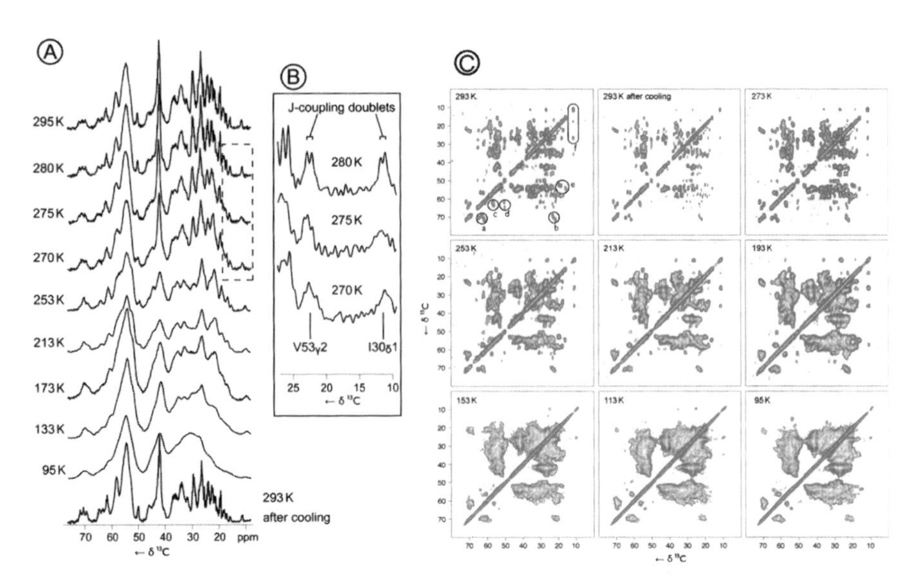

Fig. 16 (**a**) 1D ^{13}C CPMAS NMR spectra of the SH3 at temperatures between 295 and 95 K. The sample does not contain any cryoprotectant or radical. (**b**) The zoomed isoleucine 30 $C_{\delta1}$ and valine 53 $C_{\gamma2}$ region of the 1D spectra to represent the removal of scalar couplings upon cooling the sample. (**c**) The 2D ^{13}C–^{13}C NMR spectra of the SH3 sample in the same temperature range to observe the site specific changes. Reproduced from Linden et al. with permission from Springer [49]

hydration shell solvent molecules, and the solvent-independent thermal motions of the protein itself [49, 171]. Nevertheless, similar high-resolution protein structures were determined at ambient and cryogenic temperature by X-ray crystallography [172]. Moreover, high resolution NMR spectra of favorable systems led to determination of peptide conformations at ~100 K. This situation was demonstrated on a water-free tri-peptide N-*f*-MLF-OH at ~90 K [143, 173], Furthermore, secondary chemical shifts from spectra recorded at ~180 K on the GB1 protein revealed protein conformations [174].

A recent study by Linden et al. analyzed the effect of slow temperature changes on the NMR spectra of a microcrystalline SH3 domain sample between 95 and 295 K [49]. 1D and 2D NMR spectra were recorded to follow resolution and lineshape in those spectra (Fig. 16). No sudden resolution changes were observed at a particular temperature in the range of 295–95 K, not even below and above the protein glass transition temperature of ~200 K [175, 176]. In general, line broadening continued until each resonance reached a coalescence temperature after which multiple signals appeared. Nevertheless, at low temperatures some of the individual peaks showed a linewidth similar to that of the motionally averaged peaks at 293 K. In these spectra, different scenarios may still be observed concerning the line broadening/splitting effects. The threonine signals were split into many different peaks at low temperatures, and cross peaks were always observed. For proline residue P54, a coalescence phenomenon appeared. The signal was observed until 193 K, which then disappeared at lower temperatures, and multiple peaks reappeared at ~113 K. Alanine residue A55 similarly showed a coalescence behavior, which was not observed between 153 and 113 K, but was present again at 95 K. For a complete description of the resonance-specific linewidth changes, the reader is referred to the work of Linden et al. [49].

It has been demonstrated that it is possible to record NMR spectra at cryogenic temperatures without compromising spectral resolution. Barnes et al. reported linewidths of ~0.4–1 ppm in the temperature range of 82–273 K for a crystalline model peptide *N-f*-MLF-OH at 9.4 T [144]. However, distinct chemical shifts observed at 75 K represent two coexisting conformations which are slightly different from the room temperature conformation. In this study, it was shown that for the APG peptide, high resolution is obtained at 75 K, though with some minor heterogeneous contribution appeared as signal splitting. Remarkably, resolved homonuclear carbon–carbon scalar couplings were observed in a 2D ^{15}N-^{13}C correlation spectrum. For this sample, similar structures were determined by solid-state NMR and X-ray crystallography, the latter at cryogenic temperatures.

Another promising example of the application of DNP enhanced NMR to proteins has been presented by Linden et al. A high-resolution NMR spectrum of ^{13}C labeled neurotoxin II (NT-II) bound to the nicotinic acetylcholine-receptor (nAChR) in native membranes was observed at ~100 K [49]. Same TOTAPOL radical was found to be located at the surfaces of the membrane, compromising resolution due to the PRE effect (see Fig. 17). However, the total concentration of the active biradical decreased while storing the sample at -20 °C for 1 month. In this time, these TOTAPOL radicals localized close to the membrane were inactivated, while active TOTAPOL remained immobilized in the frozen solvent glass. This led to well-resolved NMR spectra with several resolvable residues, though with a lower DNP enhancement ($\varepsilon = 12$). The experiments on the acetylcholine-receptor system in its native membrane demonstrate that DNP-enhanced NMR of extensively labeled preparations is possible with good resolution. Several spin systems such as leucines and prolines could be assigned. In this case, it was possible to obtain a well resolved 2D spectrum in ~14 h instead of the 10 days of measurement time required without DNP. However, the underlying mechanism of achieving narrow lines in protein NMR needs further investigations to open the way to routine high resolution DNP enhanced protein NMR spectroscopy.

Application of DNP enhanced NMR to materials is less affected by cryogenic broadening effects, in part due to the relatively broad resonances observed for the material samples already at room temperature as a result of many different conformations simultaneously present in powder samples. At ~100 K, spectra with a resolution very similar to that at room temperature can be obtained in studies of, silica based materials [170, 177, 178].

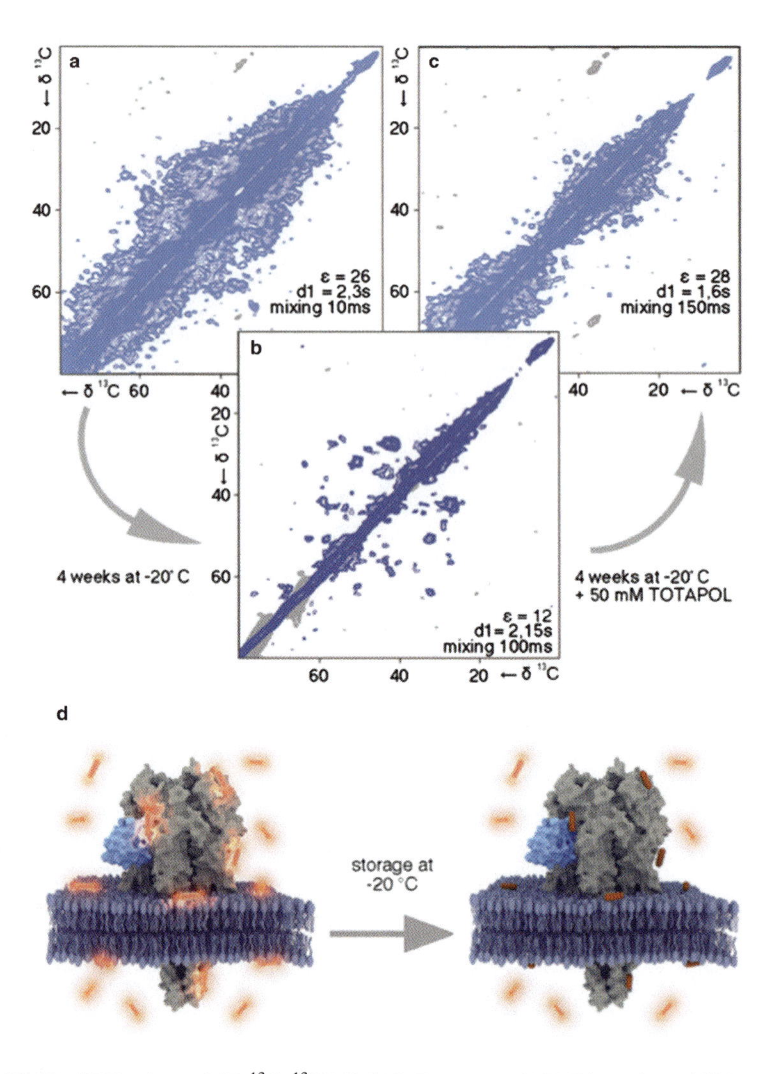

Fig. 17 The DNP enhanced 2D ^{13}C–^{13}C MAS NMR spectra of NT-II bound to nAChR recorded at 9.4 T, ~100 K and 8 kHz MAS by using TOTAPOL radical. (**a**) Spectrum with fresh sample preparation with 50 mM radical. (**b**) Spectrum recorded after keeping the sample at −20°C for 4 weeks. (**c**) Spectrum recorded with fresh radical addition. (**d**) The representation of the hypothesis explaining the observed DNP enhancement reduction and increase in the spectral resolution. The active radical layer present in the newly prepared sample (*left*), and the inactivated radical layer present with fresh radical addition (*right*). Reprinted with permission from Linden et al., Copyright (2011) American Chemical Society [50]

5.3 DNP Enhancement Profiles

The DNP enhancement profiles must be known for optimizing DNP efficiency for a specific polarization mechanism and sample. There are well-defined resonance conditions at which the enhancement reaches maxima and minima. These

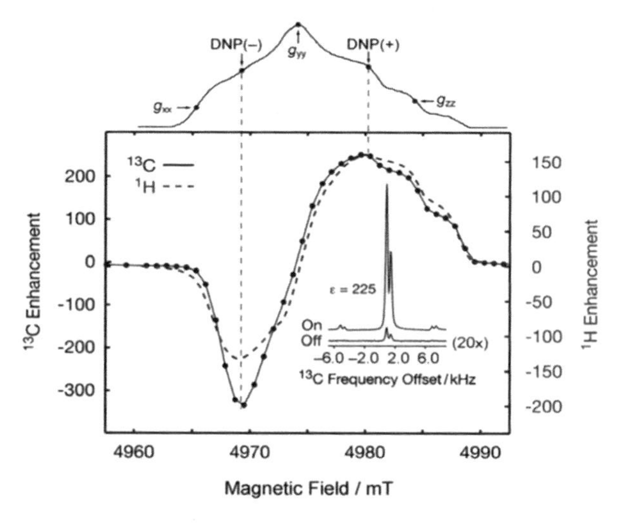

Fig. 18 The representation of the magnetic field dependent DNP enhancements for ^{1}H and ^{13}C DNP utilizing TOTAPOL. The EPR spectrum is recorded at 90 K and 5 T. The *inset* shows the ^{13}C direct excitation DNP enhanced NMR spectra with and without MW, representing an enhancement of ~225. Reprinted from Maly et al. with permission from Wiley [165]

resonance conditions for MAS DNP NMR experiments are matched when the MW frequencies are at ~$\omega_e \pm \omega_n$ for SE, TM, and CE DNP. The width and separation of the resonance conditions at which polarization transfer occurs provide valuable information about the nature of the active transfer mechanism. DNP enhancement profiles may be recorded in either a frequency- or a magnetic field-dependent way. Practically, enhancement profiles are usually obtained with a fixed MW frequency by sweeping the magnetic field at which the NMR experiment is performed, utilizing an additional coil in the NMR magnet.

A comparison of field-dependent ^{1}H and ^{13}C DNP enhancements utilizing TOTAPOL as the polarizing agent was presented recently by Maly et al. [165] (Fig. 18). The magnetic field dependency represents a classical pattern without well-resolved features, typical of TEMPO-based polarizing agents. Maximum positive DNP ("+" DNP) and maximum negative DNP ("−" DNP) enhancements are observed at two different magnetic field strengths equivalent to two different MW frequencies at $\omega_e \pm \omega_n$. The asymmetry of the enhancement profiles is due to the asymmetric EPR spectrum which leads to different enhancements at those two peak positions.

The DNP condition is satisfied when irradiating on different sides of the TOTAPOL EPR spectrum ("+" and "−" DNP conditions) for different types of nuclei due to the asymmetric lineshape [165, 179]. ^{1}H DNP is achieved best at the "+" DNP condition, whereas the ^{13}C DNP is achieved best at the "−" DNP condition. Maly et al. reported ^{1}H enhancements of a $^{13}C_3$-labeled glycerol sample of 130 and −108 for the "+" and "−" DNP conditions, respectively. ^{13}C enhancements of the same sample were determined to be 225 and −281 at the "+" and "−" DNP conditions, respectively (Fig. 18). Similarly, field-dependent ^{1}H and ^{13}C DNP enhancement profiles were recorded on protonated and deuterated SH3 samples containing TOTAPOL [179]. The results with SH3 agree with the observations by Maly et al., though significantly larger DNP enhancements were observed for the deuterated protein. For example, the largest reported ^{13}C DNP enhancement of approximately ~400 was observed at 5 T at the "−" DNP condition.

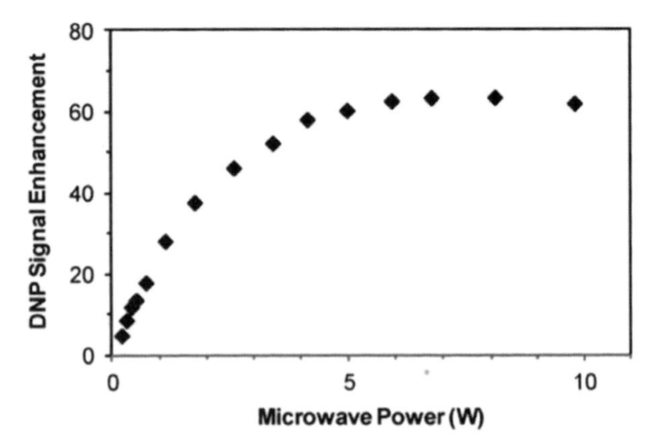

Fig. 19 The MW power dependent ^{1}H DNP enhancements recorded for a uniformly labeled proline sample in a water/glycerol mixture. The enhancements were obtained with 8 kHz MAS, at ~100 K and 9.4 T. Reprinted from Rosay et al. with permission of the PCCP Owner Societies [41]

The Larmor frequency of the nucleus to be polarized may fit to either side of the EPR spectrum with different efficiency [165]. For example, the ^{1}H nuclear Larmor frequency is large and fits better to the broader side of the EPR spectrum. As a result, larger ^{1}H DNP enhancements are observed at the "+" DNP condition (which is at the broader side of the EPR spectrum). However, the ^{13}C Larmor frequency is much smaller compared to the ^{1}H frequency, and fits efficiently into the narrower side of the EPR spectrum. As a result, larger ^{13}C DNP enhancements are observed at the "−" DNP condition (which is at the narrower side of the EPR spectrum). It is thus important to determine the optimum magnetic field for DNP NMR depending on the nucleus type to be polarized [164, 165].

5.4 *Effect of MW Power and Magnetic Field*

The extent of electron excitation, as well as the DNP efficiency, changes with applied MW power (nutation frequency, B_{1e}), see formulas in (1) and (2). For both the CE and SE, a quadratic increase of the DNP enhancement is expected with increasing B_{1e} [42, 107]. The optimal MW power to achieve maximum DNP enhancement has previously been investigated [41, 42, 147] (Fig. 19). From very low levels up to ~6–7 W, the DNP enhancement increased. At higher MW strengths (~8–9 W), a saturation phenomenon was observed and the DNP enhancements were stable. At ~10 W of MW power, the enhancement became slightly lower than the maximum value [41].

The observation of reduced DNP enhancements at excess MW however may be a result of destructive saturation over a large EPR frequency range, involving the simultaneous irradiation of "+" and "−" DNP conditions simultaneously (explained in Sect. 5.4). As a result, cancellation of positive and negative NMR signals could lead to the decreased DNP enhancements. The high MW powers could also have led to sample heating, resulting in reduced DNP enhancements.

The magnetic field strength may change nuclear and electron relaxation behaviors. This may modify the DNP mechanism, since relaxation times play an important role in the electron to nuclear polarization transfer. Generally, with increasing magnetic field, the SE DNP enhancement scales quadratically, whereas the CE DNP enhancement scales linearly. At higher magnetic fields, it becomes more difficult to saturate electron spin transitions, due to a broader EPR spectrum of heterogeneously broadened radicals such as TEMPO. This is particularly a problem when the MW strength cannot be increased linearly with the increased magnetic field.

Investigations concerning the effect of increasing magnetic field strengths were performed on TEMPO-containing model systems. A ^{1}H DNP enhancement of ~50 was achieved at 40 mM radical concentration at 5 T. Increasing the magnetic field led to a reduction in enhancement to ~17 at ~9 T under very similar conditions [97, 119]. Similar results were obtained when using TOTAPOL [179]. Proton DNP enhancements of 31 and 120 were observed for the protonated and deuterated SH3 samples, respectively, at 9.4 T and ~100 K [99]. The DNP enhancements of the same samples were also measured at 5 T and ~80 K under otherwise similar conditions, and DNP enhancements of 115 and 200 were observed for the protonated and deuterated proteins, respectively. A twofold increase in the magnetic field from 5 up to 9.4 T results in a nearly 2–4-fold decrease in the observed DNP enhancement for proton DNP experiments.

This reduction in enhancement at higher magnetic fields may put a boundary for high-field DNP NMR experiments, though the effect might be mitigated by improving the efficiency of the polarization transfer mechanisms through stronger MW irradiation, using optimized radicals with reduced anisotropy which results in smaller EPR linewidth.

5.5 *Effect of Magic Angle Spinning*

To achieve high resolution in solid state DNP NMR experiments, sufficiently fast MAS is required. The effects of MAS on the different DNP mechanisms are still not fully understood. Furthermore, effects of MW penetration efficiency and the effective sample temperature at different sample spinning frequencies might also result in the observation of MAS-dependent DNP enhancements. Moreover, modification of the static EPR lineshape may occur under MAS. In a first study by Rosay et al. [41], different DNP enhancements were observed at different MAS frequencies in the range 0–14 kHz using TOTAPOL as the polarizing agent and a field of 9.4 T (Fig. 20). A maximum DNP enhancement of ~80 was found at ~3 kHz MAS, which decreased to ~25 under static conditions, and to ~40 at 14 kHz spinning. To investigate the effect of MAS and MW on the sample temperature measurements, a KBr temperature calibration setup was done [180]. For the sapphire rotors filled with KBr a ~4 K temperature increase was observed in the static case due to heating by MW (6 W). With MAS and MW a total increase of 1–2 K in temperature was

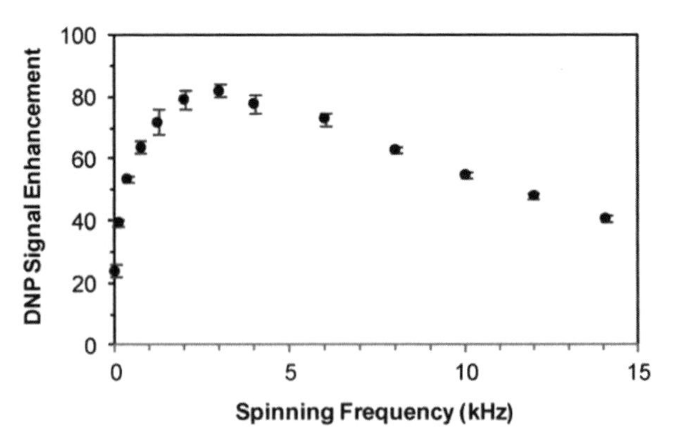

Fig. 20 The MAS frequency dependent ^{1}H DNP enhancements recorded for a uniformly labeled proline sample in a water/glycerol mixture. The enhancements were obtained at ~100 K and at 9.4 T. Reproduced from Rosay et al. with permission of the PCCP Owner Societies [41]

observed at 1.5 kHz MAS (which is less than in the static case), and below 1 K at 8 kHz MAS. Zirconia rotors led to ~12 K sample heating when the sample was static, and ~4 K at 12.5 kHz MAS and with MW on. In general, sample spinning ensures a more uniform MW penetration and lower sample temperature. While temperature variations over the sample may partly contribute to the observed maximum ε at ~3 kHz of MAS, it cannot fully account for the phenomenon. The MW heating effect is expected to be less severe at higher MAS, due to increased cold nitrogen gas flow and less restricted exposure of rotor surface. As a result, higher DNP enhancements should be observed at higher MAS. However, this was not the case; instead a decrease is obtained at MAS >3 kHz in the study by Rosay et al. (Fig. 20), hinting at a principle spectroscopic reason.

Lafon et al. [178] demonstrated the effect of MAS on direct ^{29}Si DNP. The enhancements at 5 and 10 kHz were found to be similar. Around 10% less DNP enhancement was observed at 2.5 kHz MAS compared to the enhancements observed at 5 and 10 kHz. This fits well with the observation of Rosay et al., assuming that the heating effects are more pronounced at low MAS frequencies [41].

Mentink-Vigier et al. observed similar MAS-dependent DNP enhancements on model compounds and biological samples, and provided a theoretical background of the phenomenon [181]. A similar theoretical treatment of the influence of MAS was given by the work of Thurber et al. [182]. In these studies, the differences between DNP mechanisms active in the static case and under MAS were explained. In the static case, a limited amount of electrons are excited by the MW field, not only because of insufficient MW penetration, but also due to insufficient excitation connected to their orientations. However, under MAS, the orientations and therefore frequencies of individual electron spins change, enabling simultaneous excitation of many more electrons. Furthermore, at the fixed MW irradiation frequency changing electron and nuclear frequencies lead to periodically time-dependent polarization-transfer matching conditions. As a result, polarization transfer occurs by fast MW passages through CE resonance conditions (on-resonance irradiation situation on a very short time scale). This phenomenon is termed “fast-passage DNP” on rotating solids. In the work by Mentink-Vigier et al. and Thurber et al., the effects of MAS on the observed DNP enhancement were simulated applying powder averaging. It was demonstrated that the CE DNP enhancements changed by MAS, depending on the relative magnitudes of ω_R and T_{1e} [181, 182].

5.6 Polarization Buildup Times

During the DNP process, electron spin polarization is transferred to the surrounding core nuclei and subsequently to the bulk nuclei (described in Sect. 2.3). The DNP-enhanced signal reaches its maximum intensity after a certain period of MW irradiation. Every polarized nucleus shows a characteristic exponential polarization signal buildup, with a time constant τ_B. Understanding and determining this time constant can be informative about the DNP process. τ_B depends on some of the intrinsic properties of the nucleus and its surroundings including nuclear and electron relaxation, electron concentration, MW power, gyromagnetic ratio, overall proton concentration in the system, and spin-diffusion rates. The electron concentration and relaxation times determine the polarization transfer from electrons to the core nuclei. Spin diffusion then mediates the polarization distribution from the core to the bulk nuclei, which is also important for the polarization transfer from the bulk nuclei to the site of interest.

Experimental determination of τ_B is achieved using a saturation recovery NMR experiment under continuous MW irradiation (Fig. 21). All initial polarization is removed during a saturation period, followed by signal detection either directly or indirectly after a CP transfer step. τ_B is determined from the signal intensity change as a function of polarization time.

For a protonated SH3 sample containing 20 mM TOTAPOL, the measured proton polarization buildup $\tau_B{}^{1H}$ was $\approx$1 s [179]. Carbon signals were an order of magnitude slower, with $\tau_B{}^{13C} \approx 10$ s. Nitrogen signals show the longest DNP buildup times, with $\tau_B{}^{15N} \approx 260$ s. The different relaxation times for the proton, carbon and nitrogen nuclei contribute to these observed differences in τ_B. The observation of very long τ_B for ^{15}N may be a result of inefficient spin diffusion due to the lack of a dense nitrogen network in a uniformly labeled protein sample. Further systematic investigations are required on this topic to understand the polarization buildup process in more detail.

5.7 Effect of Deuteration and Exchangeable Proton Concentration

In DNP-enhanced NMR spectroscopy, deuterated glass-forming solvents have been used extensively, including d_6-DMSO or a 60:30:10 % v:v:v d_8-glycerol/D_2O/H_2O mixture [41, 98, 156]. These solvent mixtures were optimized to include sufficient

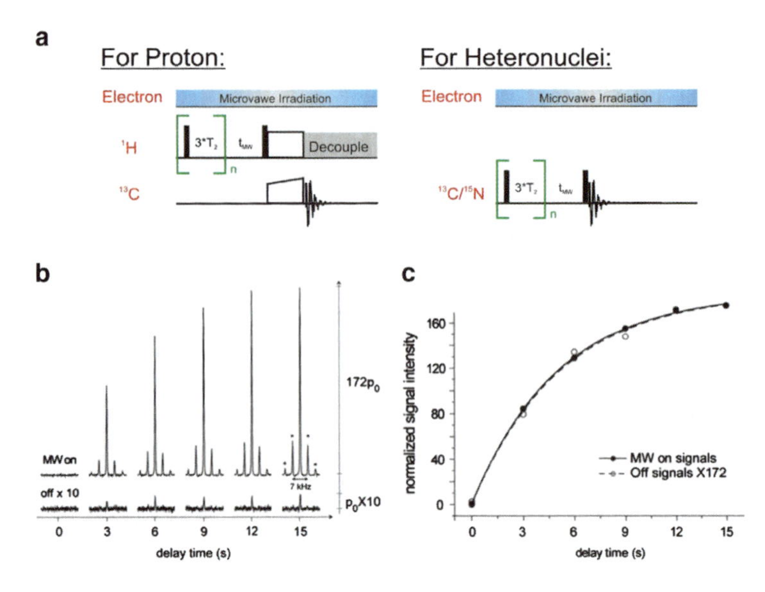

Fig. 21 (**a**) Pulse programs used to obtain the polarization buildup behavior for a CP and direct excitation types of DNP enhanced NMR experiments. (**b**) The polarization time dependent ^{1}H DNP enhanced NMR signal recorded under continuous MW irradiation. Spectra were recorded at 5 T and ~90 K using the BT2E radical on a ^{13}C labeled urea sample. (**c**) The polarization build-up time constant can be calculated by fitting the signal growth. Reprinted with permission from Song et al. [98]. Copyright (2008), American Institute of Physics

protons for efficient ^{1}H-^{1}H spin diffusion to distribute the enhanced polarization uniformly throughout the sample, while the reduced proton content ensures longer T_{1n} values that are favorable for DNP. The effects of solvent proton content on the DNP enhancement were recently shown with samples containing proline and 10 mM TOTAPOL radical dissolved in two different solvents: d_8-glycerol-D_2O-H_2O with 60:30:10 volume ratios and h_8-glycerol-H_2O with 60:40 volume ratios. The sample with the ~85% deuterated solvent had a proton DNP enhancement of ~63, whereas the sample with the protonated solvent has only ~42. These results show that the proton content of the solvent needs to be carefully optimized for maximum DNP enhancement [41].

It was also shown that protein deuteration can significantly increase DNP enhancements by factors of ~4 and ~19 for proton and carbon DNP, respectively, compared to protonated proteins under similar conditions (Fig. 22) [99]. SH3 samples fully deuterated at the non-exchangeable sites were used to study the effect of protein deuteration. The deuterated proteins were crystallized from appropriate H_2O/D_2O buffers to adjust the ^{1}H/^{2}H ratio at the exchangeable sites. Finally, the protein was dispersed in a glass forming d_8-glycerol/D_2O/H_2O matrix containing the radical.

A comparison of one-dimensional ^{13}C CPMAS spectra of the protonated and deuterated SH3 samples, as well as ^{13}C direct excitation spectra using different recycle delays, are shown in Fig. 22. The ^{1}H DNP enhancement observed in the

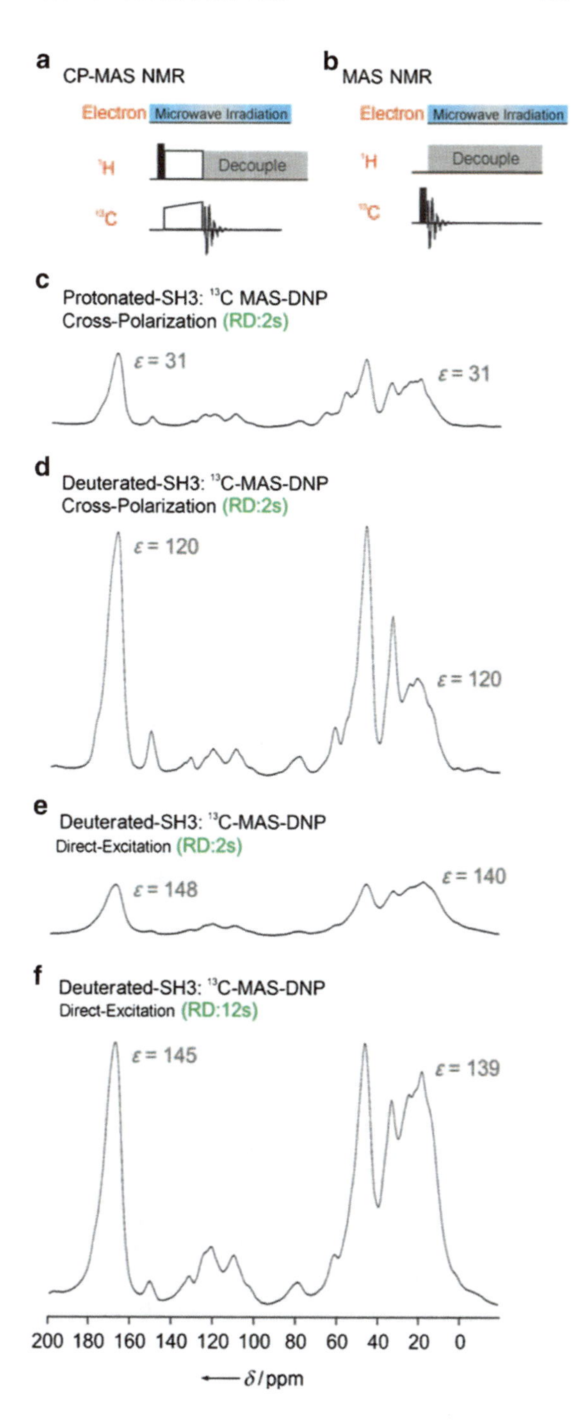

Fig. 22 The representation of the effect of protein deuteration on the DNP. (**a**, **b**) The pulse sequences used to record the ^{13}C CP and direct excitation (MAS) NMR experiments. ^{13}C CPMAS spectrum of the (**c**) protonated SH3, (**d**) deuterated SH3. ^{13}C direct excitation spectra of the deuterated SH3 with (**e**) 2 s and (**f**) 12 s of relaxation delays. The protonated and deuterated SH3 samples were dissolved in glycerol/water mixtures with ~15 and ~50 % of overall exchangeable proton content. The spectra were recorded at 9.4 T, ~100 K and 8 kHz MAS. Reproduced from Akbey et al. with permission from Wiley [99]

CPMAS spectrum from a protonated and fully $^{13}C/^{15}N$ labeled sample is ~31 at 9.4 T, ~100 K, and containing 20 mM TOTAPOL, similar to the enhancements previously reported. In the spectra of the deuterated protein, the proton enhancement increases up to $\varepsilon \approx 120$. The efficiency of the direct-excitation ^{13}C DNP efficiency increases more dramatically by protein deuteration, and an enhancement of $\varepsilon \approx 148$ is observed compared to the carbon enhancement of ~8 (not shown in the figure) obtained using fully protonated SH3. Similarly, from a direct ^{15}N DNP experiment, an enhancement of $\varepsilon \approx 207$ is obtained when deuterated SH3 is used. The $^{13}C/^{15}N$ enhancements are larger compared to the ^{1}H enhancement, although the much shorter T_1 relaxation times of protons still make them the nuclei to polarize when highest overall sensitivity is desired. Moreover, direct ^{2}H polarization resulted in large DNP enhancements of ~700 employing trityl, which has a narrow EPR spectral width in the order of ^{2}H Larmor frequency at 5 T [164].

The effect of proton concentration on the SE DNP mechanism was studied by Corzilius et al. [110]. It was found that increasing the proton concentration in solvents containing trityl or Gd^{3+} had opposite effects and can influence two separate processes: first, the initial electron to nuclear polarization transfer and second, the subsequent polarization distribution to the bulk nuclei. When using trityl as the polarizing agent, increased proton concentration decreased the DNP enhancements, due to a slow rate-determining initial polarization transfer step which is worsened by increased proton content. When using Gd^{3+} as the polarizing agent which undergoes a rapid initial polarization transfer step, increased proton concentration leads to larger DNP enhancements up to a plateau value at high proton concentrations.

5.8 High-Temperature DNP

The requirement of using cryogenic temperatures for DNP enhanced solid-state NMR spectroscopy imposes so far a boundary for applications in structural biology, due to resolution losses upon sample cooling (see Sect. 5.2 for details) [49, 171]. One possibility to overcome the detrimental effects of cooling is to record DNP-enhanced NMR spectra at increased temperatures, >100 K. This approach can lead to better resolved spectra, though at the expense of maximum DNP enhancements [167]. There can be a compromise temperature which supplies both sufficient DNP enhancement and resolution.

Temperature-dependent DNP enhancement studies on deuterated and protonated SH3 samples at temperatures between 98 and 200 K (Fig. 23) showed the feasibility of this approach. Starting from ~100 K, an increase of ~20 K results in a ~30–40% decrease in the observed DNP enhancement. Nevertheless, the deuterated protein with optimized exchangeable proton level still shows significant DNP enhancements of ~11 and ~18 at ~180 K in ^{13}C cross-polarization and direct-excitation DNP NMR experiments, respectively. These enhancements correspond

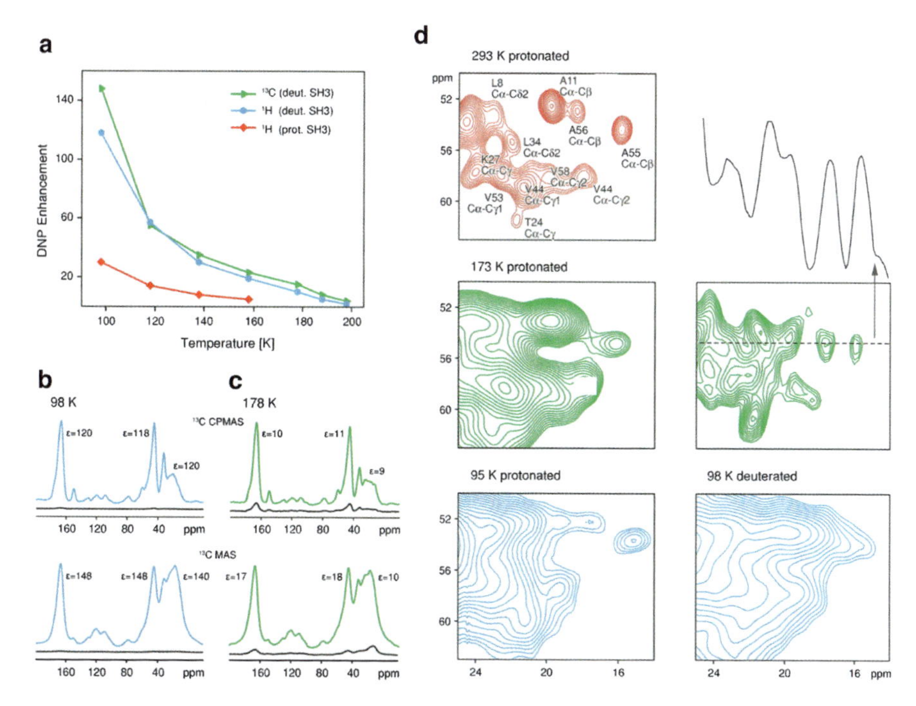

Fig. 23 (**a**) The effect of temperature on the ^{1}H and ^{13}C DNP enhancements. Results for the protonated and perdeuterated SH3 are compared. (**b**, **c**) 1D spectra representing the temperature effect to the DNP enhancements. The spectra recorded with MW are scaled to the same height. (**d**) The 2D spectra of protonated and deuterated SH3 at different temperatures. Reproduced from Akbey et al. with permission from Springer [167]

to a ~120- to 320-fold decrease in experiment time. This method was termed "High-Temperature DNP" [167] and utilizes the DNP efficiency increase by protein deuteration. High sensitivity 2D ^{13}C-^{13}C spectra were successfully recorded at 178 K in less than 10 min. The temperature area of good compromise is close to the previously mentioned protein glass transition temperature, ~200–240 K [175] and the freezing point of the protein hydration shell, ~170 K [176].

Compared to experiments at ~98 K, twofold narrower linewidths were obtained at ~178 K. The DNP enhanced 2D and 3D NMR spectra can now be recorded in reasonable experimental times at 178 K via DNP with sufficient resolution. We believe that this will facilitate unambiguous assignments of biological systems under DNP conditions. Further experimental parameters can be optimized that may lead to even larger DNP enhancements at these high temperatures, including utilizing sapphire rotors and optimized radicals. This may lead to further elevations in the operational DNP temperature with improved spectral resolution.

6 Summary

With the commercialization of DNP enhanced MAS NMR instruments, many laboratories focusing on functional materials and biomolecules have started to utilize the technique. The increasing number of applications in the past five years has driven further developments of the methodology with promising examples. In particular, resolved DNP spectra of protein samples were published, enhancements were increased by making use of sample deuteration, and new radicals have been developed. Further research on improving enhancements and spectral quality is ongoing, including DNP at higher fields and tunable MW source – NMR combinations, for example. This chapter summarizes the current state-of-the-art, concentrating on the major factors determining overall DNP efficiency with an emphasis on the benefits and the drawbacks of specific methods. Particularly, the effects of the cryogenic temperatures usually employed for DNP are explored, especially with regards biological samples optimized for room temperature experiments, as well as the loss of signals due to the bleaching effect of the biradical dopants and its effect on the sensitivity of the experiment [49, 50, 169]. The potential of diluting protons by extensive deuteration to increase enhancements both for biological [99] and inorganic materials [183] is discussed with considerable focus.

Together with the expected further rapid development of DNP, new experimental approaches are already emerging, such as the application to whole cells and extracts [50, 184–186], and in the future atomic-level in vivo structure determination of biomolecules under physiologically relevant conditions. DNP can also be expected to play a continuing role in the study of in-cell metabolomics and drug discovery, where the relevant concentrations of the molecules of interest are within the picomolar range [187, 188]. Further development of polarizing agents with longer relaxation lifetimes should allow for more efficient signal enhancements and DNP experiments at higher temperatures [162, 167].

Besides the use of DNP in structural biology and biomedical studies, DNP has proven itself an important technique for materials science investigations. Particularly the detection of low-γ nuclei may benefit. These type of nuclei are inherently difficult to detect by conventional NMR techniques, and might give important information, e.g., for catalysis and their surface chemistry, as well as reaction intermediates [189–195]. Continuing development of polarizing agents designed to enhance directly signals from low gamma nuclei should also be expected.

Acknowledgements We gratefully acknowledge very helpful discussions and proof reading of Shimon Vega, Akiva Feintuch, Yonatan Hevov, and Thorsten Maly. UA and HO acknowledge funding from the European Union Seventh Framework programs (FP7/2007–2013 under the grant agreements 261863 (Bio-NMR) and Deutsche Forschungsgemeinschaft (grant 05106/12-1 of the DIP program).

References

1. Griffin RG (1998) Dipolar recoupling in MAS spectra of biological solids. Nat Struct Biol 5:508–512
2. McDermott AE (2004) Structural and dynamic studies of proteins by solid-state NMR spectroscopy: rapid movement forward. Curr Opin Struct Biol 14:554–561
3. Jaroniec CP, MacPhee CE, Bajaj VS, McMahon MT, Dobson CM, Griffin RG (2004) High-resolution molecular structure of a peptide in an amyloid fibril determined by magic angle spinning NMR spectroscopy. Proc Natl Acad Sci USA 101:711–716
4. Wasmer C, Lange A, Van MH, Siemer AB, Riek R, Meier BH (2008) Amyloid fibrils of the HET-s(218-289) prion form a beta solenoid with a triangular hydrophobic core. Science 319:1523–1526
5. Tycko R (2006) Characterization of amyloid structures at the molecular level by solid state nuclear magnetic resonance spectroscopy. Methods Enzymol 413:103–122
6. Castellani F, van Rossum B, Diehl A, Schubert M, Rehbein K, Oschkinat H (2002) Structure of a protein determined by solid-state magic-angle-spinning NMR spectroscopy. Nature 420:98–102
7. Nieuwkoop AJ, Wylie BJ, Franks WT, Shah GJ, Rienstra CM (2009) Atomic resolution protein structure determination by three-dimensional transferred echo double resonance solid-state nuclear magnetic resonance spectroscopy. J Chem Phys 131:095101
8. Renault M, Bos MP, Tommassen J, Baldus M (2011) Solid-state NMR on a large multidomain integral membrane protein: the outer membrane protein assembly factor BamA. J Am Chem Soc 133:4175–4177
9. Lange V, Becker-Baldus J, Kunert B, van Rossum BJ, Casagrande F, Engel A, Roske Y, Scheffel FM, Schneider E, Oschkinat H (2010) A MAS NMR study of the bacterial ABC transporter ArtMP. Chembiochem 11:547–555
10. Franks WT, Linden AH, Kunert B, van Rossum BJ, Oschkinat H (2012) Solid-state magic-angle spinning NMR of membrane proteins and protein–ligand interactions. Eur J Cell Biol 91:340–348
11. Shahid SA, Bardiaux B, Franks WT, Krabben L, Habeck M, van Rossum BJ, Linke D (2012) Membrane-protein structure determination by solid-state NMR spectroscopy of microcrystals. Nat Methods 9:1212–1217
12. Park SH, Das BB, Casagrande F, Tian Y, Nothnagel HJ, Chu MN, Kiefer H, Maier K, De Angelis AA, Marassi FM et al (2012) Structure of the chemokine receptor CXCR1 in phospholipid bilayers. Nature 491:779–783
13. Griffin RG (2010) Spectroscopy clear signals from surfaces. Nature 468:381–382
14. Bargon J, Fischer H, Johnsen U (1967) Kernresonanz-Emissionslinien Wahrend Rascher Radikalreaktionen .I. Aufnahmeverfahren und Beispiele 27. Z Naturforsc Part A Astrophys Phys Physikalisc Chem A 22:1551–1555
15. Ward HR, Lawler RG (1967) Nuclear magnetic resonance emission and enhanced absorption in rapid organometallic reactions. J Am Chem Soc 89:5518–5519
16. Closs GL, Closs LE (1969) Induced dynamic nuclear spin polarization in reactions of photochemically and thermally generated triplet diphenylmethylene. J Am Chem Soc 91:4549–4550
17. Daviso E, Diller A, Alia A, Matysik J, Jeschke G (2008) Photo-CIDNP MAS NMR beyond the T(1) limit by fast cycles of polarization extinction and polarization generation. J Magn Reson 190:43–51
18. Daviso E, Janssen GJ, Alia A, Jeschke G, Matysik J, Tessari M (2011) A 10 000-fold nuclear hyperpolarization of a membrane protein in the liquid phase via solid-state mechanism. J Am Chem Soc 133:16754–16757
19. Matysik J, Diller A, Roy E, Alia A (2009) The solid-state photo-CIDNP effect. Photosynth Res 102:427–435

20. Bouchiat MA, Carver TR, Varnum CM (1960) Nuclear polarization in He-3 gas induced by optical pumping and dipolar exchange. Phys Rev Lett 5:373–375
21. Eisenschmid TC, Kirss RU, Deutsch PP, Hommeltoft SI, Eisenberg R, Bargon J, Lawler RG, Balch AL (1987) Para hydrogen induced polarization in hydrogenation reactions. J Am Chem Soc 109:8089–8091
22. Bowers CR, Weitekamp DP (1986) Transformation of symmetrization order to nuclear-spin magnetization by chemical-reaction and nuclear-magnetic-resonance. Phys Rev Lett 57:2645–2648
23. Lampel G (1968) Nuclear dynamic polarization by optical electronic saturation and optical pumping in semiconductors. Phys Rev Lett 20:491–493
24. Bifone A, Song YQ, Seydoux R, Taylor RE, Goodson BM, Pietrass T, Budinger TF, Navon G, Pines A (1996) NMR of laser-polarized xenon in human blood. Proc Natl Acad Sci USA 93:12932–12936
25. Pietrass T, Bifone A, Room T, Hahn EL (1996) Optically enhanced high-field NMR of GaAs. Phys Rev B Condens Matter 53:4428–4433
26. Overhauser AW (1953) Polarization of nuclei in metals. Phys Rev 92:411–415
27. Carver TR, Slichter CP (1953) Polarization of nuclear spins in metals. Phys Rev 92:212–213
28. Carver TR, Slichter CP (1956) Experimental verification of the overhauser nuclear polarization effect. Phys Rev 102:975–980
29. Bennett LH, Torrey HC (1957) High negative nuclear polarizations in a liquid. Phys Rev 108:499–500
30. Hausser KH, Stehlik D (1968) Solid DNP on metals. Adv Magn Reson 3:79
31. Song CS, Hu KN, Joo CG, Swager TM, Griffin RG (2006) TOTAPOL: a biradical polarizing agent for dynamic nuclear polarization experiments in aqueous media. J Am Chem Soc 128:11385–11390
32. Gunther UL (2013) Dynamic nuclear hyperpolarization in liquids. Top Curr Chem 335:23–69
33. Ardenkjaer-Larsen JH, Fridlund B, Gram A, Hansson G, Hansson L, Lerche MH, Servin R, Thaning M, Golman K (2003) Increase in signal-to-noise ratio of >10,000 times in liquid-state NMR. Proc Natl Acad Sci USA 100:10158–10163
34. Joo CG, Hu KN, Bryant JA, Griffin RG (2006) In situ temperature jump high-frequency dynamic nuclear polarization experiments: enhanced sensitivity in liquid-state NMR spectroscopy. J Am Chem Soc 128:9428–9432
35. Wind RA, Duijvestijn MJ, Vanderlugt C, Manenschijn A, Vriend J (1985) Applications of dynamic nuclear-polarization in C-13 NMR in solids. Prog Nucl Magn Reson Spectrosc 17:33–67
36. Afeworki M, McKay RA, Schaefer J (1992) Selective observation of the interface of heterogeneous polycarbonate polystyrene blends by dynamic nuclear-polarization C-13 NMR-spectroscopy. Macromolecules 25:4084–4091
37. Afeworki M, Vega S, Schaefer J (1992) Direct electron-to-carbon polarization transfer in homogeneously doped polycarbonates. Macromolecules 25:4100–4105
38. Afeworki M, McKay RA, Schaefer J (1993) Dynamic nuclear-polarization enhanced nuclear-magnetic-resonance of polymer-blend interfaces. Mater Sci Eng A Struct Mater 162:221–228
39. Becerra LR, Gerfen GJ, Temkin RJ, Singel DJ, Griffin RG (1993) Dynamic nuclear-polarization with a cyclotron-resonance maser at 5-T. Phys Rev Lett 71:3561–3564
40. Hall DA, Maus DC, Gerfen GJ, Inati SJ, Becerra LR, Dahlquist FW, Griffin RG (1997) Polarization-enhanced NMR spectroscopy of biomolecules in frozen solution. Science 276:930–932
41. Rosay M, Tometich L, Pawsey S, Bader R, Schauwecker R, Blank M, Borchard PM, Cauffman SR, Felch KL, Weber RT et al (2010) Solid-state dynamic nuclear polarization at 263 GHz: spectrometer design and experimental results. Phys Chem Chem Phys 12:5850–5860

42. Maly T, Debelouchina GT, Bajaj VS, Hu KN, Joo CG, Mak-Jurkauskas ML, Sirigiri JR, van der Wel PCA, Herzfeld J, Temkin RJ et al (2008) Dynamic nuclear polarization at high magnetic fields. J Chem Phys 128:052211–052219
43. Rosay M, Lansing JC, Haddad KC, Bachovchin WW, Herzfeld J, Temkin RJ, Griffin RG (2003) High-frequency dynamic nuclear polarization in MAS spectra of membrane and soluble proteins. J Am Chem Soc 125:13626–13627
44. Mak-Jurkauskas ML, Bajaj VS, Hornstein MK, Belenky M, Griffin RG, Herzfeld J (2008) Energy transformations early in the bacteriorhodopsin photocycle revealed by DNP-enhanced solid-state NMR. Proc Natl Acad Sci USA 105:883–888
45. Bajaj VS, Mak-Jurkauskas ML, Belenky M, Herzfeld J, Griffin RG (2009) Functional and shunt states of bacteriorhodopsin resolved by 250 GHz dynamic nuclear polarization-enhanced solid-state NMR. Proc Natl Acad Sci USA 106:9244–9249
46. van der Wel PCA, Hu KN, Lewandowski J, Griffin RG (2006) Dynamic nuclear polarization of amyloidogenic peptide nanocrystals: GNNQQNY, a core segment of the yeast prion protein Sup35p. J Am Chem Soc 128:10840–10846
47. Rosay M, Weis V, Kreischer KE, Temkin RJ, Griffin RG (2002) Two-dimensional C-13-C-13 correlation spectroscopy with magic angle spinning and dynamic nuclear polarization. J Am Chem Soc 124:3214–3215
48. Debelouchina GT, Bayro MJ, van der Wel PCA, Caporini MA, Barnes AB, Rosay M, Maas WE, Griffin RG (2010) Dynamic nuclear polarization-enhanced solid-state NMR spectroscopy of GNNQQNY nanocrystals and amyloid fibrils. Phys Chem Chem Phys 12:5911–5919
49. Linden AH, Franks WT, Akbey U, Lange S, van Rossum BJ, Oschkinat H (2011) Cryogenic temperature effects and resolution upon slow cooling of protein preparations in solid state NMR. J Biomol NMR 51:283–292
50. Linden AH, Lange S, Franks WT, Akbey U, Specker E, van Rossum BJ, Oschkinat H (2011) Neurotoxin II bound to acetylcholine receptors in native membranes studied by dynamic nuclear polarization NMR. J Am Chem Soc 133:19266–19269
51. Lingwood MD, Han S (2011) Chapter 3 – solution-state dynamic nuclear polarization. In: Graham AW (ed) Annual reports on NMR spectroscopy, Elsevier Academic, 73:83–126
52. Mullerwarmuth W, Meisegresch K (1983) Molecular motions and interactions as studied by dynamic nuclear-polarization (DNP) in free-radical solutions. Adv Magn Reson 11:1–45
53. Griesinger C, Bennati M, Vieth HM, Luchinat C, Parigi G, Höfer P, Engelke F, Glaser SJ, Denysenkov V, Prisner TF (2012) Dynamic nuclear polarization at high magnetic fields in liquids. Prog Nucl Magn Reson Spectrosc 64:4–28
54. Turke MT, Tkach I, Reese M, Hofer P, Bennati M (2010) Optimization of dynamic nuclear polarization experiments in aqueous solution at 15 MHz/9.7 GHz: a comparative study with DNP at 140 MHz/94 GHz. Phys Chem Chem Phys 12:5893–5901
55. Turke MT, Parigi G, Luchinat C, Bennati M (2012) Overhauser DNP with 15N labelled Fremy's salt at 0.35 Tesla. Phys Chem Chem Phys 14:502–510
56. Wind RA, Ardenkjaer-Larsen JH (1999) (1)H DNP at 1.4 T of water doped with a triarylmethyl-based radical. J Magn Reson 141:347–354
57. Wind RA, Bai S, Hu JZ, Solum MS, Ellis PD, Grant DM, Pugmire RJ, Taylor CMV, Yonker CR (2000) H-1 dynamic nuclear polarization in supercritical ethylene at 1.4 T. J Magn Reson 143:233–239
58. Villanueva-Garibay JA, Annino G, van Bentum PJM, Kentgens APM (2010) Pushing the limit of liquid-state dynamic nuclear polarization at high field. Phys Chem Chem Phys 12:5846–5849
59. Loening NM, Rosay M, Weis V, Griffin RG (2002) Solution-state dynamic nuclear polarization at high magnetic field. J Am Chem Soc 124:8808–8809
60. Prandolini MJ, Denysenkov VP, Gafurov M, Lyubenova S, Endeward B, Bennati M, Prisner TF (2008) First DNP results from a liquid water-TEMPOL sample at 400 MHz and 260 GHz. Appl Magn Reson 34:399–407

61. Denysenkov VP, Prandolini MJ, Krahn A, Gafurov M, Endeward B, Prisner TF (2008) High-field DNP spectrometer for liquids. Appl Magn Reson 34:289–299
62. Denysenkov V, Prandolini MJ, Gafurov M, Sezer D, Endeward B, Prisner TF (2010) Liquid state DNP using a 260 GHz high power gyrotron. Phys Chem Chem Phys 12:5786–5790
63. Sezer D, Gafurov M, Prandolini MJ, Denysenkov VP, Prisner TF (2009) Dynamic nuclear polarization of water by a nitroxide radical: rigorous treatment of the electron spin saturation and comparison with experiments at 9.2 Tesla. Phys Chem Chem Phys 11:6638–6653
64. Sezer D, Prandolini MJ, Prisner TF (2009) Dynamic nuclear polarization coupling factors calculated from molecular dynamics simulations of a nitroxide radical in water. Phys Chem Chem Phys 11:6626–6637
65. Cheng CY, Wang JY, Kausik R, Lee KY, Han S (2012) An ultrasensitive tool exploiting hydration dynamics to decipher weak lipid membrane–polymer interactions. J Magn Reson 215:115–119
66. Lingwood MD, Sederman AJ, Mantle MD, Gladden LF, Han S (2012) Overhauser dynamic nuclear polarization amplification of NMR flow imaging. J Magn Reson 216:94–100
67. Armstrong BD, Choi J, Lopez C, Wesener DA, Hubbell W, Cavagnero S, Han S (2011) Site-specific hydration dynamics in the nonpolar core of a molten globule by dynamic nuclear polarization of water. J Am Chem Soc 133:5987–5995
68. Franck JM, Pavlova A, Han S (2012) Quantitative cw Overhauser DNP analysis of hydration dynamics. In: Abstracts of papers of the American Chemical Society, p 241. arXiv:1206.0510 [cond-mat.soft]
69. Kausik R, Han S (2011) Dynamics and state of lipid bilayer-internal water unraveled with solution state (1)H dynamic nuclear polarization. Phys Chem Chem Phys 13:7732–7746
70. Reese M, Lennartz D, Marquardsen T, Hofer P, Tavernier A, Carl P, Schippmann T, Bennati M, Carlomagno T, Engelke F et al (2008) Construction of a liquid-state NMR DNP shuttle spectrometer: first experimental results and evaluation of optimal performance characteristics. Appl Magn Reson 34:301–311
71. Reese M, Turke MT, Tkach I, Parigi G, Luchinat C, Marquardsen T, Tavernier A, Hofer P, Engelke F, Griesinger C et al (2009) (1)H and (13)C dynamic nuclear polarization in aqueous solution with a two-field (0.35 T/14 T) shuttle DNP spectrometer. J Am Chem Soc 131:15086–15087
72. Krahn A, Lottmann P, Marquardsen T, Tavernier A, Turke MT, Reese M, Leonov A, Bennati M, Hoefer P, Engelke F et al (2010) Shuttle DNP spectrometer with a two-center magnet. Phys Chem Chem Phys 12:5830–5840
73. McCarney ER, Armstrong BD, Lingwood MD, Han S (2007) Hyperpolarized water as an authentic magnetic resonance imaging contrast agent. Proc Natl Acad Sci USA 104:1754–1759
74. Lingwood MD, Siaw TA, Sailasuta N, Ross BD, Bhattacharya P, Han SG (2010) Continuous flow overhauser dynamic nuclear polarization of water in the fringe field of a clinical magnetic resonance imaging system for authentic image contrast. J Magn Reson 205:247–254
75. McCarney ER, Han S (2008) Spin-labeled gel for the production of radical-free dynamic nuclear polarization enhanced molecules for NMR spectroscopy and imaging. J Magn Reson 190:307–315
76. Golman K, Ardenaer-Larsen JH, Petersson JS, Mansson S, Leunbach I (2003) Molecular imaging with endogenous substances. Proc Natl Acad Sci USA 100:10435–10439
77. Giraudeau P, Shrot Y, Frydman L (2009) Multiple ultrafast, broadband 2D NMR spectra of hyperpolarized natural products. J Am Chem Soc 131:13902–13903
78. Mishkovsky M, Frydman L (2008) Progress in hyperpolarized ultrafast 2D NMR spectroscopy. Chemphyschem 9:2340–2348
79. Frydman L, Blazina D (2007) Ultrafast two-dimensional nuclear magnetic resonance spectroscopy of hyperpolarized solutions. Nat Phys 3:415–419
80. Kurhanewicz J, Bok R, Nelson SJ, Vigneron DB (2008) Current and potential applications of clinical C-13 MR spectroscopy. J Nucl Med 49:341–344

81. Kurhanewicz J, Vigneron DB, Brindle K, Chekmenev EY, Comment A, Cunningham CH, DeBerardinis RJ, Green GG, Leach MO, Rajan SS et al (2011) Analysis of cancer metabolism by imaging hyperpolarized nuclei: prospects for translation to clinical research. Neoplasia 13:81–97
82. Nelson SJ, Vigneron D, Kurhanewicz J, Chen A, Bok R, Hurd R (2008) DNP-hyperpolarized C-13 magnetic resonance metabolic imaging for cancer applications. Appl Magn Reson 34:533–544
83. Kohler SJ, Yen Y, Wolber J, Chen AP, Albers MJ, Bok R, Zhang V, Tropp J, Nelson S, Vigneron DB et al (2007) In vivo (13)carbon metabolic imaging at 3T with hyperpolarized C-13-1-pyruvate. Magn Reson Med 58:65–69
84. Park I, Larson PEZ, Zierhut ML, Hu S, Bok R, Ozawa T, Kurhanewicz J, Vigneron DB, VandenBerg SR, James CD et al (2010) Hyperpolarized (13)C magnetic resonance metabolic imaging: application to brain tumors. Neuro Oncol 12:133–144
85. Gallagher FA, Bohndiek SE, Kettunen MI, Lewis DY, Soloviev D, Brindle KM (2011) Hyperpolarized (13)C MRI and PET: in vivo tumor biochemistry. J Nucl Med 52:1333–1336
86. Gallagher FA, Kettunen MI, Brindle KM (2011) Imaging pH with hyperpolarized (13)C. NMR Biomed 24:1006–1015
87. Gallagher FA, Kettunen MI, Day SE, Hu DE, Ardenkjaer-Larsen JH, In't Zandt R, Jensen PR, Karlsson M, Golman K, Lerche MH et al (2008) Magnetic resonance imaging of pH in vivo using hyperpolarized (13)C-labelled bicarbonate. Nature 453:940–943
88. Golman K, In't Zandt R, Thaning M (2006) Real-time metabolic imaging. Proc Natl Acad Sci USA 103:11270–11275
89. Dementyev AE, Cory DG, Ramanathan C (2008) Dynamic nuclear polarization in silicon microparticles. Phys Rev Lett 100
90. Dementyev AE, Cory DG, Ramanathan C (2008) Rapid diffusion of dipolar order enhances dynamic nuclear polarization. Phys Rev B 77:024413
91. Brindle KM, Bohndiek SE, Gallagher FA, Kettunen MI (2011) Tumor imaging using hyperpolarized (13)C magnetic resonance. Magn Reson Med 66:505–519
92. von Morze C, Larson PEZ, Hu S, Keshari K, Wilson DM, Ardenkjaer-Larsen JH, Goga A, Bok R, Kurhanewicz J, Vigneron DB (2011) Imaging of blood flow using hyperpolarized [(13)C] urea in preclinical cancer models. J Magn Reson Imaging 33:692–697
93. Lau AZ, Chen AP, Hurd RE, Cunningham CH (2011) Spectral-spatial excitation for rapid imaging of DNP compounds. NMR Biomed 24:988–996
94. Krummenacker JG, Denysenkov VP, Terekhov M, Schreiber LM, Prisner TF (2012) DNP in MRI: an in-bore approach at 1.5T. J Magn Reson 215:94–99
95. Marjanska M, Iltis I, Shestov AA, Deelchand DK, Nelson C, Ugurbil K, Henry PG (2010) In vivo (13)C spectroscopy in the rat brain using hyperpolarized [1-(13)C]pyruvate and [2-(13)C]pyruvate. J Magn Reson 206:210–218
96. Becerra LR, Gerfen GJ, Bellew BF, Bryant JA, Hall DA, Inati SJ, Weber RT, Un S, Prisner TF, McDermott AE et al (1995) A spectrometer for dynamic nuclear-polarization and electron-paramagnetic-resonance at high-frequencies. J Magn Reson A 117:28–40
97. Bajaj VS, Farrar CT, Hornstein MK, Mastovsky I, Vieregg J, Bryant J, Elena B, Kreischer KE, Temkin RJ, Griffin RG (2003) Dynamic nuclear polarization at 9T using a novel 250 GHz gyrotron microwave source. J Magn Reson 160:85–90
98. Hu KN, Song C, Yu HH, Swager TM, Griffin RG (2008) High-frequency dynamic nuclear polarization using biradicals: a multifrequency EPR lineshape analysis. J Chem Phys 128:052302
99. Akbey U, Franks WT, Linden A, Lange S, Griffin RG, van Rossum BJ, Oschkinat H (2010) Dynamic nuclear polarization of deuterated proteins. Angew Chem Int Ed 49:7803–7806
100. Hu KN (2011) Polarizing agents and mechanisms for high-field dynamic nuclear polarization of frozen dielectric solids. Solid State Nucl Magn Reson 40:31–41
101. Abragam A (1955) Overhauser effect in nonmetals. Phys Rev 98:1729–1735

102. Jeffries CD (1957) Polarization of nuclei by resonance saturation in paramagnetic crystals. Phys Rev 106:164–165
103. Pound RV (1950) Nuclear electric quadrupole interactions in crystals. Phys Rev 79:685–702
104. Hwang CF, Hill DA (1967) MAS DNP at MIT – cross effect DNP. Phys Rev Lett 19:1011–1013
105. Wollan DS (1976) Dynamic nuclear-polarization with an inhomogeneously broadened ESR line. 1. Theory. Phys Rev B 13:3671–3685
106. Abragam A, Goldman M (1978) Principles of dynamic nuclear-polarization. Rep Prog Phys 41:395–467
107. Farrar CT, Hall DA, Gerfen GJ, Inati SJ, Griffin RG (2001) Mechanism of dynamic nuclear polarization in high magnetic fields. J Chem Phys 114:4922–4933
108. Kozhushner MA, Provotorov BN (1964) On the theory of dynamic nuclear polarization in crystals. Soviet Phys Solid State 6:1152–1154
109. Ramanathan C (2008) Dynamic nuclear polarization and spin diffusion in nonconducting solids. Appl Magn Reson 34:409–421
110. Corzilius B, Smith AA, Griffin RG (2012) Solid effect in magic angle spinning dynamic nuclear polarization. J Chem Phys 137:054201
111. Smith AA, Corzilius B, Barnes AB, Maly T, Griffin RG (2012) Solid effect dynamic nuclear polarization and polarization pathways. J Chem Phys 136:015101
112. Hovav Y, Feintuch A, Vega S (2010) Theoretical aspects of dynamic nuclear polarization in the solid state – the solid effect. J Magn Reson 207:176–189
113. Hovav Y, Feintuch A, Vega S (2011) Dynamic nuclear polarization assisted spin diffusion for the solid effect case. J Chem Phys 134:074509
114. Shimon D, Hovav Y, Feintuch A, Goldfarb D, Vega S (2012) Dynamic nuclear polarization in the solid state: a transition between the cross effect and the solid effect. Phys Chem Chem Phys 14:5729–5743
115. Karabanov A, van der Drift A, Edwards LJ, Kuprov I, Kockenberger W (2012) Quantum mechanical simulation of solid effect dynamic nuclear polarisation using Krylov–Bogolyubov time averaging and a restricted state-space. Phys Chem Chem Phys 14:2658–2668
116. Kessenikh AV, Lushchikov VI, Manenkov AA, Taran YV (1963) Proton polarization in irradiated polyethylenes. Soviet Phys Solid State 5:321–329
117. Hovav Y, Feintuch A, Vega S (2012) Theoretical aspects of dynamic nuclear polarization in the solid state – the cross effect. J Magn Reson 214:29–41
118. Hovav Y, Feintuch A, Vega S (2013) Theoretical aspects of dynamic nuclear polarization in the solid state – spin temperature and thermal mixing. Phys Chem Chem Phys 15:188–203
119. Hu KN, Bajaj VS, Rosay M, Griffin RG (2007) High-frequency dynamic nuclear polarization using mixtures of TEMPO and trityl radicals. J Chem Phys 126:044512
120. Borghini M (1968) Spin-temperature model of nuclear dynamic polarization using free radicals. Phys Rev Lett 20:419–421
121. Atsarkin V, Kessenikh A (2012) Dynamic nuclear polarization in solids: the birth and development of the many-particle concept. Appl Magn Reson 43:7–19
122. Redfield AG (1955) Nuclear magnetic resonance saturation and rotary saturation in solids. Phys Rev 98:1787–1809
123. Provotorov BN (1962) Magnetic resonance saturation in crystals. Soviet Phys Jetp Ussr 14:1126–1131
124. Gadian DG, Panesar KS, Perez Linde AJ, Horsewill AJ, Kockenberger W, Owers-Bradley JR (2012) Preparation of highly polarized nuclear spin systems using brute-force and low-field thermal mixing. Phys Chem Chem Phys 14:5397–5402
125. Jannin S, Comment A, van der Klink JJ (2012) Dynamic nuclear polarization by thermal mixing under partial saturation. Appl Magn Reson 43:59–68
126. Lumata L, Jindal AK, Merritt ME, Malloy CR, Sherry AD, Kovacs Z (2011) DNP by thermal mixing under optimized conditions yields >60 000-fold enhancement of (89)Y NMR signal. J Am Chem Soc 133:8673–8680

127. Lumata L, Ratnakar SJ, Jindal A, Merritt M, Comment A, Malloy C, Sherry AD, Kovacs Z (2011) BDPA: an efficient polarizing agent for fast dissolution dynamic nuclear polarization NMR spectroscopy. Chem Eur J 17:10825–10827
128. Emsley L (2009) Spin diffusion for NMR crystallography. In: Encyclopedia of magnetic resonance. Wiley
129. Bloembergen N (1949) On the interaction of nuclear spins in a crystalline lattice. Physica 15:386–426
130. Goldman M (1970) Spin temperature and nuclear magnetic resonance in solids. Clarendon, Oxford
131. Thurber KR, Yau WM, Tycko R (2010) Low-temperature dynamic nuclear polarization at 9.4 T with a 30 mW microwave source. J Magn Reson 204:303–313
132. Granatstein VL, Parker RK, Armstrong CM (1999) Vacuum electronics at the dawn of the twenty-first century. Proc IEEE 87:702–716
133. Felch KL, Danly BG, Jory HR, Kreischer KE, Lawson W, Levush B, Temkin RJ (1999) Characteristics and applications of fast-wave gyrodevices. Proc IEEE 87:752–781
134. Nusinovich GS (2004) Introduction to the physics of gyrotrons. Johns Hopkins University Press, Baltimore
135. Kartikeyan MV, Borie E, Thumm MKA (2004) Gyrotrons: high power microwave and millimeter wave technology. Springer, Berlin
136. Kartikeyan MV, Borie E, Thumm M (2007) A 250 GHz, 50 W, CW second harmonic gyrotron. Int J Infrared Millimeter Waves 28:611–619
137. Joye CD, Griffin RG, Hornstein MK, Hu KN, Kreischer KE, Rosay M, Shapiro MA, Sirigiri JR, Temkin RJ, Woskov PP (2006) Operational characteristics of a 14-W 140-GHz gyrotron for dynamic nuclear polarization. IEEE Trans Plasma Sci 34:518–523
138. Han ST, Griffin RG, Hu KN, Joo CG, Joye CD, Mastovsky I, Shapiro MA, Sirigiri JR, Temkin RJ, Torrezan AC et al (2006) Continuous-wave submillimeter-wave gyrotrons. Proc Soc Photo Opt Instrum Eng 6373:63730C
139. Hornstein MK, Bajaj VS, Griffin RG, Temkin RJ (2006) Continuous-wave operation of a 460-GHz second harmonic gyrotron oscillator. IEEE Trans Plasma Sci 34:524–533
140. Hunter RI, Cruickshank PAS, Bolton DR, Riedi PC, Smith GM (2010) High power pulsed dynamic nuclear polarisation at 94 GHz. Phys Chem Chem Phys 12:5752–5756
141. Kryukov EV, Newton ME, Pike KJ, Bolton DR, Kowalczyk RM, Howes AP, Smith ME, Dupree R (2010) DNP enhanced NMR using a high-power 94 GHz microwave source: a study of the TEMPOL radical in toluene. Phys Chem Chem Phys 12:5757–5765
142. Leggett J, Hunter R, Granwehr J, Panek R, Perez-Linde AJ, Horsewill AJ, McMaster J, Smith G, Kockenberger W (2010) A dedicated spectrometer for dissolution DNP NMR spectroscopy. Phys Chem Chem Phys 12:5883–5892
143. Matsuki Y, Takahashi H, Ueda K, Idehara T, Ogawa I, Toda M, Akutsu H, Fujiwara T (2010) Dynamic nuclear polarization experiments at 14.1 T for solid-state NMR. Phys Chem Chem Phys 12:5799–5803
144. Barnes AB, Corzilius B, Mak-Jurkauskas ML, Andreas LB, Bajaj VS, Matsuki Y, Belenky ML, Lugtenburg J, Sirigiri JR, Temkin RJ et al (2010) Resolution and polarization distribution in cryogenic DNP/MAS experiments. Phys Chem Chem Phys 12:5861–5867
145. Woskov PP, Bajaj VS, Hornstein MK, Temkin RJ, Griffin RG (2005) Corrugated waveguide and directional coupler for CW 250-GHz gyrotron DNP experiments. IEEE Trans Microw Theory Tech 53:1863–1869
146. Weis V, Bennati M, Rosay M, Bryant JA, Griffin RG (1999) High-field DNP and ENDOR with a novel multiple-frequency resonance structure. J Magn Reson 140:293–299
147. Barnes AB, De Paepe G, van der Wel PCA, Hu KN, Joo CG, Bajaj VS, Mak-Jurkauskas ML, Sirigiri JR, Herzfeld J, Temkin RJ et al (2008) High-field dynamic nuclear polarization for solid and solution biological NMR. Appl Magn Reson 34:237–263

148. Wind RA, Anthonio FE, Duijvestijn MJ, Smidt J, Trommel J, Devette GMC (1983) Experimental setup for enhanced C-13 NMR-spectroscopy in solids using dynamic nuclear-polarization. J Magn Reson 52:424–434
149. Cho H, Baugh J, Ryan CA, Cory DG, Ramanathan C (2007) Low temperature probe for dynamic nuclear polarization and multiple-pulse solid-state NMR. J Magn Reson 187:242–250
150. Nanni EA, Barnes AB, Matsuki Y, Woskov PP, Corzilius B, Griffin RG, Temkin RJ (2011) Microwave field distribution in a magic angle spinning dynamic nuclear polarization NMR probe. J Magn Reson 210:16–23
151. Potapov A, Thurber KR, Yau WM, Tycko R (2012) Dynamic nuclear polarization-enhanced H-1-C-13 double resonance NMR in static samples below 20 K. J Magn Reson 221:32–40
152. Maly T, Cui D, Griffin RG, Miller AF (2012) 1H Dynamic nuclear polarization based on an endogenous radical. J Phys Chem B 116:7055–7065
153. Ardenkjaer-Larsen JH, Laursen I, Leunbach I, Ehnholm G, Wistrand LG, Petersson JS, Golman K (1998) EPR and DNP properties of certain novel single electron contrast agents intended for oximetric imaging. J Magn Reson 133:1–12
154. Koelsch CF (1957) Syntheses with triarylvinylmagnesium bromides – alpha, gamma-bisdiphenylene-beta-phenylallyl, a stable free radical. J Am Chem Soc 79:4439–4441
155. Duijvestijn MJ, Wind RA, Smidt J (1986) A quantitative investigation of the dynamic nuclear-polarization effect by fixed paramagnetic centra of abundant and rare spins in solids at room-temperature. Physica B C 138:147–170
156. Hu KN, Yu HH, Swager TM, Griffin RG (2004) Dynamic nuclear polarization with biradicals. J Am Chem Soc 126:10844–10845
157. Matsuki Y, Maly T, Ouari O, Karoui H, Le Moigne F, Rizzato E, Lyubenova S, Herzfeld J, Prisner T, Tordo P et al (2009) Dynamic nuclear polarization with a rigid biradical. Angew Chem Int Ed 48:4996–5000
158. Salnikov E, Rosay M, Pawsey S, Ouari O, Tordo P, Bechinger B (2010) Solid-state NMR spectroscopy of oriented membrane polypeptides at 100 K with signal enhancement by dynamic nuclear polarization. J Am Chem Soc 132:5940–5941
159. Dane EL, Corzilius B, Rizzato E, Stocker P, Maly T, Smith AA, Griffin RG, Ouari O, Tordo P, Swager TM (2012) Rigid orthogonal bis-TEMPO biradicals with improved solubility for dynamic nuclear polarization. J Org Chem 77:1789–1797
160. Kiesewetter MK, Corzilius B, Smith AA, Griffin RG, Swager TM (2012) Dynamic nuclear polarization with a water-soluble rigid biradical. J Am Chem Soc 134:4537–4540
161. Ysacco C, Rizzato E, Virolleaud MA, Karoui H, Rockenbauer A, Le Moigne F, Siri D, Ouari O, Griffin RG, Tordo P (2010) Properties of dinitroxides for use in dynamic nuclear polarization (DNP). Phys Chem Chem Phys 12:5841–5845
162. Zagdoun A, Casano G, Ouari O, Lapadula G, Rossini AJ, Lelli M, Baffert M, Gajan D, Veyre L, Maas WE et al (2012) A slowly relaxing rigid biradical for efficient dynamic nuclear polarization surface-enhanced NMR spectroscopy: expeditious characterization of functional group manipulation in hybrid materials. J Am Chem Soc 134:2284–2291
163. Corzilius B, Smith AA, Barnes AB, Luchinat C, Bertini I, Griffin RG (2011) High-field dynamic nuclear polarization with high-spin transition metal ions. J Am Chem Soc 133:5648–5651
164. Maly T, Andreas LB, Smith AA, Griffin RG (2010) (2)H-DNP-enhanced (2)H-(13)C solid-state NMR correlation spectroscopy. Phys Chem Chem Phys 12:5872–5878
165. Maly T, Miller AF, Griffin RG (2010) In situ high-field dynamic nuclear polarization-direct and indirect polarization of (13)C nuclei. Chemphyschem 11:999–1001
166. Iijima T (1998) Thermal analysis of cryoprotective solutions for red blood cells. Cryobiology 36:165–173
167. Akbey U, Linden A, Oschkinat H (2012) High-temperature dynamic nuclear polarization enhanced magic-angle-spinning NMR. Appl Magn Reson 43:81–90

168. Akbey U, Rossum BJ, Oschkinat H (2012) Practical aspects of high-sensitivity multidimensional (13)C MAS NMR spectroscopy of perdeuterated proteins. J Magn Reson 217:77–85
169. Lange S, Linden AH, Akbey U, Franks WT, Loening NM, van Rossum BJ, Oschkinat H (2012) The effect of biradical concentration on the performance of DNP-MAS-NMR. J Magn Reson 216:209–212
170. Lesage A, Lelli M, Gajan D, Caporini MA, Vitzthum V, Mieville P, Alauzun J, Roussey A, Thieuleux C, Mehdi A et al (2010) Surface enhanced NMR spectroscopy by dynamic nuclear polarization. J Am Chem Soc 132:15459–15461
171. Siemer AB, Huang KY, McDermott AE (2010) Protein–ice interaction of an antifreeze protein observed with solid-state NMR. Proc Natl Acad Sci USA 107:17580–17585
172. Franks WT, Wylie BJ, Schmidt HLF, Nieuwkoop AJ, Mayrhofer RM, Shah GJ, Graesser DT, Rienstra CM (2008) Dipole tensor-based atomic-resolution structure determination of a nanocrystalline protein by solid-state NMR. Proc Natl Acad Sci USA 105:4621–4626
173. Barnes AB, Mak-Jurkauskas ML, Matsuki Y, Bajaj VS, van der Wel PCA, DeRocher R, Bryant J, Sirigiri JR, Temkin RJ, Lugtenburg J et al (2009) Cryogenic sample exchange NMR probe for magic angle spinning dynamic nuclear polarization. J Magn Reson 198:261–270
174. Franks WT, Zhou DH, Wylie BJ, Money BG, Graesser DT, Frericks HL, Sahota G, Rienstra CM (2005) Magic-angle spinning solid-state NMR spectroscopy of the beta 1 immunoglobulin binding domain of protein G (GB1): N-15 and C-13 chemical shift assignments and conformational analysis. J Am Chem Soc 127:12291–12305
175. Doster W (2010) The protein-solvent glass transition. Biochim Biophys Acta 1804:3–14
176. Tompa K, Bánki P, Bokor M, Kamasa P, Lasanda G, Tompa P (2009) Interfacial water at protein surfaces: wide-line NMR and DSC characterization of hydration in ubiquitin solutions. Biophys J 96:2789–2798
177. Lelli M, Gajan D, Lesage A, Caporini MA, Vitzthum V, Mieville P, Heroguel F, Rascon F, Roussey A, Thieuleux C et al (2011) Fast characterization of functionalized silica materials by silicon-29 surface-enhanced NMR spectroscopy using dynamic nuclear polarization. J Am Chem Soc 133:2104–2107
178. Lafon O, Rosay M, Aussenac F, Lu XY, Trebosc J, Cristini O, Kinowski C, Touati N, Vezin H, Amoureux JP (2011) Beyond the silica surface by direct silicon-29 dynamic nuclear polarization. Angew Chem Int Ed 50:8367–8370
179. Akbey U, Corzilius B, Griffin RG, Oschkinat H. In preperation
180. Thurber KR, Tycko R (2009) Measurement of sample temperatures under magic-angle spinning from the chemical shift and spin-lattice relaxation rate of 79Br in KBr powder. J Magn Reson 196:84–87
181. Mentink-Vigier F, Akbey U, Hovav Y, Vega S, Oschkinat H, Feintuch A (2012) Fast passage dynamic nuclear polarization on rotating solids. J Magn Reson 224:13–21
182. Thurber KR, Tycko R (2012) Theory for cross effect dynamic nuclear polarization under magic-angle spinning in solid state nuclear magnetic resonance: the importance of level crossings. J Chem Phys 137:084508
183. Zagdoun A, Rossini AJ, Conley MP, Grüning WR, Schwarzwälder M, Lelli M, Franks WT, Oschkinat H, Copéret C, Emsley L et al (2013) Improved dynamic nuclear polarization surface-enhanced NMR spectroscopy through controlled incorporation of deuterated functional groups. Angew Chem Int Ed 52:1222–1225
184. Jacso T, Franks WT, Rose H, Fink U, Broecker J, Keller S, Oschkinat H, Reif B (2012) Characterization of membrane proteins in isolated native cellular membranes by dynamic nuclear polarization solid-state NMR spectroscopy without purification and reconstitution. Angew Chem Int Ed 51:432–435
185. Takahashi H, Ayala I, Bardet M, De Paëpe G, Simorre JP, Hediger S (2013) Solid-state NMR on bacterial cells: selective cell wall signal enhancement and resolution improvement using dynamic nuclear polarization. J Am Chem Soc 135:5105–5110

186. Renault M, Pawsey S, Bos MP, Koers EJ, Nand D, Tommassen-van BR, Rosay M, Tommassen J, Maas WE, Baldus M (2012) Solid-state NMR spectroscopy on cellular preparations enhanced by dynamic nuclear polarization. Angew Chem Int Ed Engl 51:2998–3001
187. Lerche MH, Meier S, Jensen PR, Hustvedt SO, Karlsson M, Duus JO, Ardenkjaer-Larsen JH (2011) Quantitative dynamic nuclear polarization-NMR on blood plasma for assays of drug metabolism. NMR Biomed 24:96–103
188. Dafni H, Ronen SM (2010) Dynamic nuclear polarization in metabolic imaging of metastasis: common sense, hypersense and compressed sensing. Cancer Biomark 7:189–199
189. Rossini AJ, Zagdoun A, Hegner F, Schwarzwälder M, Gajan D, Copéret C, Lesage A, Emsley L (2012) Dynamic nuclear polarization NMR spectroscopy of microcrystalline solids. J Am Chem Soc 134:16899–16908
190. Rossini AJ, Zagdoun A, Lelli M, Canivet J, Aguado S, Ouari O, Tordo P, Rosay M, Maas WE, Copéret C et al (2012) Dynamic nuclear polarization enhanced solid-state NMR spectroscopy of functionalized metal – organic frameworks. Angew Chem Int Ed 51:123–127
191. Rossini AJ, Zagdoun A, Lelli M, Canivet J, Aguado S, Ouari O, Tordo P, Rosay M, Maas WE, Coperet C et al (2012) Dynamic nuclear polarization enhanced solid-state NMR spectroscopy of functionalized metal-organic frameworks. Angew Chem Int Ed Engl 51:123–127
192. Rossini AJ, Zagdoun A, Lelli M, Gajan D, Rascon F, Rosay M, Maas WE, Coperet C, Lesage A, Emsley L (2012) One hundred fold overall sensitivity enhancements for silicon-29 NMR spectroscopy of surfaces by dynamic nuclear polarization with CPMG acquisition. Chem Sci 3:108–115
193. Lafon O, Thankamony ASL, Kobayashi T, Carnevale D, Vitzthum V, Slowing II, Kandel K, Vezin H, Amoureux JP, Bodenhausen G et al (2012) Mesoporous silica nanoparticles loaded with surfactant: low temperature magic angle spinning 13C and 29Si NMR enhanced by dynamic nuclear polarization. J Phys Chem C 117:1375–1382
194. Vitzthum V, Mieville P, Carnevale D, Caporini MA, Gajan D, Coperet C, Lelli M, Zagdoun A, Rossini AJ, Lesage A et al (2012) Dynamic nuclear polarization of quadrupolar nuclei using cross polarization from protons: surface-enhanced aluminium-27 NMR. Chem Commun (Camb) 48:1988–1990
195. Lafon O, Thankamony ASL, Rosay M, Aussenac F, Lu X, Trebosc J, Bout-Roumazeilles V, Vezin H, Amoureux JP (2013) Indirect and direct 29Si dynamic nuclear polarization of dispersed nanoparticles. Chem Commun (Camb) 49:2864–2866

Top Curr Chem (2013) 338: 229–300
DOI: 10.1007/128_2013_427

Published online: 14 May 2013

Photo-CIDNP NMR Spectroscopy of Amino Acids and Proteins

Lars T. Kuhn

Abstract Photo-chemically induced dynamic nuclear polarization (CIDNP) is a nuclear magnetic resonance (NMR) phenomenon which, among other things, is exploited to extract information on biomolecular structure via probing solvent-accessibilities of tryptophan (Trp), tyrosine (Tyr), and histidine (His) amino acid side chains both in polypeptides and proteins in solution. The effect, normally triggered by a (laser) light-induced photochemical reaction in situ, yields both positive and/or negative signal enhancements in the resulting NMR spectra which reflect the solvent exposure of these residues both in equilibrium and during structural transformations in "real time". As such, the method can offer – qualitatively and, to a certain extent, quantitatively – residue-specific structural and kinetic information on both the native and, in particular, the non-native states of proteins which, often, is not readily available from more routine NMR techniques. In this review, basic experimental procedures of the photo-CIDNP technique as applied to amino acids and proteins are discussed, recent improvements to the method highlighted, and future perspectives presented. First, the basic principles of the phenomenon based on the theory of the radical pair mechanism (RPM) are outlined. Second, a description of standard photo-CIDNP applications is given and it is shown how the effect can be exploited to extract residue-specific structural information on the conformational space sampled by unfolded or partially folded proteins on their "path" to the natively folded form. Last, recent methodological advances in the field are highlighted, modern applications of photo-CIDNP in the context of biological NMR evaluated, and an outlook into future perspectives of the method is given.

L.T. Kuhn (✉)
DFG Research Center Molecular Physiology of the Brain (CMPB), European Neuroscience Institute Göttingen (ENI-G) and EXC 171 Microscopy at the Nanometer Range, Göttingen, Germany
e-mail: lars.kuhn@gwdg.de

Keywords CIDNP · NMR · Nuclear spin polarization · Photosensitizer · Protein folding

Contents

Abbreviations

ADC	Analogue-to-digital converter
BLA	Bovine α-lactalbumin
CD	Circular dichroism
CHESS	Chemical shift selective excitation
CIDEP	Chemically induced dynamic electron polarization
CIDNP	Chemically induced dynamic nuclear polarization
COSY	Correlation spectroscopy
CSA	Chemical shift anisotropy
CSD	Chemical shift deviation
CW	Continuous wave
DBPO	Dibenzoylperoxide

DNA	Deoxyribonucleic acid
DNP	Dynamic nuclear polarization
DP	2,2′-Dipyridyl
DPFGSE	Double pulsed field gradient spin echo
DRYSTEAM	Drastic reduction of water signals in spectroscopy based on the stimulated echo acquisition mode
EPR	Electron paramagnetic resonance
FAD	Flavin adenine dinucleotide
FID	Free induction decay
FMN	Flavin mononucleotide
FT	Fourier transform
HEWL	Hen egg-white lysozyme
HFC	Hyperfine coupling constant
HMQC	Heteronuclear multiple quantum coherence
HSQC	Heteronuclear single quantum correlation
INEPT	Insensitive nuclei enhanced by polarization transfer
IR	Infrared
MFE	Magnetic field effect
NMR	Nuclear magnetic resonance
NOE	Nuclear Overhauser effect
NOESY	Nuclear Overhauser effect spectroscopy
PDB	Protein data bank
PEEK	Polyetheretherketone
PPII	Polyproline II helix
Ppm	Parts per million
PTFE	Polytetrafluoroethylene
R_1	Spin–lattice relaxation rate constant
R_2	Spin–spin relaxation rate constant
RPM	Radical pair mechanism
RT	Real time
SNR	Signal-to-noise ratio
T_1	Spin–lattice relaxation time constant
T_2	Spin–spin relaxation time constant
TOCSY	Total correlation spectroscopy
VAPOR	Variable pulse power and optimized relaxation delays
VCD	Vibrational circular dichroism
WATERGATE	Water suppression through gradient tailored excitation
WET	Water suppression enhanced through T_1 effects

1 Introduction

It is a remarkable fact that a nuclear property called "spin" can alter the yields of free radical reactions in solution and the effect can be detected by NMR. The accidental observation of anomalously emissive NMR resonances during the thermally induced decomposition reaction of the small organic compound dibenzoylperoxide (DBPO) (Fig. 1) made by Bargon et al. around 45 years ago [1] marked the discovery of this phenomenon which shortly afterwards became known as chemically induced dynamic nuclear polarization, or simply CIDNP.[1] The effect, which was independently observed by Ward and Lawler at about the same time [3], arises from the ability of magnetic nuclei to modulate the electronic spin state of a radical pair and hence its reactivity. It manifests itself by emission or enhanced absorption in the nuclear magnetic resonance (NMR) signals of the products of free radical reactions in solution occurring in the presence of a strong magnetic field.

Early theories, trying to interpret and explain CIDNP qualitatively and quantitatively [4–6], were based on the mechanisms which, in those days, were already known to be responsible for the nuclear Overhauser effect (NOE) and dynamic nuclear polarization (DNP). However, they all failed to explain both the magnitude of the detected enhancements as well as the occurrence of simultaneous emission and absorption lines, i.e., the multiplet effect (see below), within the same multiplet of signals for a given nucleus observed in many CIDNP spectra. A few years later, however, an alternative explanation, which accounted much more convincingly for these experimental observations, was proposed by Closs [7] and independently by Kaptein and Oosterhoff, the so-called CKO model [8]. Their theory placed the emphasis on the ability of magnetically active nuclei to alter the electronic spin state of a radical pair and hence modulate its reactivity. This theoretical approach subsequently became known as the radical pair mechanism (RPM) and led to the application of both thermally or photochemically induced CIDNP to mechanistic problems in organic chemistry as the NMR enhancements observed could be used to identify and elucidate the pathways of reactions which proceed via one or more radical pair intermediates [9–12]. Of particular importance in the context of this review was the development made by Kaptein et al. of a CIDNP technique in which the polarization is generated in reversible photochemical reactions between an excited photosensitizer and certain aromatic amino acid side chains present on the surface of a protein [13], thereby monitoring their solvent accessibility or "exposure". Ever since its first application, this "photo-CIDNP" technique has proved to be a powerful probe of protein structure and of the wide variety of factors that modify the accessibility of the polarizable amino acid residues.

[1] The acronym "CIDNP" was suggested to Bargon et al. for the first time by Sir Rex E. Richards from the Physical Chemistry Laboratory (PCL) at Oxford University, Oxford, UK, who, long before the discovery of the CIDNP effect itself, had set out a thought experiment on how to induce the – already at that time – well-known Overhauser effect in a chemical way, i.e., by breaking a chemical bond [2].

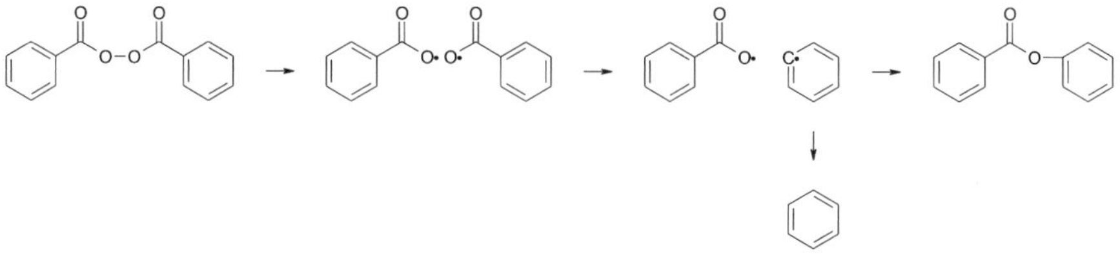

Fig. 1 Thermal decomposition of dibenzoylperoxide (DBPO) during which the first observation of the CIDNP effect was made [1]. Magnetic nuclei of recombination (ester) and escape product (benzene) exhibit opposite polarization patterns

Applications include studies of the interactions of proteins with cofactors, inhibitors, nucleic acids, lipids, and other proteins, comparison of related proteins, and investigation of conformational changes and denaturation both in equilibrium and in "real time" (RT).

2 Chemically Induced Dynamic Nuclear Polarization

2.1 *Radical Pairs*

Radicals are chemical species which possess an unpaired electron and are hence paramagnetic. When two radicals are closely associated they can form a so-called *spin-correlated* radical pair, which means that the overall spin state of the pair is well defined. The value of the total electron spin angular momentum quantum number (S) of this radical pair is given by a Clebsch–Gordon series:

$$S = s_1 + s_2,\ s_1 + s_2 - 1, \ldots, |s_1 - s_2|,$$

where s_j is the electron spin angular momentum of radical j in the pair. Hence, for a radical pair S can take the values 0 or 1. For $S = 0\,(M_s = 0)$ the radical pair resides in a singlet state (S), whereas for $S = 1\ (M_s = 0 \pm 1)$ there are three triplet states labeled T_-, T_0, and T_+. The spin parts of the singlet and triplet electronic wavefunctions of a radical pair can be defined in terms of the products of the individual electronic wavefunctions $|\alpha\rangle$ and $|\beta\rangle$:

$$\mathrm{S} = \frac{1}{\sqrt{2}}(\alpha_1\beta_2 - \beta_1\alpha_2);\quad \mathrm{T}_- = \beta_1\beta_2;\quad \mathrm{T}_0 = \frac{1}{\sqrt{2}}(\alpha_1\beta_2 + \beta_1\alpha_2);\quad \mathrm{T}_+ = \alpha_1\alpha_2.$$

In the liquid state, the radical pair and its component radicals tumble rapidly.[2] Hence, anisotropic terms of the spin Hamiltonian are averaged to zero and the spin

[2] Tumbling frequencies of around 10^{11} Hz for free amino acids and around 10^9 Hz for proteins are usually observed.

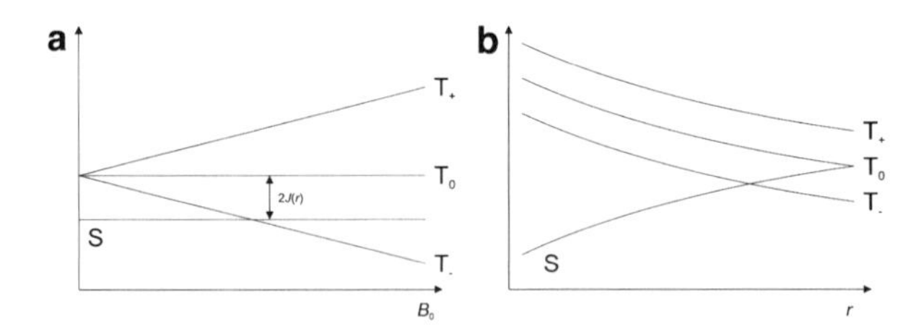

Fig. 2 Relative energies (*vertical axes*) of the singlet and triplet states (**a**) as a function of the applied magnetic field strength and (**b**) as a function of the radical separation r (*horizontal axes*). $J(r)$ is the electron exchange interaction (see text)

Hamiltonian required to describe a radical pair needs to contain only isotropic terms; in particular, the electron Zeeman interaction, the exchange interaction, and the isotropic component of the hyperfine interaction need to be included (cf. Appendix).

In the absence of an applied magnetic field, the three triplet states of a radical pair are degenerate at all separations of the two radicals, whereas the singlet and triplet states only become degenerate at large separations, i.e., when the exchange interaction is negligibly small. In the presence of a static magnetic field, however, the degeneracy of the triplet states is being lifted, which changes the overall properties of the radical pair dramatically. This behavior is illustrated in Fig. 2, where the energies of the singlet and triplet states at a given applied magnetic field strength in the absence of any magnetic nuclei are shown as a function of the electron separation r. When the interactions responsible for the mixing of electronic states are much weaker than the energy difference separating them, the efficiency (or extent) of mixing between the different states is – amongst other things – inversely proportional to this energy difference. In the presence of strong magnetic fields larger than approximately 0.1 T, the T_-, and T_+ states are so far removed from the singlet state (S) that mixing is restricted to the S and T_0 states provided that the separation of the two electrons of the radical pair is sufficiently large. It is exactly this mixing process between the S and the T_0 state, the so-called singlet–triplet mixing, which accounts for the generation of CIDNP in a spin-correlated radical pair under high-field conditions.

2.2 The Spin Hamiltonian

The spin dynamics of a spin-correlated pair of radicals can be described using a spin Hamiltonian ($\hat{H}$) which comprises terms for the Zeeman interaction ($\hat{H}_z$), the electron-nuclear hyperfine interaction ($\hat{H}_{\mathrm{hf}}$), and the exchange interaction ($\hat{H}_J$) that exist between electrons. As in solution the radical pair is allowed to undergo free tumbling and diffusion, these spin interactions are assumed to be fully isotropic, i.e., they do not exhibit any spatial or orientational dependence, with the only exception being the exchange interaction ($\hat{H}_J$) which is spatially dependent on the distance r between the two radicals. The individual terms that contribute to the overall spin Hamiltonian are explained in more detail below.

The interaction of the spin angular momentum with the applied magnetic field is known as the Zeeman interaction and its magnitude is given by the following Hamiltonian:

$$\hat{H}_z = \left(g_1\hat{S}_{1z} + g_2\hat{S}_{2z}\right)B_0\frac{\mu_B}{\hbar},$$

where $\hat{S}_{jz}$ is the z-component of the electron spin angular momentum operator $\hat{S}_j$ on radical j. Furthermore, g_j represents the respective g-values of the component radicals and B_0 is the magnetic field strength; μ_B is the Bohr magneton of the free electron. In general, the effect of the nuclear Zeeman interaction is assumed to be negligibly small as compared to its electronic equivalent.

The hyperfine Hamiltonian represents the interaction of electronic and nuclear spin angular momentum and is usually described using the following expression:

$$\hat{H}_{\mathrm{hf}} = \sum_i a_i\hat{I}_i \cdot \hat{S}_1 + \sum_j a_j\hat{I}_j \cdot \hat{S}_2$$

where a_i and a_j are the isotropic hyperfine coupling constants of nuclei i and j on radical 1 and 2 which arise from the Fermi contact interaction. According to the theory of the RPM explained below, it is this interaction which gives rise to the CIDNP effect. Hyperfine interactions between the unpaired electron on one radical and the magnetically active nuclei on the other radical are usually considered to be negligibly small and hence do not contribute to the CIDNP effect.

In addition, the two unpaired electrons present in a spin-correlated radical pair can interact with each other via the quantum mechanical exchange interaction which is due to the overlap of the electronic orbital wavefunctions. It is described using the following Hamiltonian:

$$\hat{H}_J = -J(r)\left(\frac{1}{2} + \hat{S}_1 \cdot \hat{S}_2\right)$$

where $J(r)$ is a distance-dependent function describing the strength of the exchange interaction. The exact form of this function is unknown, but it is assumed that its decay is approximately exponential with increasing radical separation r; it can normally be neglected for separations greater than 1 nm. The manifestation of the exchange interaction is to remove the degeneracy between the singlet and the three triplet states of the radical pair even in the absence of any applied magnetic field (see above). The representation of the spin Hamiltonian in the S, T_-, T_0, and T_+ basis set is given, in angular frequency units, by

$$\hat{H} = \begin{array}{c} \\ T_+ \\ S \\ T_0 \\ T_- \end{array} \begin{array}{c} \begin{array}{cccc} T_+ & S & T_0 & T_- \end{array} \\ \begin{pmatrix} \omega - J(r) & 0 & 0 & 0 \\ 0 & J(r) & Q & 0 \\ 0 & Q & -J(r) & 0 \\ 0 & 0 & 0 & -\omega - J(r) \end{pmatrix} \end{array}$$

where $\omega = 1/2(\omega_1 + \omega_2)$ and $Q = 1/2(\omega_1 - \omega_2)$. Moreover, ω_1 and ω_2 are given by the following relationship where m_i and m_j are the magnetic quantum numbers of nuclei i and j, respectively:

$$\omega_1 = g_1 B_0 \frac{\mu_B}{\hbar} + \sum_i a_i m_i,$$

$$\omega_2 = g_2 B_0 \frac{\mu_B}{\hbar} + \sum_j a_j m_j.$$

2.3 The Vector Model and Singlet–Triplet Mixing

The intersystem-crossing process between the singlet state (S) and the triplet state (T_0), i.e., the singlet–triplet mixing, is most simply described using a vector model approach. Within this semiclassical representation, the magnetic moments associated with the electron spin states s_1 and s_2 of the two paramagnetic components of a radical pair can be represented by vectors precessing about the direction of the applied static magnetic field B_0 as shown in Fig. 3. In this representation, the S and T_0 states differ only in the phase difference between the two vectors whose precession frequencies are equal to ω_1 and ω_2. As shown schematically, the conversion between the S and T_0 states will thus occur at a frequency determined by the difference in precession frequencies of the two vectors. In the absence of any hyperfine interactions, ω_1 and ω_2 are determined solely by the g-factors of the two radicals. If, however, the precession frequencies of the radicals are modulated by hyperfine interactions, the intersystem crossing rate of the radical pair and hence its recombination probability will be nuclear spin state-dependent.

2.4 The Radical Pair Mechanism (RPM)

The CIDNP phenomenon can readily be interpreted in terms of the RPM which accounts for almost all phenomena observed during the analysis of thermally or photochemically induced reactions of spin-correlated radical species in solution

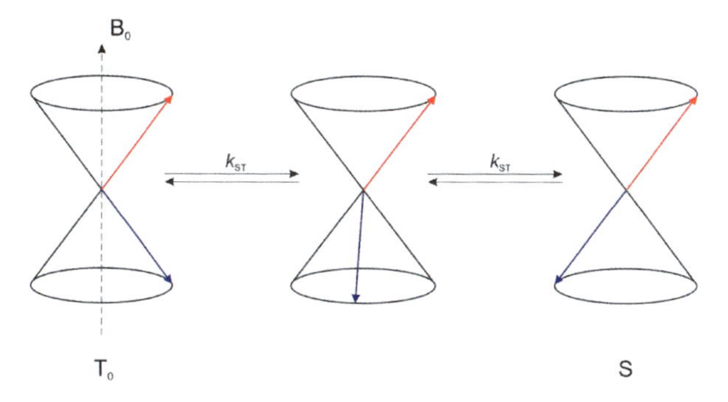

Fig. 3 Vector model representation of singlet–triplet mixing. The *cones* represent the individual electron spin states and the *arrows* the electron spins. The intersystem crossing rate (k_{ST}) is determined by the precession frequency difference of the two radical electrons

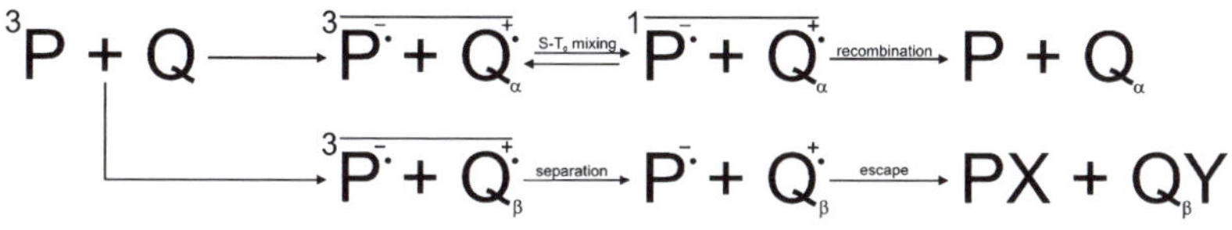

$$^3\mathrm{P} + \mathrm{Q} \longrightarrow \overline{^3\mathrm{P}^{\cdot-} + \mathrm{Q}^{\cdot+}}_{\alpha} \overset{\text{S-T}_0\text{ mixing}}{\rightleftharpoons} \overline{^1\mathrm{P}^{\cdot-} + \mathrm{Q}^{\cdot+}}_{\alpha} \xrightarrow{\text{recombination}} \mathrm{P} + \mathrm{Q}_{\alpha}$$

$$\longrightarrow \overline{^3\mathrm{P}^{\cdot-} + \mathrm{Q}^{\cdot+}}_{\beta} \xrightarrow{\text{separation}} \mathrm{P}^{\cdot-} + \mathrm{Q}^{\cdot+}_{\beta} \xrightarrow{\text{escape}} \mathrm{PX} + \mathrm{QY}_{\beta}$$

Fig. 4 Schematic representation of the radical pair mechanism (RPM) and the spin sorting process. In this example, singlet–triplet mixing is faster when the proton on Q is in the "α" state. The *overbar* indicates the spin-correlation between the two radical partners

using NMR or EPR spectroscopy.[3] To understand how the nuclear spin dependence of singlet–triplet mixing leads to CIDNP, the reaction scheme shown in Fig. 4 can be considered. In a first step, a spin-correlated radical pair in its triplet state is formed as a result of an electron abstraction reaction between the triplet excited electron acceptor P and the ground state electron donor Q. Held together by a cage of solvent molecules, this triplet radical pair then has two alternative ways to react: (1) the two partner radicals can either diffuse apart and get scavenged at a later stage to form so-called *escape products* (here PX and QY) or, alternatively, (2) react via a back electron transfer reaction to yield so-called *recombination products*, i.e., the initial diamagnetic starting compounds. In the absence of any hyperfine interactions between the unpaired electron and magnetically active nuclei of the same compound, both pathways have an almost equal chance to occur. If, however, the radical electron's precession frequency is modulated by these hyperfine interactions, the probability of the radical pair to recombine will be *spin-dependent*.

The recombination likelihood of a radical pair depends strongly on the electronic spin state of the pair, with reaction usually only possible from a singlet state as

[3] The RPM is also responsible for the ESR equivalent effect of CIDNP called chemically induced dynamic electron polarization (CIDEP) [14, 15], and for the magnetic field effect (MFE) observed during radical pair recombination reactions [16].

the Pauli principle has to be obeyed. Hence, the triplet radical pair must first convert to a singlet state in order to proceed via the recombination route. If singlet–triplet mixing is faster when a nucleus on Q is in its α state, then a nuclear spin sorting process will take place with triplet radical pairs containing a nucleus in its α state more likely to interconvert into a singlet state and then recombine. On the other hand, triplet pairs comprising nuclei of opposite spin quantum number will have a greater probability of diffusing apart and react via the escape route depicted in Fig. 4. As a consequence of this nuclear spin selective chemistry, recombination products will contain an excess of α spin state nuclei and hence show an absorptive, i.e., positive, NMR enhancement upon application of a $\pi/2$-pulse and, conversely, escape products will possess an excess of β spin state nuclei leading to an emissive, i.e., negative, enhancement in the NMR spectrum. In exactly the same way, when singlet–triplet mixing is slower for radical pairs comprising α spin nuclei, i.e., Δg and a of opposite sign, or for an initial radical pair formed from a singlet precursor, the situation is reversed, leading to positive enhancements in the magnetic nuclei of the escape products and negative enhancements in those of the recombination products [17, 18].

2.5 Sign Rules

The observations mentioned in the previous section have been conveniently summarized in a multiplicative sign rule proposed by Kaptein [19, 20] in which the phase of the enhancement (Γ), i.e., positive or negative, for a given nucleus i is given by the product of four signs:

$$\Gamma(i) = \mu \cdot \varepsilon \cdot \Delta g \cdot a_i,$$

where a_i is the sign of the hyperfine coupling constant for nucleus i and Δg is the sign of the difference of the two g-factors of the two radicals, i.e., $g_1 - g_2$, where g_i is the g-factor of the radical carrying nucleus i; μ is the expression for the spin state of the precursor which can either be in a singlet (negative sign) or a triplet (positive sign) state and ε describes the way of the reaction route and is positive for recombination products and negative for escape products. If $\Gamma(i)$ is of positive sign, Kaptein's sign rule predicts a positive (absorptive) NMR enhancement for the observed nucleus and a negative (emissive) NMR enhancement is expected for a negative sign of $\Gamma(i)$. Kaptein's sign rule may be transformed to yield the sign of the precursor, the hyperfine coupling constant, or the reaction route if $\Gamma(i)$ is known from the CIDNP spectrum. In a similar manner, g-factor differences can be estimated if the experiment is repeated with different radicals of known g-factors combined with the radical to be studied. Hence, this rule is of great value in the study of radical reactions.

It is worth noting that the sign rule described above only applies to CIDNP spectra recorded at very high magnetic fields which are usually present when using contemporary state-of-the-art NMR spectrometers. In this case, the hyperfine interaction is small compared to $\Delta g \cdot B_0$ and hence every multiplet component for a given nucleus has the same polarization. This type of CIDNP is known as the *net effect*. At fields much lower than those used in photo-CIDNP experiments of amino acids and proteins or for small values of Δg (or large a_i), a so-called *multiplet effect*, in which simultaneous emission and absorption occurs within the same multiplet for signals of a single nucleus, is often observed together with the net effect. For this case, Kaptein has formulated a separate sign rule to predict emission and absorption polarization patterns [19, 20].[4]

2.6 Spin Dynamics: The Concept of Secondary Recombination

The discussion of the origins of the CIDNP effect has so far largely ignored both the dynamic nature of radical pairs as well as the effects of the aforementioned distance dependent exchange interaction. As was described in Sect. 2.1 and shown schematically in Fig. 2, the S and T_0 states of a radical pair only become degenerate once the component radicals have separated by a few nanometers and hence S–T_0 mixing is precluded within the initially formed primary pair of radicals. Hence, for recombination to occur from a triplet precursor, the component radicals must first diffuse apart before they re-encounter to form a so-called secondary radical pair. The timescale of this re-encounter step (ca. 10^{-10} to 10^{-7} s) is normally such that singlet–triplet mixing (ca. 10^{-9} to 10^{-8} s) can take place while the radicals are separated and experience a very small and hence negligible exchange interaction. Normally, the relative timescales of the spin dynamics and the motional behavior of the component radicals can be described in greater detail within the quantitative framework of a CIDNP diffusion model [20, 24, 25]. Spin relaxation times are typically on the order of 10^{-6} s and hence loss of spin correlation between the two radicals of the pair does not interfere with the somewhat faster processes involved in generating CIDNP, i.e., spin evolution and diffusion processes. Once CIDNP has been generated within the diamagnetic products of the radical pair reaction, the time window for observation is significantly extended due to slow nuclear spin relaxation which normally takes around 0.1–10 s.

[4] The Kaptein sign rules for CIDNP, however, do not apply in all cases [21, 22]. For example, an exception to the CIDNP sign rule for the net effect can occur when the dominant relaxation process in an electron-nucleus spin system is transverse ΔHFC–Δg cross-correlation. This phenomenon has been shown to lead to an inversion in the geminate fluorine-19 CIDNP phase provided the molecule of interest has a large rotational correlation time [23].

2.7 *CIDNP of Biomolecules: The Cyclic Reaction Scheme*

The application of photo-CIDNP to the study of biological systems involves the generation of the radical pair comprising the aromatic side chains of tyrosine (Tyr), tryptophan (Trp), or histidine (His) and a suitable photosensitizer, usually the tricyclic isoalloxazine derivative flavin mononucleotide (FMN). Methionine (Met), the only non-aromatic CIDNP-active amino acid side chain, also shows weak polarization at its γ-CH position [26]; however, it is not of general interest. From an experimental point of view, the basic photo-CIDNP experiment as applied to amino acids and proteins involves adding a small amount of a suitable photosensitizer to an amino acid or protein sample [18, 27]. Spectra are then recorded with (light spectrum) and, subsequently, without (dark spectrum) prior irradiation of the sample in situ using an NMR spectrometer that has been adapted to allow light from a suitable light source, e.g., an argon ion laser, to pass into the sample within the probe.

The generation of polarization in biomolecules proceeds as follows (Fig. 5). The production of CIDNP in the aromatic amino acid side chain begins with the photochemical excitation of the FMN photosensitizer yielding a flavin molecule in its first excited singlet state (^{1}F). The fluorescence quantum yield of this state is rather low and approximately half of the molecules rapidly undergo an intersystem crossing process to a longer lived excited triplet state (^{3}F). Triplet excited flavin has a high electron affinity [28] and hence, in a second step, the triplet state photoexcited sensitizer reacts reversibly with the amino acid side chain (A) in a diffusion-controlled electron transfer or a hydrogen abstraction reaction (see below) to form a spin-correlated triplet electron transfer state, i.e., a radical pair.[5] The generated nuclear polarization is most easily detected by subtracting the spectrum without laser irradiation ("dark") from the spectrum obtained with laser irradiation ("light"). Photo-CIDNP experiments on amino acids and proteins using an argon ion laser as the light source and FMN as the photosensitizer exhibit proton NMR signal enhancements of up to one and a half orders of magnitude. Heavier spin-1/2 nuclei, e.g., ^{13}C, ^{15}N, and ^{19}F, often show larger polarizations. For certain fluorinated aromatic systems, for example, it was shown that a hyperpolarization of approximately two orders of magnitude can be generated and sustained for several seconds [29, 30].

The reaction scheme mentioned above is essentially identical to the RPM described in Sect. 2.4, with the only exception being that the idealized reaction is *cyclic*. This means that, unlike in Fig. 4, where recombination and escape products represent chemically different species, amino acid radicals that escape from the radical pair are no longer scavenged, but react with independently formed photosensitizer radicals to regenerate the original amino acid and the flavin in its

[5] Generally speaking, the photo-CIDNP phenomenon is not restricted to the magnetic nuclei of the three aromatic amino acids but the radical pair can be formed between many different excited sensitizers and various electron donors provided the ionization potential of the latter is sufficiently low.

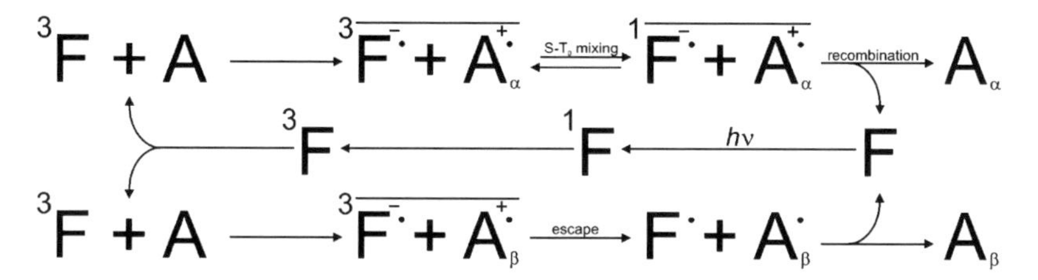

Fig. 5 Schematic of the cyclic photochemical reaction scheme responsible for the production of CIDNP in amino acids (A) and proteins – via the radical pair mechanism (RPM) – using a flavin photosensitizer (F). An electron transfer mechanism is shown; a similar set of reactions takes place if the initial step is hydrogen abstraction. The *overbar* indicates spin-correlation. Recombination and escape products are chemically identical. Thus, the net polarization depends on the amount lost due to spin–lattice relaxation in the longer lived escape radicals

ground state. The cyclic nature of the mechanism is extremely important in ensuring that, in the case of proteins, polarization is exclusively observed in the intact macromolecule rather than in a chemically modified form.

In summary, CIDNP in amino acids arises from the remarkable fact that nuclear spin modulates chemical reactivity. The effect is not very pronounced and the spin sorting process described above is not very efficient, but it does not need to be. A difference in recombination rates of only 0.01% for radicals containing protons in both spin states, i.e., "α" and "β", will lead to an order of magnitude enhancement in the NMR intensity, on the basis that the equilibrium population difference is 1 part in 10^5. More detailed, quantitative quantum mechanical descriptions of the origin of CIDNP can be found in the literature [11, 27].

2.8 *Biological Photo-CIDNP and the Protein Folding Problem*

2.8.1 Photo-CIDNP of the Amino Acids

The basic photo-CIDNP technique, as applied to amino acids and proteins, is very selective and involves laser irradiation of the sample solution in the presence of a suitable photosensitizer followed by detection of the nuclear polarization. Among the 20 naturally occurring amino acids, however, only the magnetic nuclei of the three aromatic residues tyrosine (Tyr), tryptophan (Trp), and histidine (His), and, to a certain extent, the non-aromatic residue methionine (Met) are CIDNP-active and hence polarizable (Fig. 6).[6] In the following section, the polarization pattern

[6] Given the photochemistry of the naturally occurring amino acids, it is not likely that additional polarization routes exist which could extend the observation of the photo-CIDNP effect beyond tryptophan, tyrosine, histidine, and methionine. Nonetheless, a number of amino acid derivatives are, under certain circumstances, polarizable [18]. Also, CIDNP can be induced in nucleic acids and oligophenols. However, the application of the photo-CIDNP technique to nucleic acids has received relatively little attention, compared to proteins, and is hence poorly understood. In

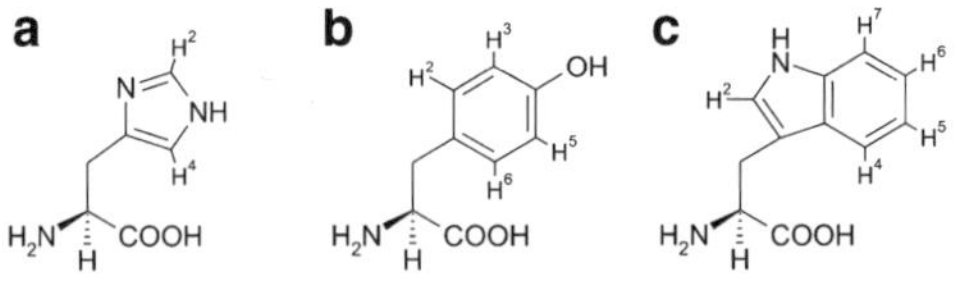

Fig. 6 Chemical structure of the three (aromatic) CIDNP-active amino acids (**a**) histidine, (**b**) tyrosine, and (**c**) tryptophan including the numbering scheme for the aromatic side chain protons

occurring in the photo-CIDNP spectra of these three side chains will be described and the reaction mechanisms in generating a radical pair with one of the most common CIDNP dyes – flavin mononucleotide (FMN) – will be discussed.

Tyrosine

The CIDNP spectrum of tyrosine exhibits a strong emissive resonance for the 3,5 protons of the aromatic ring along with weaker absorptive enhancements for the aromatic 2,6 and the aliphatic HB protons.[7] The lack of polarization for the HA proton suggests that it is too far removed from the unpaired electron's spin density – mainly located on the six-membered ring of the tyrosyl radical – to experience an appreciable hyperfine interaction.

Earlier, the reaction between tyrosine and the excited triplet state of the flavin was believed to occur via a transfer of the phenolic hydrogen atom to the FMN yielding a neutral pair comprising a tyrosyl radical with a *g*-value of 2.0041 and a flavosemiquinone radical with a *g*-value of 2.0030 [34]. Evidence for hydrogen rather than an electron abstraction mechanism was based on a number of experimental observations, most convincingly the observation of a lack of CIDNP in a tyrosine derivative in which the abstractable phenolic hydrogen had been replaced by a methyl group [18]. More recently, however, it was shown that the initial step of radical pair formation involves electron abstraction from the tyrosine to the FMN in the first place yielding a spin-correlated pair of two radical ions [35]. The electron transfer is followed by swift deprotonation of the tyrosine radical cation to give a pair comprising two neutral radical partners. The rate constant of quenching of the triplet flavin by the tyrosine is on the order of 10^9 M^{-1} s^{-1} and is independent of the pH value of the solution [35]; this, in turn, is indicative of electron transfer reactions as it happens at the diffusion controlled limit. The hyperfine coupling constants for the tyrosyl radical have been determined

particular, it is not yet clear what the conditions are for detecting polarization from the nucleotide bases in double-stranded nucleic acids [18]. In addition, photo-CIDNP spectra of linear and cyclic *ortho*-methylene-bridged oligophenols show significant polarization for the phenolic units with the largest number of *ortho* and *para* alkyl substituents [31–33].

[7] The emissive enhancement of the 3,5 protons allows an easy identification of exposed tyrosine residues in the CIDNP spectrum of a protein since neither histidine nor tryptophan produce a similar, i.e., negative, CIDNP enhancement in this region of the spectrum.

experimentally and are $a_{3,5} = -6.15$ G, $a_{2,6} = 1.5$ G, and $a_\beta = 6.7$ G. If Kaptein's sign rule for the net effect is applied to the 3,5 protons, it can be confirmed that the triplet state of the flavin molecule is involved in the formation of the radical pair:

$$\Gamma(i) = \mu \cdot \varepsilon \cdot \Delta g \cdot a_{3,5} = + \cdot + \cdot + \cdot - = -(E).$$

Histidine

Under similar experimental considerations, histidine shows absorptive enhancements for the protons at the H2 and H4 positions of the aromatic ring and an emissive resonance for the HB protons. In this case, a hydrogen abstraction mechanism is believed to occur and is supported by similar experimental evidence as for tyrosine whose radical pair formation process was initially assumed to occur via hydrogen abstraction as well [18, 35]. The polarizations observed in the CIDNP spectrum are opposite in phase to those observed for tyrosine, suggesting that Δg is smaller than zero and hence that the g-value of the neutral histidine radical is smaller than that of the flavosemiquinone radical.

Tryptophan

Tryptophan CIDNP spectra show absorptive enhancements for the 2, 4, and 6 protons of the aromatic ring and emission for the HB protons. In aqueous solutions containing only small amounts of deuterated water (D_2O), the indole NH proton is also visible and shows a weak absorptive CIDNP enhancement. In solutions having a high D_2O content, however, the indole resonance is normally not observable due to hydrogen–deuterium exchange.[8] The polarization of the tryptophan indole proton is useful as an identifier for exposed tryptophan residues in a protein as there are hardly any other signals observable in the region of the NMR spectrum where the indole proton resonates. In contrast to histidine, the reaction between triplet excited flavin and tryptophan proceeds most probably via an electron abstraction mechanism to yield a radical pair ion comprising a tryptophan cation radical and a flavosemiquinone anion with a g-value of 2.0034 [18].

The obtained polarizations are again opposite in phase to those observed for tyrosine, suggesting that $\Delta g < 0$ and hence that the g-value of the tryptophan radical cation is lower than that of the flavosemiquinone anion. The lack of polarization for both the 5 and 7 ring protons suggests negligible spin densities at these positions in the radical cation. Evidence for an electron abstraction mechanism is mainly based on the observation of CIDNP in 1-methyltryptophan and on the absorptive enhancement observed for the indole NH proton in both

[8] In NMR spectroscopy, protein sample solutions normally exhibit a D_2O content between 5% and 10% which allows the detection of both the Trp indole NH proton as well as the deuterium lock reference signal.

water and dimethylsulfoxide (DMSO) which is consistent with its position on a tryptophan radical cation formed by electron transfer, i.e., negative Δg and negative NH coupling constant [18]. Moreover, the rate constant for quenching of the triplet flavin by the tryptophan is – as in the case of tyrosine – on the order of 10^9 M^{-1} s^{-1} and hence happens at the diffusion controlled limit, thereby further corroborating the assumption of an electron transfer reaction rather than hydrogen abstraction.

Methionine

Aside from histidine, tryptophan, and tyrosine, methionine is the only commonly occurring amino acid susceptible to the photo-CIDNP technique [36, 37]. Irradiation of methionine in the presence of flavins yields a tiny absorptive enhancement for the γ-CH_2 protons consistent with a sulfur-centered radical having a g-value larger than that of the flavin radical [37]. The weakness of the effect probably owes more to a large value of Δg, due to the strong spin–orbit coupling at the sulfur atom, rather than to a slow reaction with triplet flavin. It has been suggested that the mechanism of the reaction is electron transfer. The weakness of the effect makes it unlikely to be generally useful for structural studies unless other dyes can be found whose radicals have substantially higher g-values than flavins. The only report of directly polarized methionine in a protein has been in the *lac* repressor and its headpiece [38].

2.8.2 CIDNP Cross-Polarization

^{1}H photo-CIDNP spectra for tyrosine, histidine, and tryptophan exhibit enhanced absorptive and emissive signals for nuclei that experience appreciable hyperfine interactions with the unpaired electron within their radical pair precursors: H3,5 and – to a lesser extent – H2,6 in tyrosine; H2, H4, and H6 in tryptophan; H2 and H4 in histidine; and the β-CH_2 protons in all three side chains. In addition to these *direct* enhancements, there is an *indirect* polarization route which transfers magnetization from directly polarized nuclei to other, nearby nuclei [39–41]. As the generation of CIDNP produces large, non-equilibrium spin populations on certain nuclei in the amino acids, dipolar cross-relaxation partially transfers this direct polarization to other nuclei of the same side chain. The effect has exactly the same origin as the NOE, that is to say modulation of nuclear dipole–dipole interactions by molecular motion. This so-called *cross-polarization* enables nuclei which are not directly polarized by the radical pair reaction to become either weakly enhanced absorptive or weakly emissive. Thus, the H5 and H7 protons in tryptophan may be indirectly polarized, or *cross-polarized*, by the H4 and H6 protons. Similarly, the small absorptive enhancement of the H2,6 protons in tyrosine may be considerably modified by magnetization transfer from the strongly emissive H3,5 protons. Principal cross-relaxation pathways for all three CIDNP-polarizable amino acids are described in the literature [18]. By analogy with the NOE, the extent and phase of the transferred polarization depends on the distance between the ^{1}H nuclei involved in the interaction and, importantly, on the molecular tumbling rate or rotational correlation time (τ_c) of the molecule of interest [40, 41], the latter being (roughly) defined as the average time it takes for a molecule to end up at an orientation about one radian from its starting point. Hence, for the same reasons that rapidly tumbling molecules exhibit positive proton–proton NOEs, while slowly moving molecules give negative NOEs, CIDNP transfer is more likely to occur via a double quantum relaxation process ($\Delta m_I = \pm 2$) and thus with *inversion* of phase, e.g., emission to absorption, for rapidly tumbling molecules, e.g., small molecules such as free amino acids in non-viscous solutions, but with *retention* of phase for slower tumbling molecules, e.g., proteins, as, in this case, transfer tends to occur via a zero quantum relaxation pathway ($\Delta m_I = 0$). In tyrosine itself and in small peptides, the weak absorptive polarization of H2,6 is enhanced by cross polarization from H3,5. By contrast, in a larger and less mobile molecule, the H2,6 doublet becomes emissive as a result of cross-relaxation. Similarly, the H5 and H7 protons in tryptophan become absorptively polarized in a protein unless there is considerable local mobility, when they may appear in emission. By direct analogy with the NOE, the time dependence of cross polarization may be used to assist in making resonance assignments and in studying dynamics. A simple calculation shows that indirectly polarized protons become more prominent with respect to their directly polarized partners with increasing irradiation times and for longer delays between the light and radio frequency pulses [18]. This behavior makes it straightforward to distinguish between H3,5 and H2,6 in tyrosine, and, when combined with multiplicity information, provides an unambiguous assignment method for the five tryptophan aromatic protons [18, 41–43]. Assignment of resonances is also facilitated by discovering the source of the indirect polarization. This may be done either by double resonance-selective saturation of directly polarized protons during the light pulse so as to suppress any cross-polarization [43] or, less straightforwardly, but more satisfactorily, by a CIDNP analogue of the two-dimensional NOESY experiment (Sect. 3.4.4). Turning to dynamics, the relative enhancements of the H3,5 and H2,6 tyrosine resonances in various proteins have been used to provide a qualitative measure of the degree of mobility of the tyrosine side chain [44–47]. Similarly in histidine, cross polarization from the emissively polarized β-CH_2 protons to H4 cancels some of its absorptive polarization in a protein leading to a H2/H4 ratio that is indicative of the degree of mobility [45]. This effect might also explain why the CIDNP signal of H4 is often much weaker than H2, particularly in large proteins.[9]

---

[9] CIDNP-related cross-polarization in ^{19}F-labeled amino acids and proteins leads to interesting multiplet intensity patterns arising from the interaction of various relaxation pathways such as dipole–dipole interactions and relaxation arising from the chemical shift anisotropy (CSA) of the ^{19}F nucleus [30]. In all cases, directly and indirectly polarized signals can normally be distinguished by comparing the time dependence of the polarization build-up as enhancements due to

2.8.3 Factors Affecting Polarization

The cyclic nature of the mechanism responsible for the production of photo-CIDNP in aromatic amino acid side chain nuclei also introduces a number of additional features to the RPM which are discussed here.

Recombination Cancellation

As the idealized photo-CIDNP reaction scheme for biological macromolecules outlined before is cyclic, the products of the escape and recombination channels are *chemically identical* with the only difference between them being the reciprocal overpopulation of nuclear spin states. Hence, one might be inclined to think that, since the nuclear polarization produced in the recombination and escape products is equal and opposite in phase, there would be a complete cancellation of the nuclear polarization and no enhancements of the NMR signal would be observed upon subtraction of "light" and "dark" spectra. Fortunately, however, complete cancellation does not occur because product formation via the escape route (10^{-4} s) tends to be slower than via the recombination route (10^{-7} s) since it involves the kinetically slower (second order) recombination of independent free radicals at low concentration, allowing ample time for polarization in the longer lived escaped radicals to decay to a greater extent via nuclear spin relaxation. Furthermore, nuclear spin relaxation in the escaped radicals is very efficient due to the strong magnetic moment of the electron with relaxation times of magnetic nuclei in radicals being on the order of around 10^{-4} s. This results in a substantial loss of the nuclear spin polarization in the escape products; accordingly, cancellation of the total nuclear spin polarization is incomplete. This effect is known as *recombination cancellation*. As nuclear spin relaxation in free radicals occurs predominantly via the dipolar coupling to the unpaired electron, relaxation rates are related to the electron spin density at the nucleus. If the lifetimes of escaped radicals are short, nuclei with large hyperfine coupling constants will lose less recombination polarization through cancellation than nuclei with small hyperfine couplings [41, 48].

Exchange Cancellation

A second cancellation route in cyclic reactions arises if the escaped radicals are able to undergo rapid degenerate electron exchange reactions with their diamagnetic counterparts, thereby transferring polarization to the recombination products. A spin-polarized radical "A", for example, can exchange its radical electron with

---

cross-polarization tend to increase both with the length of the laser flash used to generate CIDNP and with the delay separating the generation and detection of the polarization, i.e., the acquisition delay.

a paired electron of a recombined radical showing opposite polarization.[10] Since this escape polarization is opposite in phase to that produced via the recombination channel, the exchange will result in a decrease in the recombination product polarization. This phenomenon is known as *exchange cancellation.* Exchange effectively shortens the lifetime of escaped radicals and therefore competes efficiently with nuclear spin–lattice relaxation. Hence, the CIDNP intensity for a given proton on amino acid "A" obtained in the presence of exchange is given by

$$I_{\mathrm{A}} = P_{\mathrm{A}} \frac{R_1}{k_{\mathrm{ex}}[\mathrm{A}] + R_1},$$

where P_{A} is the polarization generated per radical pair in the absence of exchange and k_{ex} is the rate constant for electron exchange. R_1 represents the nuclear spin–lattice relaxation time in the escaped radical. The rate constant k_{ex} is typically on the order of 10^7 to 10^8 mol^{-1} dm^3 s^{-1} and hence for concentrations of "A" in the range of 10^{-3} to 10^{-2} M exchange will occur on a timescale (10^{-6} to 10^{-4} s) that competes efficiently with nuclear spin relaxation (10^{-4} s). In addition, the equivalent hydrogen atom transfer of the protonated radical is much slower and hence the CIDNP intensities for amino acids may vary considerably with the pH of the solution (Sect. 3.5.1).

As proteins represent very large macromolecular species, their translational diffusion in solution is rather slow. Hence, both recombination cancellation and exchange cancellation rates are normally very much reduced as compared to small molecules. Nuclear spin–lattice relaxation times, on the other hand, are broadly similar and, as a consequence, whilst both recombination and exchange cancellation effects play an important role in determining the CIDNP intensities of polarized nuclei of the free amino acids in solution, they play a relatively minor role within the bulkier protein molecules [18].

Competition Effects

When more than one CIDNP-active side chain is able to react with the excited flavin dye at the same time, e.g., a mixture of free histidine and tryptophan amino acids, the CIDNP intensities observed for every individual amino acid type are likely to depend in a very sensitive way on its ability to compete for the small concentration of excited triplet flavin. Furthermore, triplet flavins are quenched by fluorescence or reaction with molecular oxygen (O_2) and hence the absolute CIDNP intensity (I_{A}) of a proton in an amino acid A in a binary mixture of two different amino acids A and S in the absence of degenerate exchange is determined by a competition between these three processes:

[10] Even though this phenomenon is usually referred to in the literature as an electron "exchange" it is in fact based on an electron "hopping" process rather than on a swapping of electrons.

$$I_{A} = P_{A}\frac{k_{A}[A]}{k + k_{A}[A] + k_{S}[S]},$$

where P_{A} is the polarization produced per radical pair in the absence of any competing mechanisms; k_{A} and k_{S} are the second order rate constants for the reaction of A and S with triplet excited flavin, respectively, and k is the first order rate constant for the decay of triplet flavin by fluorescence or quenching by molecular oxygen. The second order rate constants for the reaction of the three amino acids with triplet excited flavin are on the order of 10^{8} to 10^{9} mol^{-1} dm^{3} s^{-1} [49]. Hence, when using millimolar amino acid concentrations, reactions will occur on a timescale on the order of 10^{-7} to 10^{-5} s, which competes efficiently with quenching via fluorescence or by molecular oxygen (10^{-6} s) [37].

The relative magnitudes of the second order rate constants for all three CIDNP-active amino acids have been estimated by both transient absorption and photo-CIDNP competition experiments [37, 49]. In both cases, it was found that the order of reactivity for the three amino acids is such that tryptophan reacts more favorably with triplet flavin molecules than tyrosine which, on the other hand, is much more efficient than histidine. For CIDNP studies of proteins, this means that even though a histidine residue may be sufficiently exposed to undergo the hydrogen atom abstraction required to generate the nuclear polarization, it will be less able to compete efficiently for the triplet flavin when at least partially exposed tyrosine or tryptophan residues are also present. Accordingly, the CIDNP signal intensity for this histidine residue will not reflect its true relative surface accessibility with respect to other polarizable side chains. Hence, great care must be taken in interpreting the signal intensities found in protein CIDNP spectra in particular for histidine both in the "steady state" as well as under conditions where changes in the relative accessibility and hence reactivity of amino acid residues other than the one under observation are likely to occur, e.g., in protein folding or unfolding reactions observed in a time-resolved fashion (see below).

2.8.4 Photo-CIDNP of Proteins and the Protein Folding Problem

In a protein, the CIDNP spectra of Tyr, Trp, and His residues should resemble those of the free amino acids, except that the chemical shifts will differ, and the relative intensities may be affected by relaxation and other effects. The assumption underlying the photo-CIDNP technique as applied to proteins is that only residues accessible to flavin triplets are polarizable. Hence, for a given amino acid residue to exhibit a CIDNP enhancement it has to be in contact with the surrounding solvent in order to be able to react with triplet state excited flavin molecules. Hence, when applied to proteins, the photo-CIDNP technique can be used to probe the surface exposure – or surface accessibility – of tryptophan, tyrosine, and histidine residues both in the steady state as well as in "real time" (Sect. 4.2). In the case of tyrosine and histidine, which react by hydrogen abstraction, the static or dynamic structure of the protein must be such as to allow close approach of the N5 atom of the triplet flavin and the OH or NH atom to be abstracted. The electron transfer mechanism for tryptophan is perhaps less likely to suffer from stringent constraints on the orientation of the reactants and may take place at larger dye–residue separations, possibly by a tunneling process. An additional and more questionable assumption pervades the literature on CIDNP of proteins – that an increase (or decrease) in the CIDNP intensity of a residue on changing conditions, e.g., pH, concentration, temperature, ligand, etc., directly corresponds to an increase (or decrease) in accessibility. As said before, this hypothesis needs to be approached with caution.

To understand the effects of the photo-CIDNP technique as applied to proteins more visibly, one can compare the NMR and photo-CIDNP spectra of the protein bovine α-lactalbumin (BLA), which contains four tryptophan, four tyrosine, and three histidine residues. In common with numerous other proteins, these three residue types occur frequently enough in the polypeptide chain of BLA to serve as structural probes distributed throughout the three-dimensional structure of the molecule. A direct comparison of both CIDNP as well as standard ^{1}H NMR spectroscopy shows that the CIDNP approach represents a vast simplification as compared to a standard one-dimensional NMR spectrum, since only signals arising from polarized amino acid residues are detectable. Of the four tryptophan, four tyrosine, and three histidine side chains that can be found in BLA's amino acid sequence, only Trp 118, Tyr 18, and His 68 exhibit a sufficiently high protein surface accessibility in the native state and are hence polarizable [50–52]. The ability of the photo-CIDNP technique to probe the surface of a protein in a residue-specific manner has led to its application in a wide variety of different contexts. In most cases, changes in the surface accessibilities of CIDNP-active side chains caused by the addition of some form of structure or surface perturbant, e.g., a chemical denaturing agent, are being examined in equilibrium [18]. An alternative application of the photo-CIDNP technique involves monitoring the changes in the accessibilities of the amino acid side chains that occur as a protein unfolds from its native state or as it refolds from its unfolded, i.e., denatured, state. The use of an NMR in situ mixing device designed to transfer protein quickly into a refolding buffer containing a suitable amount of a CIDNP photosensitizer inside the NMR spectrometer allows the detection of protein folding/unfolding in "real time" on timescales going as low as milliseconds. Several proteins have been studied by this technique, including, for example, hen egg-white lysozyme (HEWL) and bovine α-lactalbumin, as in these cases folding to the native state is slow enough, i.e., it proceeds on the order of seconds or minutes [53–57]. Slightly changing the above-mentioned experimental mixing setup allows an extension of the method to the observation of submillisecond folding events (see below).

2.8.5 Excursus: The Protein Folding Problem: An Introduction

The protein folding process represents a crucial step in the *central dogma* of molecular biology which states that the genetic information contained within a given sequence of deoxyribonucleic acid (DNA) is converted into a proteinaceous macromolecule exhibiting biological activity [58]. A clear and unambiguous understanding of the protein folding process in which a disordered chain of amino acids is transformed into a compact and well-structured biopolymer is hence one of the main challenges of modern structural biology. A better understanding of the complex relationship between the amino acid sequence and the formation of tertiary structure, i.e., the *protein folding problem*, will not only allow the de novo design of specific protein structures but also aid in understanding the cause and origin of a large number of neurodegenerative diseases associated with the incorrect folding of certain protein species [59, 60]. As of now, however, the protein folding problem remains largely unsolved and several attempts have been undertaken so far in order to shed more light on this area of biochemical research.

A simplistic way to find an answer to the protein folding problem would be to assume that – following its synthesis on the ribosome – all possible conformations are searched by the unstructured polypeptide chain in a random fashion to find the energetically most favorable one. Early on, however, it was found that it would take around a million years for even a small protein to fold properly if the folding process were to occur via this random search of conformational space. In the majority of cases, proteins fold on a timescale of seconds or milliseconds and hence at an early stage it became clear that a certain pathway must in some way be involved in directing the search for the native structure. In fact, the essence of protein folding is rather complicated and most approaches trying to find answers to the problem put strong emphasis on the retention of partially folded intermediate structures and transition states that are being established as folding progresses [61, 62]. Two additional considerations concerning the energetics of protein folding make the problem even more challenging to answer. First, proteins are only marginally stable. The free energy difference between the folded, i.e., native, and the unfolded state of a protein of average size, i.e., ca. 100 amino acid residues in length, is only about 10 kcal mol^{-1}. Hence, the average stability per residue is about 0.1 kcal mol^{-1} which is less than the random thermal energy kT found at ambient conditions. This also means that folding intermediates containing native substructures can be lost along the folding pathway. Second, some folding intermediates have favorable energies but are not on the path to the final folded form. For these species to contribute eventually to the ensemble of correctly folded native proteins they have to escape this energy well or "kinetic trap" by reversely folding back into a less structured conformer of higher energy. It is for these reasons no great surprise that protein folding is such an intriguing problem for both theoreticians and experimentalists.

3 Experimental Methods and Instrumentation

3.1 Light Source and Coupling Methods

3.1.1 Light Source

Even though most of the early work on photo-CIDNP employed lamps of various sorts [10–12, 18], two light sources have been used predominantly for conducting photo-CIDNP experiments of biomolecules, in particular amino acids and proteins: (1) continuous wave (CW) argon ion lasers and (2) rare gas halide excimer lasers. Little or no use has been made of other suitable lasers systems, e.g., Nd–YAG, dye lasers, etc., mainly because the use of the argon-ion/flavin combination in particular has yielded excellent results.

Operating in multi-line mode, the argon-ion laser(s) delivers blue-green light with principal wavelengths (λ) of 488 and 514 nm. The output is usually between a few up to approximately 25 W, and the narrow, highly collimated beam – a millimeter or two in diameter – is directed into the NMR probe using an optical light guide and suitable (fiber) optics. These properties make it a nearly ideal choice for conducting biological photo-CIDNP experiments. In addition, the wavelength range is well suited for protein photo-CIDNP studies as it falls within the absorption bands of the flavin-based photosensitizers. FMN, for example, has absorption maxima at 375 and 450 nm, but avoids absorption via the protein, with the exception of, for example, proteins that contain visible chromophores (see below). The light is gated into pulses of typically 50 ms to 1 s in duration using a mechanical shutter controlled by the spectrometer computer. Commonly used and commercially still available (high-power) argon-ion laser(s) are, for example, systems from the "Innova" series (Coherent Inc., Santa Clara).

An alternative to the aforementioned argon-ion light source are rare gas halide – usually either XeCl or KrF – excimer lasers operating at wavelengths of 308 and 249 nm, respectively [63, 64]. The beam, usually rectangular in shape, has a significantly larger divergence than its argon ion counterpart and is hence less facile to couple efficiently into an NMR probe. No gating is necessary, although it can be advantageous to use bursts of 10–100 pulses to increase further the photo-CIDNP enhancements. The much shorter irradiation times of excimer lasers – pulse lengths of ca. 10 ns are common – can be exploited to obtain spectra with, at best, sub-microsecond resolution, free from some of the secondary factors that affect CIDNP intensities such as cross-polarization, F-pair polarization, cancellation effects, and intramolecular electron transfer [64–68]. Nevertheless, this technique has found little application in structural studies of proteins. One potential drawback is that tryptophan and tyrosine residues absorb at wavelengths below approximately 320 nm with the possibility of direct light-induced oxidation and, in addition, flavins have low absorbance at 308 nm. Also, the beam is often wide – between 2 and 3 cm in length – and inadequately collimated, making it more difficult to get the light efficiently into the NMR probe. Typical commercial units are the COMPex (Pro) Series (Coherent Inc.) or PulseMaster 860/880 Series (LightMachinery, Ottawa).

3.1.2 Light Coupling

Essentially two different methods of guiding light into the coil region of a superconducting magnet have been developed over the years (Fig. 7) – *either* a cylindrical quartz rod *or*, alternatively, an optical fiber.

A slightly old-fashioned method of directing light into an NMR sample inside a superconducting magnet is to install a cylindrical quartz rod (diameter ~5 mm) in the NMR probe, running from its base up to the sample (Fig. 7a,b). A quartz rod inside the probe can be placed centrally so that light enters the sample vertically from below (Fig. 7b), or off-center in which case a right-angle prism or an inclined mirror is placed on top of the rod (Fig. 7a), or else the rod must be cleaved so as to direct the light in a horizontal fashion through the coil and into the sample. In either case, diverging the narrow beam by means of, for example, concave lenses or convex mirrors is useful to reduce sample heating and increase the volume of sample illuminated. Alignment of the beam can be assisted by the use of an HeNe laser or a silicon solar cell. Modifications to commercial NMR probes are in most cases needed to accommodate the light guide including prisms or coated mirrors required to illuminate the NMR tube from the side or below. If placed too close to the NMR coil these optical components may compromise the spectral resolution. In addition, a centrally placed quartz rod has several drawbacks. First, it may not be compatible with the variable temperature (VT) insert of the probe. Second, restrictions are placed on the optical density of the sample, if excited dye molecules are to be generated throughout the sensitive region of the r.f. coil. Third, homogeneity optimization of the magnetic field can be awkward if the base of the, preferably flat-bottomed, NMR tube is placed close to the coil to allow light to reach the sensitive region.

A more advantageous method is to use optical fibers to guide the light beam into the NMR sample tube, particularly when using an argon ion laser as the light source (Fig. 7d–f). Coupling of argon-ion laser(s) light into a narrow-diameter fiber is relatively straightforward with appropriate launcher optics and avoids the need for prisms and mirrors. Routinely, a 1 mm diameter optical fiber (e.g., model F-MMC, Newport Optics, Irvine) with launcher optics (e.g., Newport F-915T) is used. Using this setup, the light is introduced via the probe body, but it is simpler to bring the fiber in from above through the magnet bore. The output end of the fiber, or bundle of fibers, may either be dipped into the sample – although this might lead to problems with sample contamination, excessive local heating effects, and damage to the tip of the fiber – or installed into the probe body from below. Alternatively, the fiber is mounted in a coaxial insert – e.g., Wilmad WGS 5BL – fitted into the sample tube (Fig. 7d) [69]. The latter is an attractive option, although it precludes sample spinning. Alternatively, a quartz rod with a conic-shaped tip

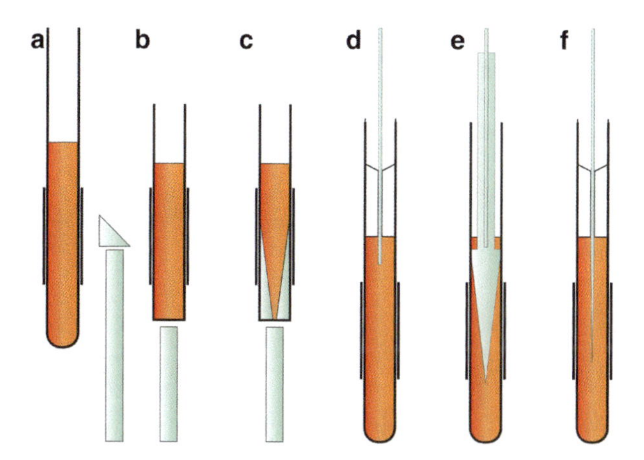

Fig. 7 Schematic drawings of NMR sample illumination methods, e.g., for photo-CIDNP experiments. (a) Illumination from the side through the receiver coil via a cylindrical quartz rod installed inside the probe body together with a prism or mirror (see text). (b) Illumination from below with a quartz light guide and a flat-bottomed NMR tube. (c) A variant of (b) using a cone-shaped NMR tube to permit more homogeneous irradiation of optically dense samples. (d) Illumination from above using an optical fiber held inside a coaxial glass insert. (e) A variant of (d) in which light is distributed by means of a pencil-shaped tip insert. (f) Another variant of (d) with a stepwise tapered optical fiber [70]

inserted into the NMR tube and reaching into the coil region can be used to give more uniform illumination of the sample (Fig. 7e; [56, 57]).[11]

Recently, a method for the illumination of optically dense NMR samples has also been presented [70]. Ideally, the entire sample solution volume should be uniformly illuminated to maximize sensitivity and to avoid concentration and temperature gradients, primarily disturbing during kinetic photo-CIDNP experiments. These problems are likely to be most severe for optically dense samples and can only be insufficiently dealt with using one of the aforementioned tip designs. To circumvent this, Hore et al. propose a straightforward and inexpensive method of illumination from above in which the laser light is distributed along the axis of the NMR tube by means of a tapered optical fiber (Fig. 7f). While illumination from above the coil (Fig. 7d) gives rise to an exponential fall in light intensity from the top of the sensitive r.f. receiver coil region to the bottom, the tapered tip – optical path length 20 mm in the case of the former vs ca. 3 mm in the case of the latter – gives almost uniform illumination. In addition, the approach requires

[11] Utilizing this tip, a coupling of two lasers operating at different wavelengths – one to release calcium ions from the photolabile ion chelator DM-Nitrophen and the other to induce CIDNP – has been designed to observe the RT refolding of calcium-depleted BLA thereby making physical mixing of solutions superfluous. The potential gain in acquiring spectra right after the start of the reaction is, however, compromised by long irradiation times (~200 ms) required to release sufficient metal ions to fold the protein and difficulties in uniformly photolysing the entire sample volume associated with low SNRs (see below).

no modification of the probe, leads to less than a 5% loss of filling factor and, as a consequence, produces minimal degradation of spectral resolution. Moreover, discontinuities in magnetic susceptibility produced by introducing the fiber are almost only perpendicular to the axis of the tube.

3.2 Photosensitizers

3.2.1 Flavins and Other Dyes

The nuclear polarization generated during a CIDNP experiment is due to a photochemical reaction between an electron or hydrogen donor molecule – in this case the CIDNP-active amino acid side chain – and a photoexcited electron or hydrogen acceptor dye molecule, i.e., the CIDNP photosensitizer. The most widely employed photoactive CIDNP dyes are flavins, e.g., lumiflavin, riboflavin, FMN, and FAD. Their general structure, based on the tricyclic isoalloxazine core, is shown in Fig. 8. The chemical, biochemical, and photochemical properties of flavin compounds are well characterized and understood. They occur widely in nature due to their ability to undergo facile, reversible one- and two electron redox reactions. The three common redox states are the (oxidized) flavin itself (F), the flavosemiquinone radical, the flavohydroquinone (FH_2), and their deprotonated forms. Flavins are commonly found as co-enzymes, often non-covalently bound to proteins, and are responsible for the electron transfer properties of flavoproteins, e.g., hydrogenases, oxidases, and mono-oxygenases. Flavins also act as co-factors in cryptochromes, a class of blue light-sensitive flavoproteins involved in animal and plant magnetoception [71] as well as the stabilization of their circadian rhythm [72].

The UV–Vis absorption spectrum of flavins is characterized by broad absorption bands occurring in the near UV and the blue and green regions of the electromagnetic spectrum at wavelengths (λ) of about 375 and 450 nm, respectively. While absorption at a wavelength $\lambda = 514$ nm is rather weak, the extinction coefficient ε is greater than 10^4 dm^3 mol^{-1} cm^{-1} at $\lambda = 488$ nm. Hence, excitation into this band leads to efficient and rapid formation of the triplet state (quantum yield ~0.5) when using argon-ion laser(s) light [73, 74]. All commonly employed flavin-derived dyes (Table 1) are water soluble and induce CIDNP in tryptophan (Trp), tyrosine (Tyr), histidine (His), and/or methionine (Met) side chains via electron or hydrogen abstraction to yield a pair of radicals in a cyclic photochemical reaction, giving rise to the observed nuclear polarization. The relative CIDNP efficiencies, however, do slightly differ depending sensitively on the charge and size of the substituents R_1 and R_2. For selective amino acid polarization, e.g., Trp and/or Tyr polarization only, xanthenes (eosin Y and fluorescein), and thiazines (thionine) can be used as alternatives to flavins. Both xanthenes and thiazines have a high solubility in water and produce strong CIDNP over a wide pH range (see below). A significant drawback of all CIDNP dies mentioned so far is that they result in photodegradation of amino acids and proteins, limiting the number of transients that can be recorded

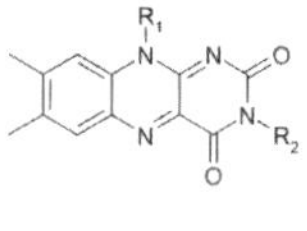

Fig. 8 Structural representation of flavin derivatives frequently used as photosensitizers in photo-CIDNP experiments of amino acids and proteins. The different chemical composition of flavin derivatives is characterized by the substituents "R_1" and "R_2" as shown in Table 1

Table 1 Chemical composition of a number of flavin derivatives used as photosensitizers in CIDNP experiments

Name	R_1	R_2
Lumiflavin	CH_3	H
3-*N*-methyllumiflavin	CH_3	CH_3
3-*N*-carboxymethyllumiflavin (Flavin 1)	CH_3	$(CH_2CO_2)^-$
3-*N*-ethylaminolumiflavin	CH_3	$(CH_2CH_2NH_3)^+$
10-*N*-carboxyethyllumiflavin	$(CH_2CH_2CO_2)^-$	CH_3
Riboflavin	$CH_2(CHOH)_3CH_2OH$	H
Flavin mononucleotide (FMN)	$[CH_2(CHOH)_3CH_2OPO_3]^{2-}$	H

for each sample. The concentration of photosensitizer used in each photo-CIDNP experiment represents a compromise between a sufficiently low optical density of the sample solution to allow its uniform illumination and, on the other hand, a high yield of triplet excited photosensitizer molecules present upon photoinduction. Flavin photosensitizers are typically used in a concentration of ca. 0.2 mM (±0.1 mM).

In addition, various dyes have been investigated as alternatives to flavins [18]. Xanthenes (fluorescein, rose bengal, or eosin), for example, polarize tyrosine but not tryptophan or histidine when irradiated with, for example, green light [75–78]. Methylene blue and several porphins also produce CIDNP in tyrosine. In addition, *p*-methoxyacetophenone, irradiated at ~250 nm, gives strong CIDNP from Tyr and Trp but little enhancement of His. Quinoxalene, excited at the same wavelength, gives strong signals for tyrosine and histidine but not tryptophan [79]. Finally, Broadhurst et al. tested a number of quinones, aromatic ketones, aza-aromatics, and a variety of dyes and sensitizers [80]. The largest enhancements were achieved with the aza-aromatic compounds 2,2′-dipyridyl (DP, 2,2′-bipyridine) and, to a lesser extent, 2,2′-bipyrazine when irradiated with UV light.[12] DP is very often employed in microsecond time-resolved CIDNP experiments using, for example, xenon-chloride excimer lasers, to investigate cancellation effects caused by recombination or degenerate exchange (e.g., [64]). Studies of the reactions of the triplet

[12] With the possible exception of the final pair, most of these sensitizers do not look very promising for biological photo-CIDNP studies as they suffer from extraneous product formation and undergo non-cyclic reactions with amino acids and proteins.

state of DP with Trp, Tyr, and His show that the rate constants and, hence, the CIDNP strongly depend on the protonation state of both DP and the amino acid side chain. Although Trp and Tyr residues are strongly enhanced, the polarization of His is weaker and varies strongly across the pH range.

3.2.2 Excursus: Photodegradation

During photo-CIDNP reactions, the flavin molecule is prone to photoreduction, leading to a substantial degradation or *bleaching* of the dye solution as the photoreaction underlying the CIDNP mechanism is not perfectly cyclic. This, in turn, can cause problems during continuous or multiple laser flash illumination – e.g., during "real time" photo-CIDNP experiments or when recording a 2D spectrum – causing a gradual loss of polarization as the concentration of oxidized flavin is depleted. For example, during the investigation of the reaction of lumiflavin with the amino acid tryptophan it was shown that the gradual loss of polarization is due to both photodegradation of the flavin molecule and degradation of the amino acid itself [81, 82]. Bleaching of the solution seems to be the more serious problem in this case because the CIDNP intensities depend sensitively on the optical density of the sample. In addition, the concentration of the flavin is an order of magnitude smaller than that of polarizable amino acid side chains in a protein. The most probable cause of this photochemically induced reduction is the disproportionation of flavosemiquinone radicals as indicated by microsecond time-resolved CIDNP studies [82, 83]. Flavin photochemistry, however, is inherently complex and hence additional factors might contribute to the degradation of the dye molecule. In particular, a number of intramolecular and intermolecular light-induced reduction, addition, and dealkylation mechanisms have been proposed, the predominant reaction pathway chosen depending on factors such as the type of solvent, pH, buffer composition, and other parameters [73]. Furthermore, primarily formed products may themselves undergo a variety of secondary photolytic or thermal reactions leading to a highly complex sample mixture. Hence, by focusing on the primary intermolecular reduction of flavins, it is apparent that only an idealized picture of flavin photochemistry can be presented.[13]

Several attempts have been made to combat the problem. (1) As bleaching occurs predominantly in the irradiated portion of the NMR tube, (manual) sample mixing between acquisitions can be used to replenish the flavin [85, 86]. (2) Manipulation of the molecular oxygen (O_2) concentration is, potentially, also helpful: O_2 efficiently re-oxidizes FH_2 to F but is consumed in the process. Hence, removal of O_2 by degassing accelerates photobleaching. A high

[13] Prolonged irradiation of a protein–flavin solution can also result in photo-damage to the protein presumably caused by irreversible oxidation of side chains by excited flavin molecules. For example, extensive photolysis of hen egg-white lysozyme in the presence of FMN can, in some cases, cause a significant reduction in its thermal denaturation temperature [80, 81, 84].

concentration of molecular oxygen, on the other hand, strongly attenuates the initial polarization due to, amongst other things, quenching of photoexcited triplet flavin [81]. (3) Another way to try to overcome the problem of photodegradation has been to spin the NMR tube rapidly between flashes so as to create a vortex in the solution [85, 86]. This reintroduces oxygen and brings fresh flavin into the irradiated region. Optical fiber illumination, however, is precluded using this approach. (4) More recently, further ways to increase the lifetime of the flavin dye have been explored and include the use of newly developed mechanical mixing devices as well as alternative chemical oxidation methods such as the use of hydrogen peroxide (H_2O_2) as an oxidizing agent [87, 88].[14] For example, a PTFE transfer line (see above) fed through the coaxial insert can be used to withdraw and then inject a portion of the solution in the NMR tube between signal acquisitions. If the solution is reintroduced sufficiently rapidly, the ensuing turbulence causes efficient mixing as well as reintroducing oxygen [88]. A more straightforward method is the addition of an oxidizing agent, e.g., hydrogen peroxide (H_2O_2), which efficiently reconverts flavosemiquinone to its fully oxidized form. Although H_2O_2 is known to act as an oxidizing agent of thioether, e.g., methionine, and thiol groups, e.g., cysteine, [89], no apparent chemical modification of the protein HEWL, containing two solvent inaccessible methionine residues, was detected using standard parameters, i.e., 298 K, pH 7, 10 mM H_2O_2 [88]. Very recently, Lee and Cavagnero introduced a novel method that decreases the extent of photodegradation and enhances photo-CIDNP in experiments requiring long-term data collection. This method enables efficient regeneration of FMN while minimizing irreversible photodegradation. Since molecular oxygen is known to be involved in many photodegradation pathways, O_2 is depleted in a first step in the NMR sample by introducing glucose and the enzymes glucose oxidase and catalase. This enzyme system is widely used to minimize photodegradation in fluorescence microscopy. The consequent decrease in photo-CIDNP – given that FMN cannot be efficiently regenerated from FH_2 (see above) – is counteracted by the additional introduction of an F_2-oxidizing enzyme. The resulting tri-enzyme system improves photo-CIDNP performance significantly while reducing the extent of irreversible sample photodegradation [90].

3.3 NMR Inserts and Sample Tubes

3.3.1 The "Steady-State" Photo-CIDNP Insert

A straightforward fashion to generate photo-CIDNP in a biological sample in situ is via one-dimensional "steady-state" CIDNP NMR spectroscopy (see below). For this purpose, the state-of-the-art means to couple (laser) light into the NMR

[14] These studies also provide a more thorough analysis of the origin of the photodegradation process of flavin photosensitizers observed during many CIDNP experiments.

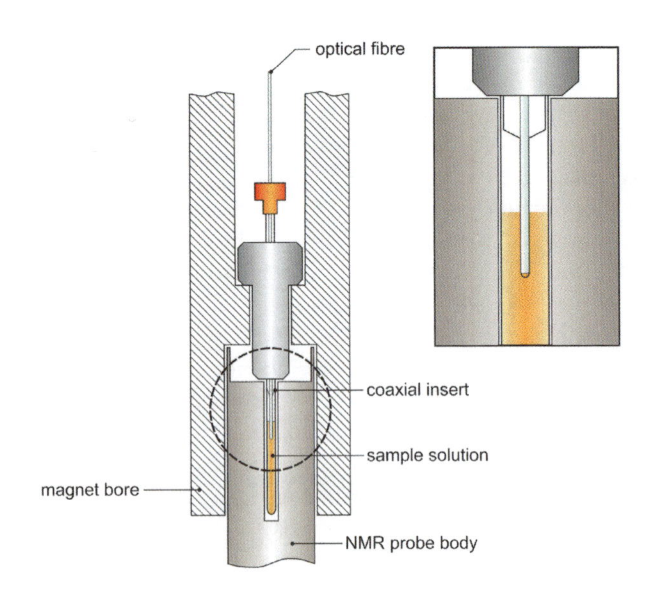

Fig. 9 Optical insert and experimental setup for conducting "steady-state" photo-CIDNP experiments of amino acids and proteins (see below). The tip of the optical fiber is placed ca. 1 mm above the upper r.f. coil threshold to minimize magnetic field (B_0) inhomogeneities in the relevant sample region of the r.f. receiver

sample region using fiber optics is presently achieved using a coaxial glass insert described above (Fig. 7d), originally designed for external chemical shift referencing. Usually, the top of the insert is held in place by a standard NMR tube cap with a small hole drilled in the center. This method of illumination was first described by Scheffler et al. [69]. The tip of the insert is placed roughly 1 mm above the top of the r.f. coil to minimize field inhomogeneities while, at the same time, maintaining reasonable CIDNP intensities [87]. The arrangement of the sample tube and insert is shown in Fig. 9. As mentioned before, the use of this setup precludes sample spinning due to the presence of the optical fiber. This, however, is not a problem with modern NMR spectrometer software/hardware as efficient automated shimming procedures are routinely available.

3.3.2 The (Stopped-Flow) "Real Time" Mixing Device

A variety of different experimental procedures designed to conduct NMR experiments in "real time" have been described in the literature [55, 88, 91–94]. While the devices used in these experiments are capable of mixing solutions inside the NMR tube with high efficiency, problems arise with most of their designs which very often cause long experimental dead times. Furthermore, it is difficult to obtain good line shapes due to decreased magnetic field homogeneities

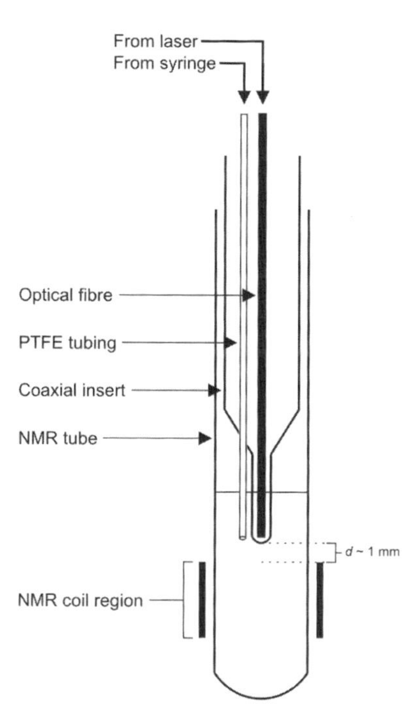

Fig. 10 Schematic drawing of the mixing device designed by Maeda et al. [88]. The positioning of the optical fiber, coaxial insert, and PTFE transfer line inside a 5 mm NMR tube is shown. Sample illumination is based on a method first described by Scheffler [69]. The PTFE line can be used, for example, to transfer rapidly in situ a small volume of a solution of unfolded protein into a refolding buffer. Benchmark experiments to assess the rapidity and uniformity of mixing have been performed on an improved version of this arrangement (Sect. 3.3.3; [55])

occurring as the tip of the mixing device which usually reside within the r.f. coil region of the probe. In addition, most of these systems are constructed from a modified optical stopped-flow apparatus and require modification of the NMR probe in order to incorporate the syringe [92, 93].

Roughly 10 years ago, a rapid-mixing device was designed and tested specifically for conducting "real-time" photo-CIDNP protein folding experiments and, in addition, photo-CIDNP "pulse-labeling" studies [55]. The system, based on an earlier photo-CIDNP stopped-flow design presented by Hore et al. [54] and an improved version thereof by Maeda et al. [88] (Fig. 10), consists of two main components: a coaxial glass insert fitted inside a 5-mm Shigemi NMR sample tube and a pneumatic triggering unit placed outside the magnet. The coaxial insert consists of glass capillary tubing with one end fused to a short length of a glass micropipette with a narrow internal diameter. The liquid to be transferred is held inside the capillary and is ejected as a collimated jet into the NMR tube. The tip of the pipette is positioned right above the top of the NMR coil, so as to cause minimal disturbance to the field homogeneity, and just below the surface of the buffer solution in the NMR tube. The top of the glass capillary is coupled to a thin PTFE transfer line which feeds injectant solution into the glass capillary. An optical fiber is inserted coaxially into the glass capillary, inside the solution, with its lower end about 2.8 cm above the micropipette piece. This arrangement makes two types of experiments possible: either (1) a solution of a protein inside the capillary tube can be irradiated prior to injection as desired during photo-CIDNP pulse-labeling experiments, or (2) sample solution inside the NMR tube can be irradiated through the micropipette after the injection. The PTFE transfer line is connected to a glass syringe outside the magnet whose action is governed by a pneumatic piston, controlled using a TTL line from the spectrometer, and driven by nitrogen gas at a pressure of 10 bar. The minimum electronic gate time needed in the pulse sequence to depress the piston and to inject 50 mL of solution is 10 ms. Extensive benchmark testing revealed that complete mixing of reactants occurs within 50 ms, and real-time protein folding experiments can be performed in H_2O, instead of D_2O, by incorporating in vivo NMR water suppression techniques in the pulse sequence [55, 95].

3.3.3 The Rapid Mixing Device

To transfer protein solutions more swiftly into the NMR tube inside the probe of the NMR spectrometer, a specific rapid mixing device – based on both the initial design presented by Maeda et al. [88] as well as the further modifications introduced by Hore et al. [55] (Sect. 3.3.2) and designed to improve both the efficiency of mixing and to reduce experimental dead times – was introduced by Mok et al. around 5 years ago (Fig. 11; [96]); it comprises two main components: a coaxial tube insert fitted inside the susceptibility-matched NMR sample tube (BMS-3; Shigemi Inc., Allison Park) and a pneumatic trigger placed outside the magnet. To the end of the coaxial tube insert is attached a narrow glass microliter pipette (20 mm in length; internal diameter 0.17 mm) using epoxy adhesive glue. The glass capillary tip allows for a highly collimated jet of sample to be injected into the NMR tube, thereby promoting rapid and homogeneous mixing of two (or more) solutions [34, 55]. Three PTFE collars, each 5 mm in length, are situated along the tube insert to ensure the coaxial orientation within the NMR tube. The insert is screw-fitted at the top of the NMR tube by a Y-connector made from PEEK polymer which is manufactured to fit the width of the NMR bore. An optical fiber (F-MSC; Newport Corporation, Irvine) is threaded through the Y-connector and is dipped into the injectant solution enclosed by the coaxial insert. The end of the optical fiber is positioned within the insert to allow a homogeneous irradiation of the specified volume of the solution to be injected which normally does not exceed 30 μL. For experiments that do not require sample illumination, it is possible to

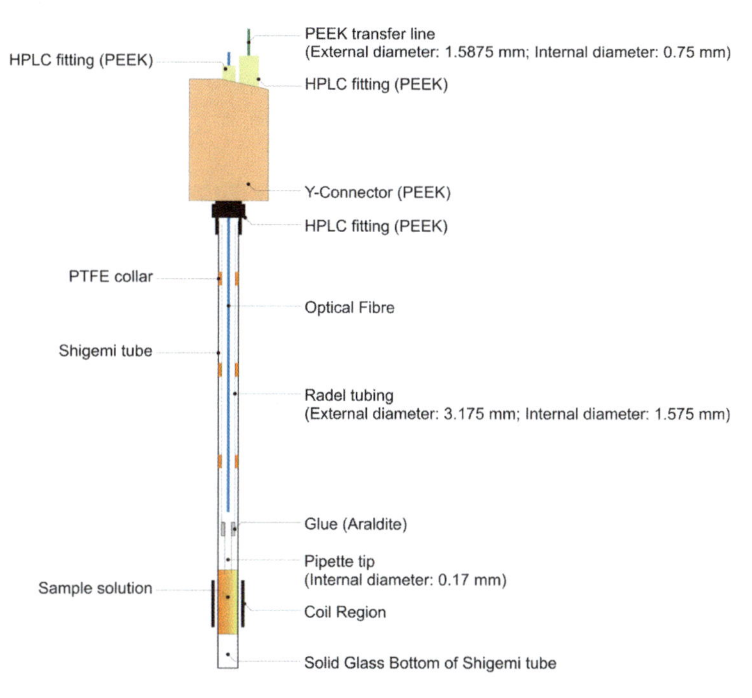

Fig. 11 Detailed schematic of the rapid mixing device used for photo-CIDNP (NOE) pulse-labeling experiments on, for example, urea-denatured proteins as described in Sect. 4 of this review

replace the optical fiber with an additional liquid transfer line, thereby permitting ternary mixing. For photo-CIDNP "real-time" as well as (NOE) pulse-labeling experiments, a small air bubble is introduced into the end of the glass capillary tip to avoid premature mixing of protein solution and buffer prior to injection. The presence of this bubble has no detectable effect on either the mixing efficiency or the signal intensity [55]. The glass capillary tip is positioned 1 mm above the top of the receiver coil to cause minimal disturbance to field homogeneities and right below the surface of the refolding buffer solution in the NMR tube (Fig. 12).

At the top of the Y-connector block, a liquid transfer line (PTFE; internal diameter: 0.75 mm), connected with an HPLC-type fitting, feeds the injectant solution into the coaxial insert. The other end of the liquid transfer tubing is connected to the glass syringe (LL-GT type; SGE International Pty. Ltd., Melbourne) via a Luer adapter, which is mounted on the pneumatic injector. The glass syringe (Model: SGE 500 R GT) mentioned above is mounted in a PTFE block in order to absorb any mechanical shock caused by the injection event. The action of the syringe is controlled by a pneumatic piston driven using nitrogen gas at 10 bar and triggered by a TTL spare control line from the NMR spectrometer. The syringe and piston assemblies are housed on a PTFE plate outside the bore of the magnet.

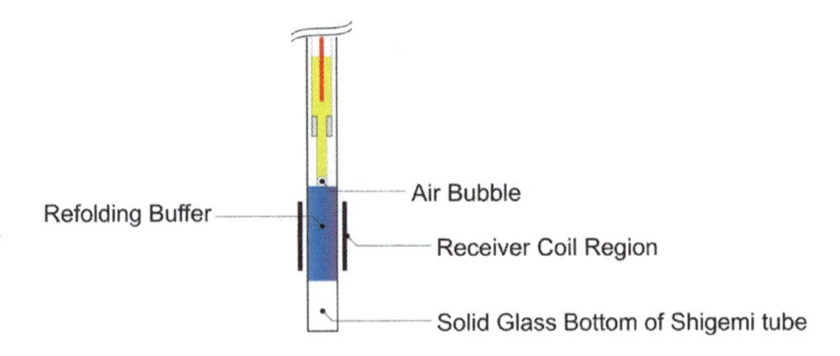

Fig. 12 Enlarged r.f. coil region of the photo-CIDNP (NOE) pulse-labeling setup prior to sample illumination. Denatured protein solution (*yellow*) and refolding buffer (*blue*) are separated by a small air bubble (see text). The optical fiber, entering the solution from above, is shown in *red*

3.3.4 Sample Tubes

Standard high-quality NMR sample tubes, usually 5 mm in diameter, are routinely used for conducting photo-CIDNP "steady-state" experiments. If fiber optics and coaxial inserts are employed, the end of the insert should be positioned close to the top of the r.f. coil region. The exact insert-tip positioning, ideally ~1 mm above the upper coil region limit, constitutes a compromise between the disturbance to the field homogeneity on the one hand, and the absorption of light by dye solution outside the NMR detection region on the other (Fig. 10). Optionally, a hole made in the shoulder of the insert can be used for the injection of reagent solutions into the NMR sample.

For CIDNP "real time" and/or CIDNP (NOE) "pulse-labeling" experiments, susceptibility-matched NMR tubes, i.e., so-called Shigemi tubes (Shigemi Inc., Allison Park, PA) – 5 mm outside diameter; magnetic susceptibility matched to D_2O – are used routinely for two reasons. First, a smaller sample volume is required when using Shigemi tubes, which is advantageous when using precious and/or isotopically labeled samples. Second, when the jet of injected solution strikes the flat bottom of the tube, greater turbulence is produced than with a round-bottomed tube, thereby improving the mixing efficiency of the (two) solutions even further [55].

3.4 CIDNP Pulse Sequences

3.4.1 One-Dimensional CIDNP

The basic photo-CIDNP experiment is relatively easy to realize as it only requires a short period of illumination to be implemented into a standard pulse-acquire NMR sequence at an appropriate position, i.e., prior to the application of the read pulse and the acquisition of the NMR spectrum. The easiest method to observe the laser light-induced photo-CIDNP effect is by difference spectroscopy [97]. As shown in Fig. 13, this involves acquiring a free induction decay (FID) with ("light") and another one without ("dark") a prior laser flash – typical irradiation times (T_L) for

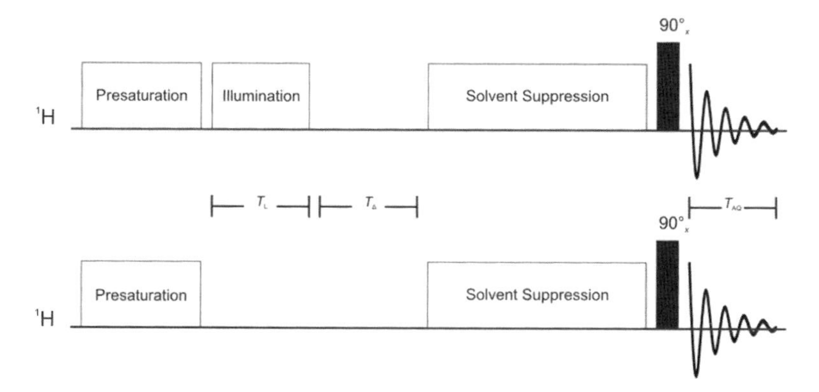

Fig. 13 The basic pulse sequence for photo-CIDNP difference spectroscopy. The "light" (*upper part*) and "dark" (*lower part*) halves of the sequence are identical but for the illumination period. The delay T_D, not shown here, is defined as the time between acquisition of "light" and "dark" spectra. Presaturation is optional, as discussed in the text

an argon–ion laser are 0.1–1.0 s – and then subtracting one from the other, resulting in a difference spectrum which exclusively exhibits resonances representing non-equilibrium spin state distributions stemming from the CIDNP effect.

Difference spectroscopy highlights the features that change between the "light" and "dark" spectra but does not provide new information and hence the signal-to-noise ratio (SNR) found in the two spectra is degraded by a factor of $\sqrt{2}$ in the difference spectrum. However, this SNR loss is, in most cases, negligible given the additional signal enhancement due to the CIDNP effect. Moreover, it can be reduced via the acquisition of more "dark" than "light" transients followed by an appropriate scaling of the "dark" spectrum prior to subtraction [85, 86]. A short delay ($T_\Delta \sim 5$ ms) is usually arranged between the light and the r.f. pulse to allow time for diffusing (free) radicals to recombine and hence avoid paramagnetic broadening of signals [75]. Also, sample heating happening during the acquisition of the "light" spectrum can lead to small chemical shift deviations (CSDs) and, as a result, subtraction artifacts in the difference spectrum. Hence, it is advisable to include a sufficiently long delay (T_D) of approximately 10–20 s between the acquisition of the two spectra, i.e., "dark" and "light", for temperature equilibration. Alternatively, one might make use of CIDNP-specific presaturation techniques outlined below.

The pulse sequence for a photo-CIDNP "real time" experiment is similar to that shown in Fig. 13, except for the inclusion of the initial injection (T_I) and the repetition of the r.f./light pulse sequence to allow spectra to be recorded at intervals after initiation of a reaction, e.g., protein folding (Fig. 14). The repetition time of these measurements is controlled by T_Δ, T_L, and the FID acquisition time (T_{AQ}). For the fastest measurements, T_Δ can be reduced to zero and T_L can be a few tens of milliseconds for an argon–ion laser or several nanoseconds for an excimer laser. The repetition time is then determined principally by T_{AQ}, i.e., the desired

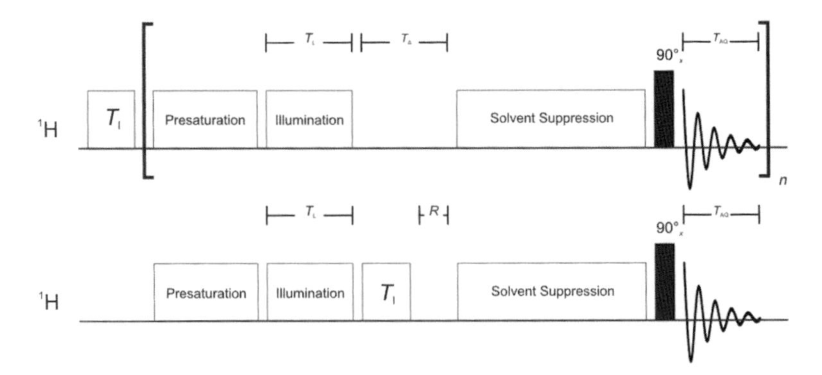

Fig. 14 Pulse sequence schemes for "non-static" photo-CIDNP experiments: *upper part* – pulse sequence for the "real time" observation of kinetic events using photo-CIDNP spectroscopy. "T_I" denotes the injection event, i.e., the voltage pulse used to trigger the mixing device, which typically takes ~20–100 ms. The delay between the injection event and the first 90° pulse can be adjusted to alter the temporal resolution of the experiment (see text); *lower part* – pulse sequence for photo-CIDNP (NOE) pulse-labeling experiments. The sample is irradiated with a laser pulse to generate nuclear spin-polarization within the sample, followed by a rapid change in the solvent conditions by in situ transfer of solutions. "R" denotes the time allowed for refolding which lies usually between 100 and 300 ms or below (cf. Sect. 4)

spectral resolution. For example, up to five spectra per second were acquired in real-time CIDNP experiments on α-lactalbumin or ca. 20 spectra per second during a "real time" CIDNP experiment of a paramagnetic protein whose short relaxation times and large chemical shift range allowed a rather short value of T_{AQ} to be used [98].

A related experiment, photo-CIDNP (NOE) pulse-labeling, involves polarizing an unfolded or partially disordered protein using a laser flash and then transferring the magnetization to the native state of the protein via fast refolding (Fig. 14; Sect. 4). This experiment gives information on side chain accessibilities of the unfolded or partially folded states of proteins whose NMR spectra are characterized by poor chemical shift dispersion and broad lines and has been applied, for example, to the acidic molten globule state of bovine α-lactalbumin [55, 99].

3.4.2 Excursus: Presaturation Methods

The generation of CIDNP results in a nuclear spin system exhibiting non-Boltzmann spin state distributions. This so-called *nuclear polarization* is detected in the NMR spectrum either as signals of enhanced absorption, opposite phase, or both. To observe these polarization signals more clearly and to provide an alternative method avoiding the problems of subtraction artifacts associated with the difference spectroscopy technique described above, it is advisable to presaturate the whole spectrum prior to the light flash [97]. The use of such a presaturation technique minimizes the signals observed from any unpolarized nuclei. The two possible ways of presaturating the whole spin system prior to the laser flash comprise either a $\pi/2$-pulse followed by a short "crush" gradient (G) pulse to dephase the coherent magnetization or, alternatively, a train of r.f. $\pi/2$-pulses of random phase with an interpulse delay determined by a converging geometric series. The former sequence is complete in approximately 10 ms as opposed to 1 s for the latter and is hence particularly attractive when a short delay between two subsequent spectra is required. Both methods effectively suppress background signals by several orders of magnitude and hence for the technique to be successful the length of the laser flash must be small compared to the respective spin–lattice relaxation times. Presaturation can replace difference spectroscopy when T_L and T_D are small compared to the shortest T_1 of the sample. The only disadvantage of the presaturation method becomes obvious when working under conditions where nuclear spins are able to return very quickly to thermal equilibrium. Since proteins are large macromolecules they tumble very slowly in solution and hence their rotational correlation times are rather long. An implication of this is that nuclear spin relaxation becomes more efficient the slower a molecule tumbles and hence, given that the length of a laser pulse applied during a CIDNP experiment is long compared to the typical relaxation time of a small protein, significant relaxation can be expected to occur during the laser pulse. This means that the effect of the presaturation applied to the spins becomes less pronounced. Since this effect can be counteracted by using difference spectroscopy methods and also because subtraction artifacts in difference spectra can be reduced by presaturation, the two techniques are commonly combined under conditions of fast spin–lattice relaxation, in particular when argon-ion laser(s) are used as a light source.

3.4.3 Excursus: Solvent Suppression Methods

Solvent suppression is an additional concern especially for samples in H_2O. Due to the high proton concentration of pure water (~10^2 M) and the comparably low amino acid or protein concentrations used in NMR experiments (~10^{-3} M), a relatively large solvent signal is usually detected which can cause problems with the dynamic range of the spectrometer's analogue-to-digital (ADC) converter. This, in turn, makes solute signals difficult to detect. To minimize these problems caused by large solvent signals, pure D_2O solutions can be used. In many cases, however, protein sample solutions containing a 10 to 1 mixture of H_2O and D_2O are more practical to observe solvent-exchangeable ^{1}H nuclei, e.g., backbone amide proton signals or the indole NH proton of tryptophan residues. Hence, solvent suppression steps are included in all pulse sequences [100]. In principle, it is possible to remove the solvent signal during subsequent data processing by deconvolution. Nonetheless, in most cases it is desirable to suppress the solvent signal prior to the acquisition of the FID. For this purpose, a double pulsed field gradient spin echo (DPFGSE) sequence is routinely incorporated into the CIDNP "steady state" experiment [101]. This sequence, which is more efficient than related pulse schemes, leads to selective dephasing of the water signal and provides uniform excitation over a wide range of offsets. The sequence is applied immediately after the final $\pi/2$-pulse which generates the detected transverse magnetization. As shown below, the DPFGSE sequence begins with a strong field gradient that phase encodes transverse magnetization. A selective π-pulse is then applied to the solvent signal, followed by a hard π-pulse applied to the whole system. Subsequently, the strong field gradient is applied again leading to a further dephasing of the solvent signal. This unit is repeated a second time using a different pair of gradients, taking care not to refocus the effects of the first one.

For photo-CIDNP "real time" and photo-CIDNP (NOE) pulse-labeling experiments a different solvent suppression method is used, since modern solvent suppression techniques that employ "excitation sculpting" such as the DPFGSE sequence described above or WATERGATE (water suppression through gradient tailored excitation) [102, 103] are not readily compatible with rapid injection experiments. Both techniques rely on a gradient echo to dephase selectively the solvent magnetization leaving the desired signals intact. Even 1 s after an injection event occurs, however, there is still enough residual bulk motion of the solution in the sample tube to prevent complete rephasing of the solute magnetization which, in turn, can result in a significant loss of NMR signal. These problems are similar to those encountered during in vivo NMR experiments, where motion is also present while acquiring data. In these cases, the problem has been successfully circumvented using methods such as WET, VAPOR, DRYSTEAM, or CHESS. Of these sequences, the CHESS (chemical shift selective excitation) experiment (or variations of it) is the easiest to implement [95]. The basic CHESS procedure consists of a single frequency-selective $\pi/2$-excitation pulse on the solvent resonance followed by a dephasing (homogeneity spoiling) gradient repeated three times using orthogonal gradients prior to the acquisition pulse. The procedure leaves the spin system in a state where no net magnetization of the unwanted component is retained while the desired component remains entirely unaffected in the form of z-magnetization. In the case of CIDNP "real time" or "pulse-labeling" experiments, CHESS solvent suppression allows the rapid transfer of a denatured protein into a refolding buffer during the experiments in protonated water, so that solvent-exchangeable protons can be observed (see below).

3.4.4 Two-Dimensional CIDNP

Photo-CIDNP spectra of proteins are far less crowded than the corresponding NMR spectra. For example, the aromatic region of the native state of HEWL contains ca. 45 resonances, only 6 of which have significant direct CIDNP polarization. Nevertheless, the assignment process is often less than trivial if, for example, more than one CIDNP-active residue of each kind, i.e., His, Trp, or Tyr, is appreciably enhanced as this often leads to overlapping of polarization signals in the aromatic region of the spectrum [43, 104]. As in conventional NMR, insufficient spectral dispersion and, as a consequence, assignment ambiguities can be alleviated by spreading resonances into a second frequency dimension. For this purpose, CIDNP versions of the basic two-dimensional NMR experiments, COSY

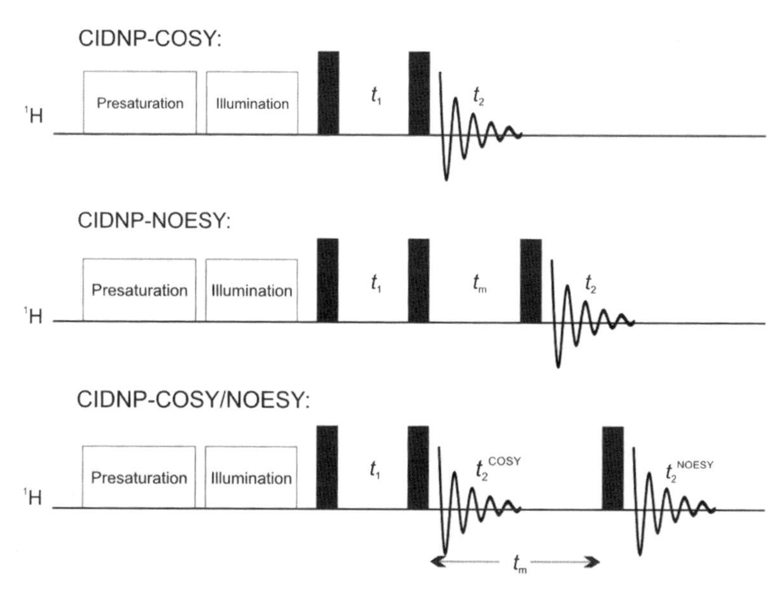

Fig. 15 Pulse sequence schematic for two-dimensional photo-CIDNP spectroscopy. Shown are schemes for CIDNP-COSY, CIDNP-NOESY, and CIDNP-COSY/NOESY experiments, where "t_m" is the NOESY mixing time. The "dark" half of each sequence is identical to the "light" part but for the absence of the laser flash and, possibly, a different number of transients (see text)

[105] and NOESY [106], were developed by Kaptein et al. [50, 85, 86, 107]. The regular two- and three-pulse sequences are preceded by a train of transmitter pulses, to destroy all z-magnetization, and a light pulse, as depicted in Fig. 15. Unfortunately, CIDNP-COSY and CIDNP-NOESY are rather less straightforward experiments than their conventional analogues because of dye exhaustion (Sect. 3.2.2). As discussed before, flavins are inactivated by various side-reactions with the result that only a limited number of transients can be recorded from a given sample, placing severe restrictions on the implementation of two-dimensional CIDNP experiments. For this reason, only one transient per value of t_1 is used – thereby making phase cycling and quadrature detection in ω_1 unnecessary – and 64 or 128 increments in t_1. To achieve acceptable digital resolution in ω_1, the transmitter frequency is placed at the low field end of the spectrum with a spectral width large enough to entail the entire aromatic region (ca. ± 2 ppm). The remainder of the spectrum, to high field, is consequently folded in ω_1. "Light" and "dark" spectra are recorded separately, and subtracted, to remove non-enhanced signals arising from z-magnetization that recovers during the light pulse as well as axial peaks resulting from relaxation during t_1. The SNR penalty associated with difference spectroscopy may be avoided by accumulating more dark spectra than light ones with appropriate scaling prior to calculation of the difference. A simultaneous photo-CIDNP COSY/NOESY pulse sequence has also been designed (Fig. 15) and is particularly attractive when protein sample is sparse [108]. In addition, CIDNP-EXSY (exchange spectroscopy) experiments can be conducted for (protein) species

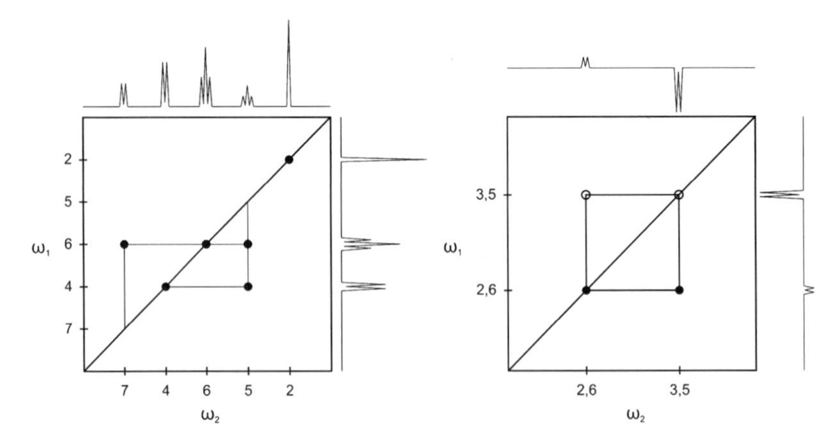

Fig. 16 Schematic two-dimensional CIDNP (COSY or NOESY) spectra of tryptophan and tyrosine side chains in a protein. Shown is only the aromatic region. The chemical shifts are arbitrary and spin–spin couplings are neglected. Cross-polarization during the light pulse is assumed to be negligible, i.e., only H2, H4, and H6 (all absorptive) in tryptophan, and H2,6 (weakly absorptive) and H3,5 (emissive) in tyrosine appear in the ω_1 dimension, and cross-polarization during the NOESY mixing period is assumed to proceed with retention of phase (Sect. 2.8.2). Emissive polarization is shown by *open circles*, absorptive by *filled circles*. Projections onto the ω_1 and ω_2 axes are shown to the right and above each two-dimensional spectrum

undergoing slow chemical exchange (roughly 0.1–10 s^{-1}) between two structurally discreet states. A two-dimensional CIDNP-EXSY spectrum recorded at the mid-point of denaturant- or temperature-induced unfolding, for example, was successfully employed to correlate chemical shifts of protein nuclei exchanging between the native and denatured state [87].

Schematic 2D CIDNP spectra for tyrosine and tryptophan residues in a protein are shown in Fig. 16. Only the aromatic region is highlighted and cross-polarization is assumed to occur with retention of phase. Moreover, cross-polarization during the light pulse is ignored. These spectra differ from their conventional counterparts in being asymmetric about the diagonal because only directly polarized protons appear in the ω_1 dimension. Thus, the projection onto the ω_1 axis gives the CIDNP spectrum *without* cross-polarization and that onto the ω_2 axis, the spectrum *with* cross-polarization (Fig. 16). The magnitudes of the spin–spin coupling constants and the internuclear distances lead to CIDNP-COSY and CIDNP-NOESY spectra of tyrosine and tryptophan that differ in the aromatic region only in the relative intensities of the peaks. The scalar and dipolar interactions between the H2 and H4 protons in histidine are, in general, too small to produce cross-peaks in either type of two-dimensional CIDNP spectrum. Each of the three amino acids has a unique pattern of cross peaks in the aromatic region, allowing unambiguous first stage assignments of chemical shifts to particular protons in individual residues. In addition to the cross peaks shown in Fig. 16, all three CIDNP-active amino acids generally show $H_\beta \rightarrow H_\alpha$ and $H_\beta \rightarrow H_\beta$ NOESY signals. Extra intra-residue peaks between the H_α or H_β and aromatic protons may appear in a CIDNP-NOESY spectrum, depending on the conformation of the side chain about the C_α–C_β and C_β–C_γ bonds. The only extra COSY peaks likely to be observed are the $H_\beta \rightarrow H_\alpha$ and $H_\beta \leftrightarrow H_\beta$, all other couplings being too small. A further advantage of the two-dimensional method is the ability to distinguish cross-relaxation pathways that would be difficult to observe in a conventional CIDNP spectrum. For example, the Trp 62 HA proton in HEWL receives absorptive polarization from H2 and emissive polarization from HB of the same residue leading to a partial cancellation of the Trp 62 HA polarization signal in a one-dimensional spectrum. The existence of the two magnetization transfer routes is immediately evident from the cross peaks in the 2D spectrum [104].

In addition to proton–proton COSY and NOESY sequences, a heteronuclear ^{15}N–^{1}H CIDNP correlation experiment has been developed by Lyon et al. [29] and produced significant signal enhancements (SE ~ 100) for polarizable Trp indole protons of ^{15}N-labeled proteins. The basic sequence involves a single reverse INEPT ^{15}N–^{1}H transfer between evolution and detection periods, i.e., t_1 and t_2. Under these conditions, the "dark" ^{15}N–^{1}H spectrum is below the noise level making subtraction of CIDNP "light" and "dark" spectra obsolete. Differences in surface accessibilities of the six Trp residues in native and urea-denatured HEWL, respectively, have been detected in this way. More recently, the sensitivity of this ^{15}N–^{1}H CIDNP experiment was further improved introducing two additional ^{15}N-photo-CIDNP-enhanced pulse sequences, namely EPIC- and CHANCE-HSQC, by Sekhar and Cavagnero [109]. Both sequences, tested on Trp and the Trp-containing protein *apo*-HmpH, were found to produce up to twofold higher sensitivity than the reference HSQC-type pulse train. In addition, Cavagnero et al. designed a series of photo-CIDNP enhanced versions of various two-dimensional ^{1}H–^{15}N heteronuclear correlation experiments, i.e., SE–HSQC, HMQC, SOFAST–HMQC, and g/s–HMQC, for the sensitivity-enhanced observation of (solvent-accessible) Trp-indole NH spin systems using low laser irradiation power of 1 W at 500 ms per increment [110]. Among these, the HPE–SOFAST–HMQC sequence has yielded the highest sensitivity, i.e., a twofold greater SNR per unit time as compared to its parent pulse scheme. Another two-dimensional CIDNP sequence introduced by the same group exploits the relatively large polarization enhancement stemming from ^{13}C CIDNP [111]. The ^{13}C-PRINT (photo-CIDNP-enhanced constant time reverse INEPT) experiment involves an initial ^{13}C nuclear spin polarization step via photo-CIDNP followed by conversion to anti-phase coherence and transfer to ^{1}H for detection. Substantial SEs, up to two orders of magnitude relative to the "dark" spectrum, are detected for resonances of both side chain and backbone CH pairs for the three aromatic residues Trp, His, and Tyr, a 32-residue peptide, and the drkN SH3 protein. The sensitivity of this experiment is unprecedented in the NMR polarization enhancement literature dealing with polypeptides in solution. In addition, data collection time is reduced up to 256-fold, thereby highlighting the advantages of ^{1}H-detected ^{13}C photo-CIDNP in solution NMR.

3.5 *Miscellaneous*

3.5.1 Additional Factors Affecting CIDNP Intensities

pH Dependence

The pH of the sample solution is an important factor in CIDNP experiments of amino acids and proteins as the second order rate constants for the reaction of the three amino acids with triplet excited flavin also depend on the proton concentration of the sample solution (see below; [18, 35, 64]). Histidine is most affected, displaying a difference in rate of more than three orders of magnitude in the range between pH 2 and 14. Also, in the pH range where most NMR experiments on proteins are conducted – roughly between pH 4 and 8 – it can be seen that histidine does not compete efficiently with tryptophan and tyrosine for the low concentration of flavin triplets [49, 112]. This is (mainly) why the His residue can only be significantly polarized in the presence of exposed Trp and Tyr when its own surface accessibility is very high. The absence of polarization for His 32 in BLA, for example, is probably a result of the competition with Trp 118 and Tyr 18. The intrinsic pH dependence of histidine polarization can be exploited by comparing CIDNP spectra acquired at low pH (no polarization) and neutral pH (polarization). Tryptophan and histidine signals, all of which are absorptive in the aromatic region, can be distinguished on the basis that the tryptophan signals do not disappear at low pH [113]. It is important, though, not to confuse the intrinsic pH dependence of CIDNP intensities with pH-induced changes in amino acid side chain accessibilities.[15]

The large differences in second order rate constants between the different CIDNP-active amino acids can been exploited in, for example, RT in situ protein refolding studies, e.g., in the case of the histidine-containing phosphocarrier protein (HPr) [116]. The sequestering of tryptophan residues to form a transient, hydrophobically collapsed intermediate has been proposed based on the appearance of polarization from relatively less competitive His residues during the early stages of single phenylalanine-to-tryptophan substituted HPr variants. As mentioned above, the Trp residues involved in the hydrophobically collapsed intermediate include the natively solvent-accessible Trp 48 side chain. Most likely, the reason for this is to minimize intermolecular aggregation by reducing the exposed hydrophobic surface area. Similar conclusions involving the early non-native sequestration

[15] For tyrosine, the fall-off of CIDNP signal intensity at high pH is believed to arise from degenerate electron exchange. The same effect is also made responsible for the lack of tryptophan polarization at low pH. Luckily, cancellation by degenerate electron exchange is much less likely to be a problem in proteins whose greater bulk reduces the electron exchange rate constant considerably. Thus, in proteins, CIDNP has been observed for tyrosine residues [114] at pH 12 and for tryptophan residues [115] at pH 2. The small His enhancements at low pH are most likely due to reduced reaction rates between triplet excited flavin and histidine [67], both of which are protonated and positively charged below approximately pH 4.5.

of hydrophobic residues have been reached for HEWL [54, 117] and for the small helical protein Im7 [118]. Accordingly, when the hydrophobic residues in question are tryptophan or tyrosine, RT CIDNP can be a valuable tool for examining early folding events.

Hydrogen Bonding

As mentioned before, the generation of CIDNP in Tyr and His side chains in the presence of FMN or other flavin derivatives is by hydrogen abstraction and, thus, requires an accessible hydroxyl (Tyr) or amide group (His), respectively. In one of the first CIDNP studies on bovine ribonuclease A, no polarization was found for Tyr 92 or His 105, both of which, on the basis of the crystal structure, were expected to be exposed to the solvent [119, 120]. In this case, the suppression of CIDNP is attributed to involvement of the OH and NH groups in hydrogen bonds with lysine and serine residues, respectively. Hydrogen bonding has since been used sometimes to explain the lack of CIDNP in tyrosine or histidine side chains, although the absence of polarized histidine signals can also be caused by insufficient competition for excited triplet flavin if Tyr and/or Trp side chains are exposed at the same time. However, there are other examples, where CIDNP is detected in residues whose anomalously high pK_a values suggest that the hydroxyl or amide groups are hydrogen-bonded [121, 122].

Charge Effects

There are studies which compare the relative CIDNP intensities for different residues in a protein in the presence of differently charged flavins. In all but one case [123] the effects are pronounced. Two closely related flavodoxins – from *M. elsdenii* and *Clostridium beijerinckii* (MP), respectively – exhibit CIDNP for two tryptophans [124–126]. For the latter, positively charged 3-*N*-ethylaminolumiflavin polarizes *both* tryptophans, while the negatively charged 3-*N*-carboxymethyllumiflavin induces polarization only in Trp 90 [124]. These observations were attributed to the carboxylate group of Glu 65 in the *C. beijerinckii* protein which hinders the approach of the negative flavin to the nearby Trp 6. In the flavodoxin of *M. elsdenii*, where the charge of the dye has a smaller effect, the glutamate is replaced by a valine. The two oppositely charged flavins also produce different spectra for *M. elsdenii* flavodoxin and the *apo*-form [125]. In addition, the positively charged dye polarizes a tyrosine only in the *holo*-protein. The opposite effect is found for the negatively charged dye. Similar charge effects have been observed for HEWL [43]. Replacing 3-*N*-carboxymethyllumiflavin with 3-*N*-ethylaminolumiflavin causes a threefold reduction in the enhancement of Trp 123 relative to Trp 62; neutral dyes, e.g., riboflavin and lumiflavin, give intermediate intensity ratios. These changes are rationalized by the distribution of positively and negatively charged groups in the proximity of the two tryptophans and, in addition, can be used for resonance assignment.

Related charge effects are observed for the Trp–Trp dipeptide [80]: at high pH, both Trp residues are strongly polarized, whereas at pH 4 very little CIDNP is seen for either side chain as a result of rapid exchange cancellation. However, at pH ~ 6, only the C-terminal tryptophan shows appreciable enhancement. The origin of this effect is unclear, but can possibly be attributed to a slower Trp radical cation formation rate near the N-terminal amino group than close to the C-terminal carboxyl group. Hence, these findings indicate that care is needed in interpreting CIDNP intensities. It is possible that changes in the local charge environment of a polarizable residue, e.g., as a result of a change in pH or binding to a (charged) ligand, affect the polarization such that it might be mistaken for changing surface accessibilities.

Flavin–Protein Binding

Although FMN and other flavin derivatives – together with their corresponding radicals – are known to form weak complexes with the amino acids tyrosine and tryptophan [127, 128], the exact effect of complexation on the extent of photo-CIDNP intensities is not clear. Weak binding most likely enhances the polarization via an increase of the radical pair's lifetime and, thus, more extensive $S–T_0$ mixing. Interactions within a radical pair, strong enough to prevent the radicals separating to a point where the exchange interaction is small, suppress polarization. Accordingly, no CIDNP is observed for flavoproteins [125, 126] in the absence of added dye despite the internal flavin being photochemically active and its proximity to aromatic amino acid residues. Similarly, the absence of CIDNP in the adenine moiety of flavin adenine dinucleotide (FAD) at neutral pH seems to be due to stacking interactions between the two aromatic rings in the biradical formed by intramolecular electron transfer. In acidic solution, where the adenine radical is protonated, the stacked conformation is less favorable, resulting in a smaller exchange interaction and substantial CIDNP [129].

Evidence as to whether flavin dye–protein interactions affect CIDNP intensities is inconclusive. It was found that binding of *N*-acetylglucosamine (NAG) to HEWL shifts and enhances the CIDNP signals of Trp 62 and also shifted the nitrogen-bound methyl resonance of 3-*N*-carboxymethyllumiflavin (Flavin 1). This was interpreted as a displacement of protein-bound flavin by competitive binding of NAG in the region of Trp 62, followed by the release of dye into solution. Similar intensity effects have been found for human lysozyme [50]. CIDNP–NOESY spectra reveal cross polarization between protons in 3-*N*-carboxymethyllumiflavin with retention of phase, an effect that disappears when NAG is fully bound, indicating slow tumbling in a flavin–protein complex [104]. In contrast, experiments on lysozyme with FMN rather than Flavin 1 show only a slight increase in the integrated CIDNP intensity of Trp 62 and no change in the signal from the other polarizable residue, Trp 123, when NAG is fully bound, suggesting that FMN may bind less strongly than 3-*N*-carboxymethyllumiflavin [80]. Addition of FMN to lysozyme solutions causes small shifts in the indole NH proton resonances of tryptophans 62 and 63, presumably due to binding. Finally, the temperature dependence of tryptophan CIDNP does not depend on the concentrations of either FMN (0.02–1.0 mM) or lysozyme (0.2–2.0 mM), which counts as evidence against significant binding effects [115]. In earlier experiments, however, it is noted that the absolute intensities dropped rapidly when the protein concentration was increased beyond 1 mM [130]. Similar effects were seen for the polarization of a small amount of added tyrosine, interpreted in terms of dye-inactivation by binding to the protein.

Denaturants and Other Factors

Highly satisfactory though they are, flavin-based CIDNP photosensitizers are not perfect. For example, the triplet-excited state of flavins is efficiently quenched by molecular oxygen, secondary and tertiary amines [50, 104], and azide. Hence, complexing agents like ethylenediaminetetraacetate (EDTA) and buffers such as "tris" have devastating effects on CIDNP intensities [36]. The co-enzymes NADH [131] and NADPH [132] are also reactive towards excited flavins, as is the hydrophobic dye 8-anilino-1-naphthalene sulfonate (ANS). As to chemical denaturants, urea is preferred over guanidine hydrochloride (Gdn-HCl) because the guanidine group attenuates the enhancements, again by competing for flavin triplets [80, 115]. Nevertheless, CIDNP has been observed for denatured and partially denatured proteins in Gdn-HCl and a concentration of up to approximately 4.0 M Gdn-HCl appears feasible [115]. Experiments on proteins with reduced disulfide bridges can be difficult, as triplet flavins are also quenched by thiols, e.g., cysteine [133]. Fully reduced HEWL, for example, has a much weaker CIDNP spectrum than the fully cysteine-protected, i.e., carboxymethylated, protein [115]. Also, continuous illumination, in principle, causes irreversible oxidation of side chains photosensitized by flavin [81], although proof of such effects is scarce.

Two other factors affecting CIDNP intensities need to be mentioned briefly, too. (1) Chromophore-containing proteins that absorb strongly at the wavelengths used to excite the flavin photosensitizer, present problems. Thus, most attempts failed to detect (flavin-sensitized) CIDNP for heme-containing proteins [36], for example, although Akutsu et al. induced CIDNP in cytochrome b_5 [134]. In addition, Day et al. demonstrated that it is possible to generate photo-CIDNP in a heme-containing protein, namely the *holo*-state of the C10A/C13A double mutant of cytochrome c_{552}, but, interestingly, not in the wild-type protein [135]. Reasons for this difference were attributed to the different electronic properties of the prosthetic heme group, which may influence quenching of the triplet-excited flavin or, alternatively, interaction of the heme group with the transiently formed radicals involved in the radical pair reaction. (2) Protein-bound paramagnetic ions have the potential to suppress CIDNP as found in a study of HEWL [18]. Here, it was found that Pr^{3+}, which binds more tightly to Trp 62 than Trp 123, attenuates the polarization of Trp 62 to a larger extent than that of Trp 123. Charge effects were ruled out by comparison with diamagnetic La(III) ions. One possible explanation is that Pr^{3+} destroys the spin-correlation in the flavin–protein radical pair via exchange interactions, and so reduces the polarization. Alternatively, triplet-excited FMN could be quenched by the bound metal ion [96, 97]. In both cases, the effect should be strongly dependent on the distance of the residue from the lanthanide binding site.

3.5.2 Time-Resolved (TR) Photo-CIDNP Techniques

Standard, i.e., continuous-wave, photo-CIDNP NMR spectroscopy detects a time-averaged, i.e., steady state, value of the photochemically generated magnetization, which yields mechanistic information but, on the other hand, precludes kinetic studies. Some of the faster physicochemical processes that can influence the observed polarization in a CIDNP experiment are obscured by the use of laser pulses longer than ca. 1 μs in duration. These include, among others, (1) the production of polarization by encounters between independently formed photosensitizer and amino acid radicals in so-called "F-pairs", (2) the cancellation of recombination and escape polarization as a result of radical recombination or degenerate electron exchange reactions between the escaped radicals and their parent molecules, and (3) intramolecular electron transfer, e.g., from tyrosine residues to the radicals of nearby tryptophan residues [63, 64]. The time-resolved (TR) photo-CIDNP technique, using a pulsed laser as the light source, is best suited for exploring kinetic events on a microsecond to millisecond timescale in photo-induced spin-selective radical reactions, allowing these processes in, for example, amino acids and dipeptides to be studied and their rate parameters to be extracted [35, 63, 64, 107, 136–139]. TR CIDNP studies of proteins, on the other hand, are scarce. Experiments on hen egg-white lysozyme, for example, suggest that intramolecular electron transfer occurs, probably in hydrophobic clusters at high temperatures in the denatured state [64]. It has yet to be established though to what extent these rapid reactions influence the CIDNP intensities observed in experiments on proteins exploiting argon-ion laser(s) illumination.

As mentioned before, TR photo-CIDNP experiments generally require a light pulse duration (much) shorter than 1 μs. In this case, the resulting incident optical power densities are not compatible with optical fibers and, thus, prohibit the use of illumination schemes used in the field of CW photo-CIDNP and/or optical NMR studies. A first description of a TR photo-CIDNP setup [140], based on an electromagnet 60-MHz FT NMR spectrometer with relatively complex, home-built, control electronics entailed straightforward light routing as the probe's r.f. coil region was easily accessible from outside the magnet [141]. Since the introduction of superconducting magnets, however, the geometry around the sample volume has become much more constrained and difficult to access, such that sample irradiation can require, at the very least, minor rearrangement of the probe's electronics, requiring sometimes even one of the r.f. coils to be removed for synchronization purposes. To this end, the method of choice for high-field TR CIDNP studies comprises the use of a cylindrical fused silica light guide (diameter 5 mm) passing off axis through the body of the NMR probe, surmounted by a prism to bring the light into the NMR sample from the side through a window in the r.f. coil. Alternatively, light can be brought in from below the sample contained in a flat-bottomed NMR tube. As both arrangements, however, also demand probe hardware modifications, it is likely that new setup solutions for TR CIDNP experiments will be produced [138].

4 Applications

4.1 Photo-CIDNP "Steady-State" Experiments

The acquisition of a standard, i.e., one-dimensional, photo-CIDNP "steady-state" NMR spectrum forms the starting point of every photo-CIDNP, e.g., "real time" or (NOE) pulse-labeling, study both to check sample conditions as well as to test the general feasibility of inducing photo-CIDNP in side chain nuclei of the protein of interest.[16] Once set up, the measurement of a basic 1D CIDNP (equilibrium) spectrum of a protein in aqueous solution is usually straightforward and fast. The protein concentration – $c \sim 1$ mM or below – can be the same as or lower than that used for conventional NMR experiments and the number of scans needed for an acceptable SNR can often be as low as 8 or 16, depending (mainly) on the sample conditions. A small amount of stock FMN solution ($c \sim 10$ mM) is added directly to the NMR sample tube yielding a final dye concentration of ca. 0.2 mM. A preliminary "steady-state" photo-CIDNP spectrum will provide general information on the accessible tryptophan, tyrosine, and histidine residues. If information on the chemical shifts is available, identification of surface-accessible residues is immediately possible as well. If further peak dispersion is necessary,

[16] As said before, the (potentially) CIDNP-active amino acid side chain must be accessible to the photosensitizer to generate the triplet-born radical pair and hence the nuclear polarization. "Native state" (static) solvent accessibilities, using either the high-resolution crystal or NMR structure, can be calculated prior to the CIDNP experiment to identify those residues that are exposed to the solvent and, thus, will most probably benefit from the CIDNP effect. The method, developed by Lee and Richards, calculates atomic accessible surfaces by rolling a probe of a given size around the outer sphere of the protein as defined by its PDB structure file [142]. When FMN is used as a photosensitizer, a probe radius of 1.4 Å is used in the calculations and the results are quoted relative to the accessibilities found for an extended conformation of the tripeptide Ala–Xaa–Ala, where Xaa is the residue of interest. Recently, it has been shown that the accessibility of the highest occupied molecular orbital (HOMO) of the aromatic side chain gives a more robust prediction of the observation of photo-CIDNP signals, particularly in the case of a limited static solvent accessibility [143].

homonuclear or, if isotopically ^{13}C- and/or ^{15}N-labelled protein sample is available, heteronuclear two-dimensional experiments can be carried out (Sect. 3.4.4).

An interesting application of (one-dimensional) "steady-state" photo-CIDNP spectroscopy involves the comparison between structurally related proteins, in particular when small differences in CIDNP intensities are interpreted carefully [18]. The principal aim of these studies is to probe both structural similarities and differences of related proteins and, in addition, to investigate the effect of structural modifications on interaction with ligands, lipids, nucleic acids, or other proteins. In addition, 1D photo-CIDNP of amino acids and proteins can be used to monitor the pH dependence of aggregation properties and changes in tertiary structure as well as providing local information about the ionization properties of photo-CIDNP polarizable side chains and neighboring groups, e.g., to distinguish between the aromatic resonances of tryptophan and histidine residues, relying on the sensitive dependence of histidine H2 and H4 chemical shifts on the ionization state of the imidazole ring [18, 55]. Most photo-CIDNP studies carried out to study the interactions between proteins and micelles or vesicles sought for information of two different kinds. (1) First, comparison of spectra taken in the presence and absence of micelles can help to identify the site of micelle binding. Those studies were either performed on enzymes that bind and hydrolyze aggregated lipid, e.g., phospholipase A_2, or on proteins to study the physical interaction with lipid bilayers involved in, for example, receptor binding. (2) Second, cross-polarization effects – usually in small proteins or peptides – have been used to explore the effect of complexation on protein mobility and the degree to which micelle binding induces structure formation in small proteins or peptides. In addition, thermal or chemical denaturation leads to the disruption of secondary and/or tertiary protein structure and the effect on CIDNP intensities – both quantitatively and qualitatively – is significant. Despite the prospect of structural information on folding intermediates and/or the denatured state of a protein, only very few detailed 1D photo-CIDNP studies at equilibrium are available in the literature, mainly due to the considerable loss of spectral dispersion and fine structure upon denaturation (see above).

4.2 Photo-CIDNP (Stopped-Flow) "Real Time" Experiments

Different methods are available to study the folding of proteins as it progresses inside the NMR tube inside the magnetic resonance spectrometer in "real time" where folding is in principle initiated following the change of solvent conditions, e.g., pH, denaturant concentration, or concentration of stabilizing agents, upon mixing of two (or more) solutions in situ [98, 144–150]. After mixing, various NMR techniques can be applied to monitor the change of different properties of the polypeptide chain during the folding reaction. The principal approach in this case is the use of one-dimensional NMR spectroscopy due to the time constraints required to acquire spectra of higher dimensions.[17] The combination of biological photo-CIDNP and "real time" NMR, which provides information on the changes in the solvent accessibilities of aromatic side chains as a function of time, has certain advantages over conventional NMR for monitoring rapid reactions [54, 55, 57, 116, 152, 153]. During a "real time" photo-CIDNP experiment, nuclear polarization is generated by the laser flash, thereby eliminating the need for relaxation delays between signal acquisitions. In addition, it is not crucial to wait for spin–lattice relaxation to establish the equilibrium nuclear polarization, when the transfer of solutions is carried out from a region of low magnetic field outside the probe of the NMR spectrometer. Accordingly, the technique allows the spectral sampling of closely spaced time points. Whereas optical spectroscopic methods report on global folding changes, the detection and investigation of individual side chain protons is feasible with "real time" photo-CIDNP NMR provided that there is sufficient chemical shift resolution. It is thus possible to monitor local structural changes which provide significant insights into protein folding at a residue-specific level. These high resolution data can later be used in computational models, for example, where comparison with experimental results serves as a valuable benchmark allowing validation of the different theoretical models underlying the calculations.

Prior to conducting photo-CIDNP "real time" unfolding or refolding experiments, it is helpful to obtain a series of "steady-state" photo-CIDNP spectra monitoring the equilibrium denaturant-, pH titration-, or temperature-dependence of the protein to be studied. Peaks observed under static conditions can then be used to differentiate any new signals that appear in the "real time" measurements. Such peaks offer potential evidence for the presence of intermediates on the way to the native state or the denatured state in the case of folding or refolding, respectively. Furthermore, if present, the molten globule state, characterized by broad peaks and poor chemical dispersion, can also be clearly identified in the early stages of folding [18, 27]. Also helpful are refolding and/or unfolding kinetic data obtained with other optical spectroscopic techniques, e.g., stopped-flow fluorescence or circular dichroism, to determine the repetition time – how frequently spectra should be acquired during the refolding reaction – and T_{AQ} (Fig. 14), i.e., the desired spectral resolution, for "real time" photo-CIDNP experiments. Kinetic constants can be extracted from high quality data. Examples include, among others, monitoring the refolding of the proteins HEWL [54], histidine-containing phosphocarrier protein (HPr) from *E. coli* [116], bovine α-lactalbumin [88], and, in addition, ribonuclease A (RNase A) [153]. Initially, these experiments were conducted employing the stopped-flow, "real time" CIDNP device. More recently, however, the rapid mixing device is used exclusively in

[17] One of the most important limitations of one-dimensional NMR experiments is the low spectral resolution and hence recent work has concentrated on the extension of "real time" NMR methods to utilize the higher resolution of multidimensional NMR spectroscopy in kinetic experiments [e.g., 151].

these measurements due to the significantly reduced dead time as well as the much more efficient mixing of solutions. (1) The in situ observation of structure formation during the refolding reaction of HEWL, possessing a total of six – potentially CIDNP-polarizable – tryptophan residues, has been studied using the photo-CIDNP “real time” approach. “Real time” spectra of the tryptophan indole proton peaks recorded during folding show a rapid decrease of the broad indole ^{1}H NMR polarization signal corresponding to the denatured state (10.1 ppm), suggesting a rapid sequestering of the tryptophan side chains that are solvent-inaccessible in the native state. Only the indole proton peaks corresponding to the two natively accessible residues (Trp 62 and Trp 123) are found to be polarizable at a later stage. (2) Histidine-containing phosphocarrier protein (HPr) from *E. coli* is a small 85-residue protein lacking Tyr or Trp residues. Site-specific single substitutions of its four phenylalanine residues to Trp yielded four variants of the protein that permit separate “real time” CIDNP analyses of the accessibilities of tryptophan side chains during refolding from the fully denatured state. In the native state, one mutant (Phe 48→Trp) is completely solvent-accessible, while two others (Phe 2→Trp and Phe 22→Trp) are partially inaccessible. Interestingly, however, it is found that, regardless of the native accessibilities, all three of these tryptophan residues are transiently inaccessible, suggesting an early hydrophobic collapse to be common. (3) Another “well-behaved” system for the study of “real time” protein folding using CIDNP is the metalloprotein bovine α-lactalbumin. For example, this protein has been studied with respect to the folding of its calcium(II)-depleted *apo*-form following the photochemical release of Ca^{2+} ions into solution. Photo-CIDNP has also been implemented as a technique to probe the structure of a non-native, i.e., partially folded, state of BLA, the so-called “A state”, exploiting the transfer of nuclear polarization from the partially folded ensemble to the native state for detection (see below). Based on these approaches, photo-CIDNP has proven successfully to have the potential of providing significant site-specific information on the various conformations adopted by a given protein as the native structure is disrupted or recreated. (4) More recently, photo-CIDNP “real time” NMR was used to monitor the refolding of ribonuclease A (RNase A) following the rapid dilution of the denatured protein in the presence of either urea-d_4 or guanidine hydrochloride-d_6 (Gdn-DCL). In general, similar results were obtained for the refolding of the protein from both the urea- and the Gdn-DCL-denatured state, revealing the existence of two distinct kinetic processes, i.e., a faster step and a slower one, the latter being attributed to the *cis*/*trans*-isomerization of one of the proline residues. In addition, the quantitative analysis of the photo-CIDNP spectra yielded time constants which are very similar for both experiments and, in addition, agree well with literature values reported by others. An intriguing difference, however, is that the distinct photo-CIDNP Tyr signal arising from an intermediate state – not observed previously using other spectroscopic techniques – was only seen in the case of refolding from Gdn-DCL. It was suggested that this discrepancy can be explained with the much higher ionic strength of the Gdn-DCL solution which serves to stabilize structurally the (metastable) folding intermediate.

4.3 The Photo-CIDNP Pulse-Labeling Experiment

Whereas the assignment of photo-CIDNP-derived NMR signals to specific residues in the well-resolved "native state" NMR spectrum is relatively simple, it becomes increasingly difficult as a protein becomes less structured and, thus, the differences in the chemical environment responsible for the chemical shift dispersion within a given residue type become less pronounced. The few photo-CIDNP studies of protein denaturation and unfolded proteins reported in the literature often relied on the poor resolution of one-dimensional ^{1}H spectra and have thus so far been limited to a somewhat qualitative description of changes in the relative exposure of different residue types, e.g., tyrosine and tryptophan, rather than of specific residues. In an attempt to quantify the exposures of the different residue types relative to those expected for a fully unfolded protein, comparisons of the relative CIDNP intensities obtained in the denatured protein to those obtained with mixtures of free amino acids have been used [115]. In addition, it became clear early on that the temporal resolution of a CIDNP "real time" experiment is normally not sufficient for observing folding or unfolding events that occur on a (sub)millisecond time-scale, e.g., the initial hydrophobic collapse typical for the early phase of folding of globular proteins, as the dead time of the experiment as well as the delays between subsequent acquisitions are well above this threshold [55]. Hence, to circumvent both the problem of a lack of signal dispersion as well as the insufficient time resolution of the "real time" approach, a more sophisticated CIDNP experiment has been developed in recent years with which structural features of non-native, i.e., partially folded or unfolded, states of proteins that refold to the native structure on a (sub)millisecond timescale, i.e., faster than nuclear spin–lattice relaxation, can be monitored. This so-called "pulse-labeling" experiment combines the two approaches of CIDNP "steady state" and CIDNP "real time".

The idea of performing a kinetic photo-CIDNP "pulse-labeling" experiment originally came from a radio frequency pulse-labeling study by Balbach et al. in which NOEs rather than CIDNP enhancements generated in the molten globule state of the protein bovine α-lactalbumin were transferred by refolding to the native state for detection [154]. The NOEs were generated somewhat unselectively by applying a 1 s r.f. excitation pulse to the aromatic envelope of the molten globule state followed by the detection of cross-polarized signals in the aliphatic region of the NMR spectrum ca. 0.8 s after the refolding was initiated. Whereas the results obtained from this approach provided a general idea of the contacts between residues within the partially folded state, it was evident that the inherent lack of both sensitivity and selectivity limits the applicability of this technique substantially. To overcome this problem, Lyon et al. designed a photo-CIDNP variant of the initial r.f. labeling technique in which solvent-accessible magnetic nuclei of CIDNP-active side chains were selectively polarized in a protein's structurally perturbed conformation, e.g., the molten globule state (M) [87, 99]. Following the refolding of the protein, these polarization signals were then observed in the well-dispersed "native state" NMR spectrum thereby making it possible to assign and identify unambiguously individual polarized side chain nuclei that were exposed to

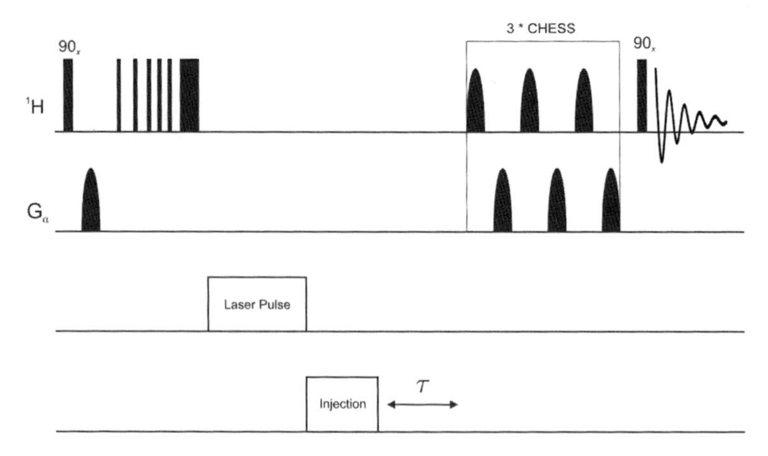

Fig. 17 Schematic of the full NMR pulse sequence used for photo-CIDNP (NOE) pulse-labeling experiments on partially folded or fully denatured proteins. The field gradient G_{α} is applied either as a strong gradient along each of the Cartesian axes (order: x, y, z) or, alternatively, along three orthogonal vectors tilted at the magic angle with respect to the B_0 field direction (see text)

the solvent in the protein's "M" state: (1) in the first step of a pulse-labeling experiment, nuclear polarization is generated in the magnetic nuclei of CIDNP-active side chains residing on the surface of the structurally perturbed protein; (2) subsequently, the protein solution is injected into an NMR tube containing the refolding buffer to trigger the rapid refolding of the protein under investigation in situ. This refolding step is followed by the application of a $\pi/2$ r.f. "read" pulse and the subsequent acquisition of the NMR spectrum (Fig. 17). The great advantage of this experiment is that solvent accessibilities of CIDNP-active side chains in a non-native environment which lacks spectral signal dispersion can be monitored in the well-resolved and hence spectrally assignable NMR spectrum of the protein's native state [99].

Lyon et al. successfully tested their methodologically unprecedented approach on both the "A-state" of bovine α-lactalbumin, a partially folded "molten globule" state of the protein characterized by extensive secondary structure and the absence of rigid side-chain packing and, in addition, the TFE-denatured state of HEWL, which represents a partially structured state of the protein exhibiting a high degree of helical structure. Interestingly, the authors managed to irradiate the sample not only in the strong field of the NMR magnet (~9.4 T) but also at lower magnetic field (4.0 T) – where ^{1}H CIDNP enhancements show a fivefold increase in intensity – followed by a transfer of the pre-polarized sample solution into refolding buffer at high field. More recent applications of the photo-CIDNP pulse-labeling technique include the structural characterization of the different equilibrium (partially folded) "molten globule" states of both bovine and human α-lactalbumin where the application of photo-CIDNP pulse-labeling made it significantly easier to differentiate patterns of hydrophobic-core surface accessibilities and, in addition, to identify multiple interactions between hydrophobic side chains capable of stabilizing the overall topology of the molecule [155].

4.4 The Photo-CIDNP NOE Pulse-Labeling Experiment

4.4.1 Concept and Setup

The photo-CIDNP NOE pulse-labeling experiment, designed to probe the presence of residual structure in the unfolded state of (ultra)fast folding proteins, further extends the concept of CIDNP pulse-labeling and combines the two aforementioned approaches of radio frequency- and photo-CIDNP "pulse-labeling", adding the feature of CIDNP-induced NOEs from selectively polarized, i.e., CIDNP-active, side chains to nearby aliphatic protons in the protein's denatured state (Fig. 18).

In the first step of a photo-CIDNP NOE pulse-labeling experiment, a solution containing the denatured protein and the flavin photosensitizer is illuminated with laser light from an argon ion laser in order to initiate the cyclic photochemical reaction responsible for the generation of CIDNP. This process leads to the initial polarization of solvent-accessible tyrosine, tryptophan, or histidine side chain nuclei. Subsequently, some of the initially generated polarization will be transferred to nearby aliphatic, i.e., CIDNP-inactive, side chain nuclei via NOE transfer processes provided that residual structure is present in the unfolded state of the protein under investigation. Since the efficiency of a polarization transfer involving cross-relaxation processes is highly distance dependent, only aliphatic side chain nuclei that are very near to the CIDNP-active side chain in the protein's denatured state will become cross-polarized.[18] Eventually, this step is followed by the transfer of the protein solution into the NMR tube containing the refolding buffer to trigger the rapid renaturation of the protein. The whole sequence is concluded by the acquisition of the NMR spectrum which exhibits additional, i.e., cross-polarized, resonances in the aliphatic region provided that these contact interactions exist in the denatured state of the protein. The advantage of the NOE driven pulse-labeling approach is that close contact interactions between CIDNP-active, i.e., aromatic, and CIDNP-inactive amino acid side chains present in the denatured state can be observed in the well-dispersed and assignable NMR spectrum of the protein's native state (Fig. 19). Hence, the technique is capable of unambiguously assigning side chain nuclei involved in the formation of non-random interactions in the unfolded state. As the build-up and the retention of additional cross-polarized NMR signal intensity competes efficiently with spin relaxation, it is clear that such an experiment can only be conducted on proteins whose refolding rate constants (k_f) – being on the order of milliseconds or below – significantly exceed the respective spin–lattice relaxation rates of cross-polarized side chain nuclei.

To perform such a photo-CIDNP-driven NOE pulse-labeling experiment, a means of rapidly transferring a solution containing the denatured protein together with the flavin photosensitizer into a refolding buffer in situ needs to be used to

[18] The distance dependence of the NOE requires that internuclear separations of more than 5 Å can normally not be observed due to the lack of a sufficiently strong mutual dipolar coupling interaction between the two potentially interacting spins.

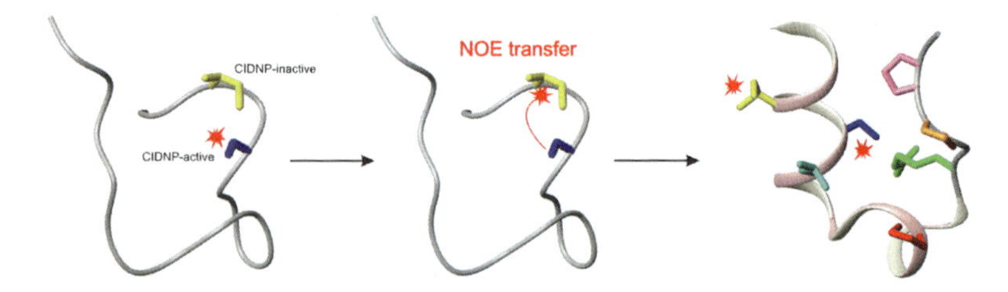

Fig. 18 Photo-CIDNP NOE pulse-labeling schematic: CIDNP generation of polarizable side chain nuclei (*left*) is followed by an NOE transfer of polarization to nearby aliphatic side chains in the protein's denatured state (*middle*). Subsequently, acquisition of the native state NMR spectrum is performed following a submillisecond refolding step (*right*)

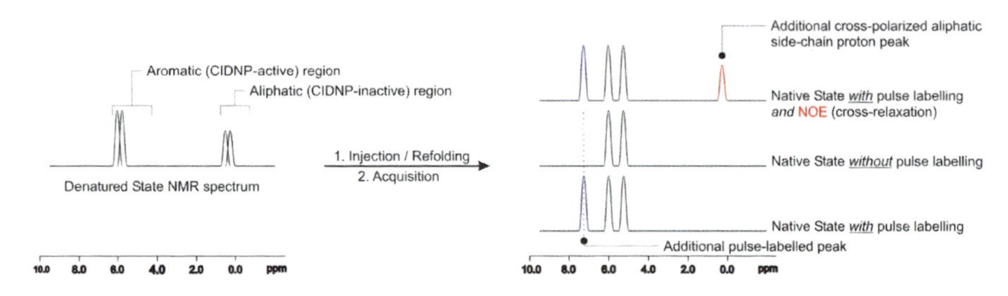

Fig. 19 Photo-CIDNP (NOE) pulse-labeling: NMR signals which overlap in the unfolded state are spread out by the greater chemical shift dispersion in the native state and additional cross-polarized signals show up in the aliphatic region. The conventional "N" state CIDNP spectrum and the result of the CIDNP pulse-labeling experiment as pioneered by Hore et al. are shown for comparison (see text)

trigger the submillisecond refolding of the protein under investigation following the CIDNP-generating illumination and excitation step described above. This fast transfer of the protein solution is achieved using the rapid mixing device described in Sect. 3.3.3 (Fig. 11). The crucial parts of this apparatus are again highlighted in Fig. 12 depicting the r.f. coil region of the NMR spectrometer immediately prior to illumination of the sample. In this representation it becomes clear that the optical fiber (shown in red) is directly immersed into the protein solution prior to the transfer to ensure an intense and homogeneous illumination of the sample solution which, in turn, will yield a high concentration of triplet-excited flavin dye molecules. A small air bubble is introduced into the capillary tip of the mixing device between refolding buffer and the protein solution in order to prevent diffusion and hence premature mixing of the two reagents prior to the transfer step.

The pulse sequence used for CIDNP pulse-labeling as well as CIDNP NOE pulse-labeling experiments is depicted in Fig. 17. The sequence consists of a presaturation period, which entails a $\pi/2$-pulse followed by a crush gradient and then a train of multiple presaturation pulses to destroy any background magnetization present prior to the illumination of the sample (see above). Since the protein solution is situated outside the r.f. detection coil region during this period, the presaturation sequence only affects the refolding buffer. This step is then immediately followed by the laser illumination – normally between 50 and 500 ms in duration – necessary to initiate the RPM responsible for the generation of nuclear polarization. The injection event takes place immediately after the illumination period and is followed by a short post-injection delay t (100 ms in length) which ensures that protein folding is completed and any remaining free, i.e., paramagnetic, radicals have recombined prior to the r.f. $\pi/2$-pulse applied for detection and subsequent collection of the FID.[19] Prior to the acquisition pulse, three CHESS pulses are applied for solvent suppression. The basic CHESS sequence consists of a selective $\pi/2$-pulse on the signal to be suppressed (H_2O) followed by a strong field gradient to dephase the transverse magnetization produced by the selective pulse. This combination of a pulse and a gradient is the so-called "crush" pulse. Lin et al. have shown that a combination of radiation damping effects and the distant dipolar field, in which long-range dipolar interactions are important, can result in a recovery of magnetization following the application of a "crush" pulse [156, 157]. Improvement of the signal suppression following a "crush" pulse is obtained if the field gradient is applied along the magic angle $\theta = 54.65^\circ$, where the dipolar interaction is zero. θ is defined as the angle between the dipolar vector and the B_0 field vector. For optimum results this selective crush pulse procedure is repeated three times, with the gradients applied along three orthogonal axes so that no accidental refocusing of the magnetization occurs. There are improved versions of this sequence in which the final selective $\pi/2$-pulse is replaced with a selective π-pulse, or the delays and flip angles in the sequence are numerically optimized to compensate for relaxation and B_1 inhomogeneity effects. In most cases, however, the delays (2 ms) and selective flip angles ($\pi/2$) are set to the same value for each CHESS pulse.

4.4.2 A Case Study: The 6 M Urea-Denatured State of the Mini-Protein "TC5b"

The TC5b Mini-Protein

The system of choice for the photo-CIDNP NOE pulse-labeling studies as pioneered by Mok et al. [96] is a designed peptide sequence comprising 20 amino acid residues (NLYIQ WLKDG GPSSG RPPPS) that folds spontaneously and cooperatively to a so-called "Trp-cage" [158]. The "Trp-cage" or TC5b (relative molecular mass ~2.2 kDa) is an iteratively sequence-optimized (de novo) mini-protein synthesized following the truncation and mutational optimization of the poorly folded helical peptide "exendin 4" [159]. Interestingly, the native state of

[19] During the processing of the data the acquired photo-CIDNP (NOE) pulse-labeling spectra are averaged over a series of several "light" and "dark" subtraction pairs – usually four or eight – to improve the signal-to-noise ratio. A new sample and a new injection event are necessary for each measurement.

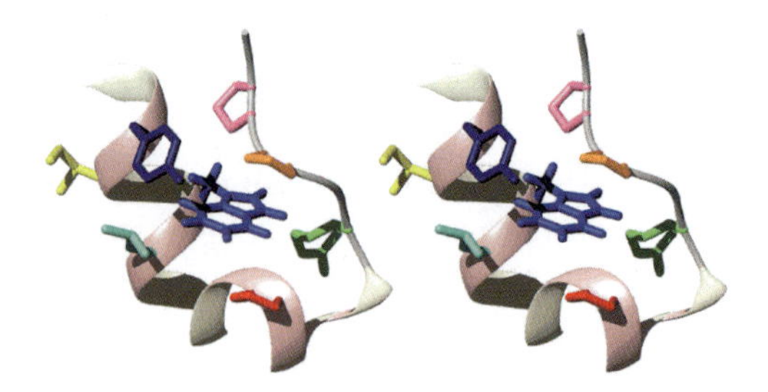

Fig. 20 Stereo view representation of the NMR-derived native state structure of TC5b (PDB entry: 1L2Y). Highlighted are the cage-forming residues (see text) together with the side chains of Ile 4 (*yellow*), Leu 7 (*light blue*), and Arg 16 (*green*). The figure was created using MOLMOL [160]

TC5b exhibits key structural features usually much more common for larger and more complex biomacromolecular systems such as secondary structure elements, a tight packing of side chains, and pronounced tertiary interactions.

The NMR-derived solution state structure (PDB entry 1L2Y) of the TC5b mini-protein (Fig. 20) reveals a compact hydrophobic core where three proline residues (Pro 12, Pro 18, and Pro 19) and a glycine (Gly 11) pack against the aromatic side chains of Tyr 3 and in particular Trp 6 which lies in the center of the hydrophobic "cage", well shielded from solvent exposure. Secondary structure elements present in the peptide's native state comprise a helix between residues 2 and 8 together with a short 3_{10}-helix comprising residues 11–14. Recent vibrational circular dichroism (VCD) studies further reveal the presence of a polyproline II helix (PPII) found at the C-terminal end of the peptide [161]. The globular fold of TC5b is structurally maintained by intramolecular hydrogen bonds and a salt bridge between Asp 9 and Arg 16 that exists in the native state. This salt bridge further interacts with the hydrophobic core of TC5b which leads to increased fold stabilization of the native state structure. Due to its highly compact and globular character, TC5b is considered to be one of the smallest fully folded amino acid sequences known to date containing not only features of secondary structure but also elements of pronounced tertiary interactions and a characteristic protein-like folding behavior. Kinetic investigations into the folding rates of TC5b have been performed and reveal an unprecedented folding speed. The spectroscopic analysis of temperature-jump experiments in combination with Trp fluorescence studies conducted by Qiu et al. showed that folding of this mini-protein is completed within a time of ca. 4.1 μs [162]. This value exceeds the fastest protein folding rates previously observed and is rather surprising given the high relative contact order of the TC5b molecule, which is more typical of slower folding proteins [163, 164]. Additionally, circular dichroism (CD) and fluorescence spectroscopy experiments reveal that folding of TC5b is largely cooperative following a simple two state transition, expected for a protein of such a small size.

In addition, NMR experiments show strong evidence for the assumption that TC5b is in a temperature-dependent equilibrium with an unfolded state that does not display random coil chemical shift values for all of the side chain nuclei due to the presence of persistent non-random interactions. Proof for that observation was mainly given by the detection of CSDs for certain side chain ^{1}H nuclei in the peptide's thermally denatured state. In an early investigation into the structurally perturbed states of TC5b, Neidigh et al. stated that even though their CD spectroscopic analysis gave no evidence for the retention of helicity after the thermal loss of tertiary structure, some unexpected negative CSDs were detectable in the NMR spectrum for the side chain protons of Gly 11 and Pro 12 [158]. These changes were as large as 1.56 and 0.86 ppm for the HA3 proton of Gly 11 and the HD3 proton of Pro 12, respectively. Following the author's statement, these increasingly negative chemical shift deviations can be understood and explained in terms of the existence of a residual high temperature hydrophobic cluster between Trp 6 and Pro 12 which places these two protons further into the shielding region of the aromatic ring current of the tryptophan side chain. As such, the photo-CIDNP NOE pulse-labeling approach was considered to be the ideal method to find definitive answers as to whether the denatured state of TC5b contains elements of residual structure or not.

Preliminary Tests

As the urea-induced unfolding behavior of TC5b was not known, an equilibrium denaturation curve of TC5b was recorded using fluorescence excitation spectrophotometry to confirm that the initial and final conditions used in the pulse-labeling CIDNP NMR experiments correspond to the truly unfolded and native states, respectively. The protein concentration of TC5b used in this experiment was 10 μmol L^{-1}, the buffer was sodium citrate (pH 7.0), and the temperature was 5 °C. Moreover, the urea-containing solutions were allowed to equilibrate overnight. Fluorescence excitation was performed at a wavelength (λ) of 295 nm (slit width: 5.0 nm) in order to measure only the tryptophan fluorescence, and emission detection was done at 360 nm (slit width: 7.5 nm). The fraction of unfolded molecules (F_u) was determined for each concentration and the data was fitted to a sigmoid function showing "two-state" denaturation behavior, albeit indicating low folding cooperativity.

Hence it was confirmed that the peptide's unfolding behavior in the presence of varying concentrations of urea is identical (or at least very similar) to the thermally induced denaturation process as determined by Andersen et al. [158]. Furthermore, the unfolding experiment revealed that at a concentration of about 5.5 M urea, denaturation of TC5b is fully completed and that at urea concentrations smaller than 0.8 M the peptide attains its fully native structure. According to these unfolding studies, the initial (6 M) and final (0.55 M) urea concentrations were carefully chosen to ensure that the pulse-labeling results are reproducible and scientifically meaningful. This change in urea concentration corresponds to an 11-fold dilution of the denaturant upon injection of the highly concentrated urea solution into an aqueous refolding buffer during the pulse-labeling experiment, thereby triggering the submillisecond refolding of the TC5b peptide. Moreover, the pulse-labeling refolding reaction was performed at a temperature of 5 °C where, according to Andersen's temperature denaturation curve in the absence of urea, almost all, i.e., approximately 95%, of the protein molecules are fully folded. In contrast, at a sample temperature of 20 °C, the unfolded state ensemble is still substantially, i.e., more than 15%, populated even in the absence of added urea [158]. Hence, it was safe to assume that essentially all of the injected unfolded molecules refolded to the native state prior to the acquisition of the spectrum during the pulse-labeling experiment. Furthermore, a temperature of 5 °C was also chosen for all other NMR equilibrium spectra, e.g., the two-dimensional TOCSY and NOESY NMR measurements, of the native and the denatured state of TC5b as described in the following section.

In a second preliminary experiment, a ^{1}H chemical shift assignment of TC5b in the presence of residual urea (0.55 M) had to be carried out. Since this low urea concentration resembles the situation after the injection step in the photo-CIDNP NOE pulse-labeling experiment, information of the chemical shift values measured using these conditions makes an unambiguous assignment of cross-polarized resonances in the aliphatic region of the NMR spectrum possible. Accordingly, a TOCSY and a two-dimensional homonuclear NOESY ^{1}H NMR experiment were performed on the native and the denatured forms of TC5b to provide definitive chemical shift assignments and to show the presence of interresidue NOEs in the denatured state. For the TOCSY experiment, a specific mixing scheme designed to reduce ROE transfers during the "spin lock" period was used with a mixing time of 80 ms, and for the NOESY experiments two mixing times of 150 and 500 ms were employed. The conditions used for determining the spectra of the native and denatured states were identical to the initial and final conditions of the pulse-labeling experiments, i.e., the temperature was set to 5 °C, the protein solutions were buffered with 0.05 M sodium citrate at pH 7.5, and the denatured state and the native state solutions contained 6.0 M urea and 0.55 M urea, respectively. The protein concentration for these non-CIDNP experiments was 2 mM. The complete protein backbone and side chain ^{1}H chemical shift resonance assignment of the native state spectrum of TC5b in the presence of 0.55 M urea was achieved analyzing the TOCSY and NOESY spectra of TC5b followed by their comparison with the reported chemical shifts of native TC5b at 0 M urea as outlined in the "Biological Magnetic Resonance Data Bank" (entry: bmr5292.str). The individual ^{1}H chemical shifts were assigned unambiguously. A close examination of these values reveals that, compared to the reported chemical shift data of TC5b in the complete absence of urea, slight changes are observable for some of the ^{1}H chemical shifts measured in the presence of 0.55 M urea, 5 °C, and pH 7.5. These minor deviations, however, can all be explained and rationalized in terms of a slight shift in the folding equilibrium from ca. 98% folded to approximately 8% unfolded with urea added according to the urea denaturation curve of the protein. In addition, the small chemical shift differences can further be attributed to slightly changing solvent conditions.

Photo-CIDNP Equilibrium Experiments

Initially, conventional photo-CIDNP "steady state" spectra of both the peptide's native as well as its urea-denatured states were acquired. As expected, the CIDNP difference spectrum of native TC5b recorded with a relatively short and intense light flash of 100 ms is considerably simpler than the NMR spectrum acquired at thermal equilibrium since only 2 of the 20 side chains present in TC5b's primary sequence react with the flavin photosensitizer and are hence potentially hyperpolarizable. Their polarization signal intensity thereby reflects the relative solvent accessibility of the two CIDNP-active side chains of Tyr 3 and Trp 6 which, in the protein's native state, was calculated to be 53.1% and 10.9%, respectively.

The aromatic region of TC5b's "native state" CIDNP spectrum displays the characteristic absorptive peaks for the side chain protons of Trp 6 (HD1, HE3, and HH2) and an emissive resonance for the aromatic HE 1,2 protons of Tyr 3. In addition, the aliphatic region contains the emissive signals of the HB ^{1}H nuclei of the tryptophan side chain. The absorptive HB CIDNP proton signals of tyrosine are obscured by the buffer resonances. The "steady state" photo-CIDNP experiment of TC5b unfolded in 6.0 M urea, on the other hand, reveals noticeable differences as compared to the native state. In this spectrum, the polarized aromatic Trp signals are stronger relative to those of tyrosine, including the aforementioned indole HE1 proton resonances clearly visible in the "low field" region, thereby confirming the increase of solvent exposure of the Trp 6 residue upon denaturation and hence disruption of the native state hydrophobic core with the tryptophan indole side chain in its center. Intuitively, this feature together with the observation of the loss of spectral resolution and the movement of signals towards random coil chemical shifts detectable in the spectrum acquired in the presence of 6 M urea suggests a less structured, i.e., more flexible, and hence more accessible protein environment in the denatured state.

Subsequently, CIDNP "steady state" spectra of native and unfolded TC5b were acquired using different laser illumination periods to examine whether cross-polarization can be observed between CIDNP-active and aliphatic CIDNP-inactive side chains. The laser output power used in these experiments was chosen with respect to the laser illumination duration in order to ensure that the total amount of light energy delivered to the sample during the laser pulse was constant. Accordingly, laser output powers of 24.0, 6.0, and 2.4 W were employed for the experiments conducted using laser illumination times of 50, 200, and 500 ms, respectively. Since it was clear even before the experiments began, that the observation of this effect in proteins constitutes a crucial prerequisite for the feasibility of the NOE-driven polarization transfer experiment described below, these studies had to be carried out prior to the acquisition of the pulse-labeling spectrum.

From the NMR-derived structure of TC5b it was known that the peptide exhibits a considerable amount of sequence-remote tertiary interactions in its native state. Accordingly, this structural arrangement of the protein represented the most likely case of observing CIDNP-induced cross-polarization visible in the aliphatic region of the NMR spectrum and hence the first CIDNP NOE equilibrium studies were performed on the fully folded state of the peptide. A comparison of the spectra recorded with increasing laser irradiation times and proportionately reduced laser powers shows that, even though the signal intensities from the directly polarized aromatic tryptophan and tyrosine protons listed above remain approximately constant, there are several resonances in the aliphatic region of the spectra acquired with 200 or 500 ms of illumination time – principally between 0.5 and 2.1 ppm – that are essentially absent in the CIDNP "steady state" spectrum acquired using a short illumination time of 50 ms. Moreover, these additional signals significantly grow in intensity as the irradiation time is increased from 200 to 500 ms. This characteristic build-up of polarization as a function of increasing laser illumination duration, i.e., mixing time, strongly suggests that the supplemental signals represent NOE peaks arising from transfer of polarization from initially polarized Tyr 3 and/or Trp 6 side chain protons to certain aliphatic nuclei.

The chemical shifts of the newly appearing aliphatic signals correlate well with cross-peaks between protons separated by ca. 2.0–3.5 Å which were also observed in two-dimensional NOESY spectra of the native state of TC5b and hence used as constraints while calculating the peptide's three-dimensional native state structure [158]. An additional feature observable in the aliphatic region of the "N" state CIDNP spectra acquired with a laser illumination time of 200 or 500 ms, respectively, relates to the overall shape of the cross-polarized signals. A closer examination of these signals shows that they exhibit both partially absorptive and emissive phase contributions. It becomes clear that this mixture of positive and negative phases reflects the opposite CIDNP enhancements of the tyrosine and tryptophan ^{1}H nuclei whose chemically induced spin state perturbations lead to cross-relaxation that occurs with retention of phase as the native state of the "Trp-cage" appears to be in the regime of "slow molecular tumbling" [18, 41]. In particular, the strongly emissive character of the cross-polarized peaks of Ile 4 HG2, HD1, and Leu 7 HD1 at a chemical shift of ca. 0.9 ppm suggests that they arise from the proximity of the (emissive) Tyr 3 HE2 protons in the native state. Likewise, NOEs to methylene groups from the side chain protons of Trp 6 yield several smaller and broader signals which are absorptive in phase, such as Pro 12 HG (ca. 2.08 ppm) and Leu 7 HG (1.6 ppm).

In a next step, the same set of experiments was conducted on the 6 M urea-denatured state of TC5b. An examination of the data reveals that no signals can be detected in the aliphatic region of the spectra acquired with laser illumination times of 50 or 200 ms. At laser illumination times of 500 ms, however, aliphatic resonances are observable in the CIDNP spectrum of TC5b's unfolded state, thereby giving evidence of NOE build-up as a function of increasing laser irradiation time in the denatured state, albeit at a lower intensity as compared to native TC5b. The observation of much less pronounced CIDNP cross-polarization in the spectra of the unfolded state is by no means surprising given the fact that the presence of a high concentration of urea significantly disrupts the structure of the peptide and hence removes most of the tertiary interactions that can be found in the native state structure. In addition, a closer examination and comparison of the "U" state CIDNP spectrum with that of the native state shows that almost all of the peaks have different chemical shifts in the unfolded state. Furthermore, most of the resonance lines representing the aliphatic as well as the amide and aromatic ^{1}H nuclei are not well-resolved in the spectrum of the denatured state, making a direct resonance assignment extremely difficult.

Hence, although denatured TC5b is relatively less compact as compared to the native structure, contacts between different amino acid residues can still be detected under denaturing conditions provided a sufficiently long laser irradiation time is used. Nonetheless, the superposition of these cross-polarized resonances in the aliphatic region of the "U" state CIDNP spectrum emphasizes the need for more resolution and signal dispersion. It is basically this observation together with the strong evidence of residual structure in TC5b's unfolded state which gave the incentive for conducting the pulse-labeling experiments described in the following section.

Photo-CIDNP NOE Pulse-Labeling Experiment

As outlined, the pulse-labeling experiment entails production of photo-CIDNP and buildup of NOEs in the denatured state, followed by a rapid injection of the protein solution into a refolding buffer in the NMR sample tube. Using the aforementioned rapid injection device, complete mixing is usually accomplished within 50 ms [55]. A FID is acquired after a post-injection delay of 100 ms, which allows ample time for complete mixing and recombination of radicals. After subtraction of the corresponding "dark" signal acquired in a separate experiment without laser irradiation, the resulting spectrum of the refolded native state contains aliphatic resonances whose intensities reflect the strength of the chemically amplified NOEs that were generated in the denatured state. TC5b is well suited to this procedure, since the refolding of the "all *trans*" proline conformers is extremely fast and occurs on the order of microseconds. As a result of the aforementioned CIDNP "steady state" experiments, it was decided to use a laser illumination time of 500 ms at a laser power of 2.4 W to ensure efficient magnetization transfer between CIDNP-active and CIDNP-inactive nuclei during the pulse-labeling experiment.

As expected, the aliphatic region of the pulse-labeling ("U" to "N") spectrum differs substantially from the conventional CIDNP spectra of native and denatured TC5b, in particular with regard to the signals that can be found in a region between 0.75 and 2.25 ppm. These cross-polarized resonances can be identified straightforwardly and unambiguously from the NMR assignments of the native state in the presence of 0.55 M urea. The resonances close to 0.9 ppm have the same chemical shift values as compared to certain aliphatic protons of Ile 4 (HG2 and HD1) and Leu 7 (HD1 and HD2) in the NMR spectrum of the native state. Furthermore, this set of signals has a pattern of absorptive enhancements that shows similarities as well as differences if it is compared to the "steady state" CIDNP spectrum of the native state. To ensure that only "true" NOE cross-polarization signals are assigned and in order to avoid the incorrect interpretation of subtraction artifacts that might potentially be present (see below), the remainder of the aliphatic NOE peaks that can be identified in the pulse-labeling spectrum was interpreted in a conservative manner, i.e., only those side chains that display two or more cross-polarized signals were taken into consideration. Following this strategy it can be seen that Ile 4 shows an additional resonance in the pulse-labeling spectrum at the chemical shift of its HB proton (1.92 ppm). In addition, a pair of resonances comprising a strong absorptive peak at 1.67 ppm and another medium-sized signal at 1.84 ppm were identified as the HB3 and HB2 proton nuclei of the side chain of Arg 16. Broad absorptive peaks, which most likely correspond to the native chemical shifts of Pro 12 HB2 at 2.0 ppm and Pro 12 HG2 at 2.1 ppm, can also be found. This completes the chemical shift assignment of the aliphatic region of the pulse-labeling spectrum. Hence, it can be seen that four side chains, i.e., Ile 4, Leu 7, Pro 12, and Arg 16, were identified straightforwardly as recipients of cross-polarization stemming from the hyperpolarized nuclei of the tryptophan residue. However, the individual distances of these side chain nuclei to their polarization source do not appear to be of equal size as their relative signal intensities differ significantly from each other. Hence, it was necessary to perform subsequently a relaxation analysis intended to extract absolute distances between the Trp 6 side chain and aliphatic cross-polarized nuclei, thereby giving the pulse-labeling data a more quantitative character.

Quantitative Analysis of the Photo-CIDNP Pulse-Labeling Data

Absolute distances between nuclei of different, i.e., CIDNP-active and CIDNP-inactive, amino acid side chains in the unfolded state of the "Trp-cage" mini-protein can be extracted using a mathematical model based on the Solomon relaxation equations which were modified to account for the generation of photo-CIDNP and a subsequent NOE-driven magnetization transfer during the laser pulse [96]. This data treatment yields absolute distances between cross-polarization donor and cross-polarization acceptor nuclei provided a few simplifications are introduced into the calculations. First, it is assumed that a single "point source" of hyperpolarization is responsible for the generation of the observed aliphatic NOEs rather than individual nuclei of the tryptophan indole side chain. This is necessary because the pulse-labeling experiment is unable to distinguish between different individual sources of polarization. Second, the CIDNP generation rate during the whole duration of the laser pulse is constant. Third, protein folding (k_f) is much faster as compared to nuclear spin–lattice relaxation (R_1) and the refolding of the protein upon injection is considered to occur instantaneously with respect to the post-injection delay (t) and the acquisition of the NMR spectrum. Subsequently, a straightforward analysis based on the Solomon relaxation equations [165] which are modified to include the production of CIDNP during the laser pulse gives expressions for the relative intensities of directly and cross-polarized signals as observed in the pulse-labeling ("U" to "N") experiment. Provided that experimentally determined ^{1}H spin–lattice relaxation times (T_1) for the native and the

Cross-polarized protons	Ile4 HB	Ile4 HG2, HD1	Leu7 HD1	Leu7 HD2	Pro12 HG2	Arg16 HB3	Arg16 HB2
Distance from Trp[6] in U state (Å)	4.2	4.5	4.0	3.4	3.3	3.8	4.2
Distance from Trp[6] in N state (Å)	8.07	8.71, 8.75	4.81	3.73	4.02	4.68	3.24

Fig. 21 Calculated absolute distances (*red*) between Trp 6 and cross-polarized side chain nuclei in unfolded TC5b. In addition, distances between Trp 6 and the same aliphatic nuclei extracted from the native state PDB file are shown in *black* for comparison

unfolded states of TC5b are available, these mathematical expressions can be used to obtain estimates of absolute intramolecular proton–proton distances. To have an internal calibration which makes the calculation of absolute distances feasible, the ensemble-averaged spatial separation of 3.41 Å between the HD1 nucleus of Trp 6 and the Arg 16 HB2 proton in the native state of TC5b was used. This value was extracted straightforwardly from the PDB file (IL2Y) of the protein's NMR-derived native state structure.

The spin–lattice relaxation times of both the native and the unfolded state necessary for this calculation were measured applying a conventional inversion recovery experiment to the signals found in the spectra of native and urea-denatured TC5b. To estimate spin–lattice relaxation times for aliphatic resonances present in the NMR spectrum of the unfolded state, signal envelope intensities found in the aliphatic region of urea-denatured TC5b were analyzed rather than individual signals due to a lack of signal dispersion and missing resonance assignments. The absolute distance estimates between the Trp 6 side chain and individual aliphatic nuclei which were extracted from these calculations are shown in Fig. 21.

A comparison of the internuclear distances extracted from the CIDNP NOE pulse-labeling data and the distances of the same side chain nuclei present in the native state of the "Trp-cage" displays an unexpected structural change upon denaturation of the peptide. Counterintuitively, the data suggest that in the 6 M urea state of TC5b non-covalent interactions between Trp 6 and aliphatic side chains that are responsible for the generation and stabilization of the native hydrophobic core, e.g., Pro 12, already exist in the peptide's unfolded state. In addition, the distances of other hydrophobic side chain protons, e.g., the HD nuclei of Leu 7 and the HB3 proton of Arg 16, seem to become smaller with respect to the side chain of Trp 6 upon the urea-induced unfolding of the molecule and only the HB2 nucleus of the side chain of Arg 16 is further removed from the tryptophan residue as compared to the native state structure. The most striking structural change, however, is observable for the amino acid side chain of Ile 4 which – in the native

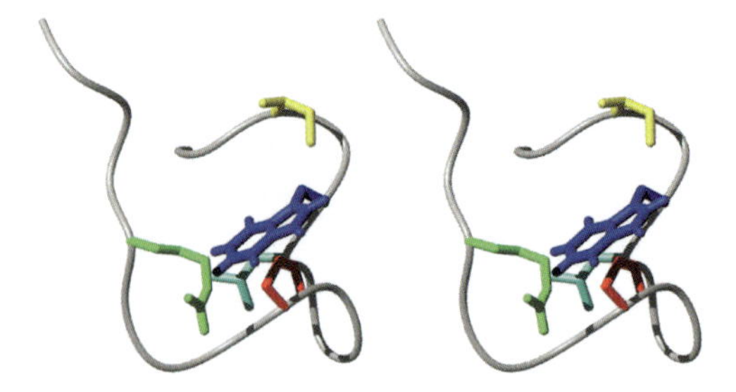

Fig. 22 Schematic representation of the 6 M urea state of the "Trp-cage" as derived from the quantitative analysis of the CIDNP pulse-labeling data. The side chains of Ile 4 (*yellow*), Leu 7 (*light blue*), Pro 12 (*red*), and Arg 16 (*green*) form a hydrophobic cluster around the tryptophan side chain (*dark blue*)

state – is far removed from the tryptophan by a distance of more than 8 Å. In the denatured state, however, the isoleucine side chain protons that exhibit signals in the aliphatic region of the pulse-labeling spectrum are strongly polarized yielding a calculated distance between them and the Trp 6 indole moiety of only 4.2 Å (HB) and 4.5 Å (HG2, HD1), respectively. Hence, this interaction can clearly be identified as a purely non-native contact which seems to exist prior to the folding and the structural rearrangement of the polypeptide chain into its native conformation.

A Pre-existing Hydrophobic Collapse

Taking the calculated internuclear distances presented in the previous section as a loose restraint it was attempted to generate a schematic representation of what the structural ensemble of the "Trp-cage" molecule present at 6 M urea might look like. The result of this attempt is shown in Fig. 22. It has to be stressed though that this schematic does not represent the result of a thoroughly performed structure calculation of an unfolded TC5b ensemble based on numerous intramolecular restraints. In fact, it should be regarded merely as a "quantitatively correct" visualization of arrangements of different amino acid side chains with respect to the Trp 6 indole ring featuring calculated distances which were extracted from the quantitative information provided by the pulse-labeling data. Furthermore, this structural representation exclusively highlights the features of those members of the unfolded ensemble which are amenable to the pulse-labeling experiment both in terms of generation and transfer of polarization as well as with respect to the ultrafast, i.e., submillisecond, refolding of the protein upon dilution of a high concentration of urea. Protein molecules that belong to other unfolded ensemble families, e.g., other proline conformers, are fully neglected and hence cannot be considered.

A close examination of this structural schematic further highlights the features of unfolded TC5b. In contrast to the general understanding that a denatured mini-protein is fully expanded, the results rather suggest a compact hydrophobic cluster around Trp 6 comprising side chains that are normally further removed from this residue in the native state. Although Leu 7, Pro 12, and Arg 16 exhibit several weak NOEs together with tryptophan protons in the native state, Ile 4 does not, nor does it have NOEs with other side chains that constitute the hydrophobic "Trp-cage". In fact, and in accordance with the native state structure, the isoleucine side chain lies on the opposite side of the N-terminal helix and hence points away from the tryptophan. In the unfolded state, however, there is clear evidence that Ile 4 is close to the tryptophan since the side chains of the two residues face each other directly. Accordingly, this interaction can be identified as an entirely non-native interaction. Moreover, NOEs between Trp 6 and the three terminal proline residues in native TC5b are fully absent in the unfolded peptide since they cannot be detected in the "U" to "N" pulse-labeling spectrum (see below). The role of Arg 16 in contributing to the formation of a hydrophobic cluster in native and unfolded TC5b is ambiguous. Although its side chain seems to enhance the stability of the compact cage structure through formation of a salt bridge with Asp 9 in the native state, it is its aliphatic part, identified by CIDNP cross-polarized HB2 and HB3 protons, that is found associated with Trp 6 in the "U" state. On the other hand, Pro 18 and Pro 19 – which seem to stabilize the native state of TC5b – do not appear to be important for the formation of residual structure in the denatured state. In contrast, and in agreement with its enhanced chemical shift deviation in thermally denatured TC5b, Pro 12 does participate significantly in the hydrophobic clustering with tryptophan. Hence, the data seem to suggest that the loss of conformational entropy due to the caging of Trp 6 by several proline side chains is too great a price to pay in the unfolded state, and that hence the formation of these contacts occurs at a later stage on the free energy landscape on the protein's way to the native state.

The assumption of a stabilizing contribution provided by a non-native interaction between isoleucine and tryptophan in the unfolded state is consistent with fluorescence correlation spectroscopic studies of a TC5b fluorophore adduct reported by Sauer et al. [166]. Their experimental results show that the substitution of Ile 4 by a glycine side chain appears to produce a larger fraction of a fully random coil structure in the denatured state ensemble as compared to "wild type" TC5b. Moreover, this paper independently reports on the presence of residual structure in denatured TC5b using a fully NMR-free methodological approach. Recently presented data obtained from UV spectroscopy suggests that the tryptophan indole side chain is actually situated within a hydrophobic environment in the thermally-denatured state, thereby further supporting the observations derived from the CIDNP NOE pulse-labeling data [167].[20]

[20] In the meantime, independently performed experiments using state-of-the-art multidimensional NMR methodology have confirmed the presence of all inter-residue contacts in the 6 M urea-denatured state of TC5b derived from CIDNP NOE pulse-labeling data. In addition, the study provides an unambiguous assignment of all interacting side chain protons in the high-temperature hydrophobic cluster [168].

5 Concluding Remarks

Biomolecular photo-CIDNP spectroscopy is a potent and very sensitive NMR method to elucidate the surface structure of polypeptides and proteins, be it in their native or in structurally ill-defined, i.e., partially folded or denatured, states. During the last 35 years or so the technique has been developed to a significant extent and has helped in finding new answers to the long-standing enigma of how proteins fold correctly into their native three-dimensional conformation following biosynthesis on the ribosome, i.e., the protein folding problem. In particular, the method – unlike optical spectroscopic techniques – offers selectivity by virtue of the chemical shift dispersion of side chain resonances in proteins containing multiple aromatic residues. This benefit has allowed its application both in equilibrium as well as in "real time" studies while a protein folds/unfolds in the NMR magnet, i.e., in situ. The more recent introduction of photo-CIDNP (NOE) pulse-labeling extends the application of the method to extract residue-specific information both on partially folded as well as the ensemble of unfolded conformers in more detail. In the light of recent experimental and theoretical results on the denatured state and, in addition, the early stages of folding, e.g., the initial collapse of hydrophobic polypeptide chains via the formation of both native and also non-native interactions [168–172], the photo-CIDNP technique appears to be a valuable source of insight into the nature of both the non-native and the transient states of a protein. There is no doubt that additional advances in the field, both methodological and experimental in nature, are to be expected in the not-too-distant future. These will surely contribute to an even further widening of its applicability.

Acknowledgements This review is dedicated to Joe Bargon, the author's former scientific mentor. Financial support from the German National Academic Foundation, the UK Biological and Biochemical Sciences Research Council (BBSRC), the European Commission, and the German Research Foundation (DFG) is gratefully acknowledged. The European Neuroscience Institute Göttingen (ENI-G) is jointly funded by the Göttingen University Medical School and the Max Planck Society.

Appendix

The radical pair mechanism (RPM) as described in Sect. 2.4 can be recast in terms of the product operator formalism [17, 18]. The triplet T_0 spin-state of the radical pair can be described by the following expression:

$$|T_0\rangle\langle T_0| = \frac{1}{4}E - S_{1z}S_{2z} + \{ZQ\}_x, \tag{1}$$

where E is the identity operator, $S_{1z}S_{2z}$ is electron two-spin order, and ZQ is electronic zero-quantum coherence:

$$\{ZQ\}_x = S_{1x}S_{2x} + S_{1y}S_{2y}, \tag{2}$$

$$\{ZQ\}_y = S_{1y}S_{2x} + S_{1x}S_{2y}. \tag{3}$$

By analogy, the singlet state is described by

$$|S\rangle\langle S| = \frac{1}{4}E - S_{1z}S_{2z} - \{ZQ\}_x. \tag{4}$$

The initial density operator of the triplet-born radical pair – ignoring the T_+ and T_- states – can be written in the following form:

$$\hat{\rho}(0) = \frac{1}{2}|T_0\alpha\rangle\langle T_0\alpha| + \frac{1}{2}|T_0\beta\rangle\langle T_0\beta| = \frac{1}{4}E - S_{1z}S_{2z} + \{ZQ\}_x, \tag{5}$$

where α and β correspond to the nuclear spin states of a spin-$\frac{1}{2}$ nucleus coupled to electron 1 with an isotropic hyperfine coupling constant a. This density operator is then allowed to evolve under the electronic Zeeman Hamiltonian ($\hat{H}_z = (\omega_{s1}S_{1z} + \omega_{s2}S_{2z})$) and the electron-nuclear hyperfine Hamiltonian ($\hat{H}_h = \pi a\, 2S_{1z}I_z$) where I is the nuclear spin:

$$\hat{\rho}(0) \xrightarrow{(\omega_{s1}S_{1z}+\omega_{s2}S_{2z})t} \begin{array}{l} \frac{1}{4}E - S_{1z}S_{2z} \\ +\{ZQ\}_x \cos(\omega_{s1} - \omega_{s2})t \\ +\{ZQ\}_y \sin(\omega_{s1} - \omega_{s2})t, \end{array} \tag{6}$$

$$\hat{\rho}(0) \xrightarrow{\pi a t 2S_{1z}I_z} \begin{array}{l} \frac{1}{4}E - S_{1z}S_{2z} \\ +\{ZQ\}_x[\cos(\omega_{s1} - \omega_{s2})t \cos(\pi a t)] \\ -2\{ZQ\}_x I_z[\sin(\omega_{s1} - \omega_{s2})t \sin(\pi a t)] \\ +\{ZQ\}_y[\sin(\omega_{s1} - \omega_{s2})t \cos(\pi a t)] \\ +2\{ZQ\}_y I_z[\cos(\omega_{s1} - \omega_{s2})t \sin(\pi a t)], \end{array} \tag{7}$$

$$= \hat{\rho}(t). \tag{8}$$

The nuclear polarization in the recombination products, which is formed through the singlet channel, is then given by the trace of $\hat{\rho}(0)$ with

$$|S\rangle\langle S|I_z = \left(\frac{1}{2}|S\alpha\rangle\langle S\alpha| + \frac{1}{2}|S\beta\rangle\langle S\beta|\right)I_z, \tag{9}$$

$$|S\rangle\langle S|I_z = \left(\frac{1}{4}E - S_{1z}S_{2z} - \{ZQ\}_x\right)I_z. \tag{10}$$

Hence, the expression for the nuclear polarization p^r found in the recombination products of the radical reaction after electron back transfer has occurred takes the following form:

$$p^r = [\hat{\rho}|\mathrm{S}\rangle\langle\mathrm{S}|I_z] = -\frac{1}{4}\sin(\omega_{s1} - \omega_{s2})t\,\sin(\pi a t). \qquad (11)$$

In addition, the nuclear polarization found in the escape products, which is equal and opposite in phase to the recombination products, is given by the trace of $\hat{\rho}(t)$ with $|\mathrm{T}_0\rangle\langle\mathrm{T}_0|I_z$:

$$p^e = -\frac{1}{4}\sin(\omega_{s1} - \omega_{s2})t\,\sin(\pi a t). \qquad (12)$$

From (11) and (12) it can easily be shown that no polarization is produced if the g-values of the two radicals are identical or if there is no hyperfine coupling. In these cases and in cases where Δg is very small, the CIDNP *multiplet effect* applies. Moreover, the phase of the polarization as predicted by Kaptein's sign rule is evident from the signs of Δg and the hyperfine coupling constant a_i of nucleus i as displayed in the respective equations.

References

1. Bargon J, Fischer H, Johnsen U (1967) Z Naturforschg A 22:1551
2. Bargon J (2006) Helv Chim Acta 89:2082
3. Ward HR, Lawler RG (1967) J Am Chem Soc 89:5518
4. Bargon J, Fischer H (1967) Z Naturforschg A 22:1556
5. Lawler RG (1967) J Am Chem Soc 89:5519
6. Fischer H, Bargon J (1969) Acc Chem Res 2:110
7. Closs GL (1969) J Am Chem Soc 91:4552
8. Kaptein R, Oosterhoff LJ (1969) Chem Phys Lett 4:214
9. Kaptein R (1975) Adv Free Radic Chem 5:319
10. Closs GL (1974) Adv Magn Reson 7:157
11. Lawler RG (1973) Progr NMR Spectrosc 9:147
12. Freed JH, Pedersen JB (1976) Adv Magn Reson 8:1
13. Kaptein R, Dijkstra K, Nicolay K (1978) Nature 274:293
14. Fessenden RW, Schuler RH (1963) J Chem Phys 39:2147
15. Kaptein R, Oosterhoff LJ (1969) Chem Phys Lett 4:195
16. Solov'yov IA, Schulten K (2012) J Phys Chem B 116:1089
17. Sørensen OW, Eich GW, Levitt MH, Bodenhausen G, Ernst RR (1983) Progr NMR Spectrosc 16:163
18. Hore PJ, Broadhurst RW (1993) Progr NMR Spectrosc 25:345
19. Kaptein R (1971) J Chem Soc D Chem Commun 732
20. Kaptein R (1972) J Am Chem Soc 94:6251
21. Salikhov KM (1982) Chem Phys 64:371
22. Hore PJ, Stob S, Kemmink J, Kaptein R (1983) Chem Phys Lett 98:409
23. Kuprov I, Craggs TD, Jackson SE, Hore PJ (2007) J Am Chem Soc 129:9004

24. Adrian FJ (1970) J Chem Phys 53:3374
25. Adrian FJ (1971) J Chem Phys 54:3912
26. Markley JL, Bax A, Arata Y, Hilbers CW, Kaptein R, Sykes BD, Wright PE, Wüthrich K (1998) Pure Appl Chem 70:117
27. Mok KH, Hore PJ (2004) Methods 34:75
28. Martin CB, Tsao ML, Hadad CM, Platz MS (2002) J Am Chem Soc 124:7226
29. Lyon CE, Jones JA, Redfield C, Dobson CM, Hore PJ (1999) J Am Chem Soc 121:6505
30. Kuprov I, Hore PJ (2004) J Magn Reson 168:1
31. Zetta L, de Marco A, Casiraghi G, Cornia M, Kaptein R (1985) Macromolecules 18:1095
32. Böhmer V, Goldmann H, Kaptein, Zetta L (1987) Chem Commun 1358
33. Zetta L, Böhmer V, Kaptein R (1988) J Magn Reson 76:587
34. Day IJ (2004) Doctoral thesis, Oxford University, Oxford
35. Tsentalovich YP, Lopez JJ, Hore PJ, Sagdeev RZ (2002) Spectrochim Acta A 58:2043
36. Kaptein R (1982) Biol Magn Reson 4:145
37. Stob S, Kaptein R (1989) Photochem Photobiol 49:565
38. Stob S, Scheek RM, Boelens R, Kaptein R (1988) FEBS Lett 239:99
39. Closs GL, Czeropski MS (1977) Chem Phys Lett 45:115
40. de Kanter FJJ, Kaptein R (1979) Chem Phys Lett 62:421
41. Hore PJ, Egmond MR, Edzes HT, Kaptein R (1982) J Magn Reson 49:122
42. Berliner LJ, Kaptein R (1981) Biochemistry 20:799
43. Hore PJ, Kaptein R (1983) Biochemistry 22:1906
44. Zetta L, Kaptein R, Hore PJ (1982) FEBS Lett 145:277
45. Zetta L, Hore PJ, Kaptein R (1983) Eur J Biochem 134:371
46. de Marco A, Zetta L, Kaptein R (1985) Eur J Biochem 11:187
47. Zetta L, Kaptein R (1984) Eur J Biochem 145:181
48. Closs GL (1975) Chem Phys Lett 32:277
49. Winder SL, Broadhurst RW, Hore PJ (1995) Spectrochim Acta A 51:1753
50. Redfield C, Dobson CM, Scheek RM, Stob S, Kaptein R (1985) FEBS Lett 185:248
51. Berliner LJ, Koga K, Nishikawa H, Scheffler JE (1987) Biochemistry 26:5769
52. Alexandrescu AT, Broadhurst RW, Wormald C, Chyan CL, Baum J, Dobson CM (1992) Eur J Biochem 210:699
53. Radford SE, Dobson CM, Evans PA (1992) Nature 358:302
54. Hore PJ, Winder SL, Roberts CH, Dobson CM (1997) J Am Chem Soc 119:5049
55. Mok KH, Nagashima T, Day IJ, Jones JA, Jones CJV, Dobson CM, Hore PJ (2003) J Am Chem Soc 125:12484
56. Kühn T, Schwalbe H (2000) J Am Chem Soc 122:6169
57. Wirmer J, Kühn T, Schwalbe H (2001) Angew Chem Intl Ed 40:4248
58. Ptitsyn OB (1995) Curr Opin Struct Biol 5:74
59. Dobson CM (2003) Nature 426:884
60. Dobson CM (2005) Nature 435:747
61. Fersht AR (1994) Curr Opin Struct Biol 5:79
62. Daggett V, Fersht AR (2003) Nat Rev Mol Cell Biol 4:497
63. Morozova OB, Yurkovskaya AV, Tsentalovich YP, Forbes MDE, Sagdeev RZ (2002) J Phys Chem B 106:1455
64. Morozova OB, Yurkovskaya OB, Tsentalovich YP, Forbes MDE, Hore PJ, Sagdeev RZ (2002) Mol Phys 100:1187
65. Turro NJ, Zimmt MB, Gould IR (1983) J Am Chem Soc 105:6347
66. Läufer M, Dreeskamp H (1984) J Magn Reson 60:357
67. Leuschner R, Fischer H (1985) Chem Phys Lett 121:554
68. Vollenweider J, Fischer H, Hening J, Leuschner R (1985) Chem Phys 97:217
69. Scheffler JE, Cottrell CE, Berliner LJ (1985) J Magn Reson 63:199
70. Kuprov I, Hore PJ (2004) J Magn Reson 171:171
71. Ahmad M, Galland P, Ritz T, Wiltschko R, Wiltschko W (2007) Planta 225:615

72. Reppert SM, Weaver DR (2002) Nature 418:935
73. Heelis PF (1982) Chem Soc Rev 11:15
74. Heelis PF, Parsons BJ, Phillips GO, McKellar JF (1978) Photochem Photobiol 28:169
75. Kaptein R, Dijkstra K, Müller F, van Schagen CG, Visser AJWG (1978) J Magn Reson 31:171
76. Muszkat KA (1977) J Chem Soc Chem Commun 872
77. Muszkat KA, Gilon C (1977) Biochem Biophys Res Commun 79:1059
78. Muszkat KA, Gilon C (1978) Nature 271:685
79. McCord EF, Boxer SG (1981) Biochem Biophys Res Commun 100:1436, http://www.ncbi.nlm.nih.gov/pubmed/7295309
80. Broadhurst RW (1990) Doctoral thesis, Oxford University, Oxford
81. Connolly PJ, Hoch JC (1969) J Magn Reson 95:165
82. Müller F (1987) Free Radic Biol Med 3:215
83. Hore PJ, Zuiderweg ERP, Kaptein R, Dijkstra K (1981) Chem Phys Lett 83:376
84. McCormick DB (1977) Photochem Photobiol 26:169
85. Scheek RM, Stob S, Boelens R, Dijkstra K, Kaptein R (1984) Faraday Discuss Chem Soc 78:245
86. Scheek RM, Stob S, Boelens R, Dijkstra K, Kaptein R (1985) J Am Chem Soc 107:705
87. Lyon CE (1999) Doctoral thesis, Oxford University, Oxford
88. Maeda K, Lyon CE, Lopez JJ, Cemazar M, Dobson CM, Hore PJ (2000) J Biomol NMR 16:235
89. Vogt W (1995) Free Radic Biol Med 18:93
90. Lee JH, Cavagnero S (2013) J Phys Chem B. doi:10.1021/jp4010168
91. Frieden C, Hoeltzli SD, Ropson IJ (1993) Prot Sci 2:2007
92. Kühne RO, Schaffhauser T, Wokaun A, Ernst RR (1979) J Magn Reson 35:39
93. Green DB, Lane J, Wing RM (1987) Appl Spectrosc 41:847
94. Hamang M, Sanson A, Liagre L, Forge V, Berthault P (2000) Rev Sci Inst 71:2180
95. Haase AJ, Frahm J, Hänicke W, Matthaei D (1985) Phys Med Biol 30:341
96. Mok KH, Kuhn LT, Goez M, Day IJ, Lin JC, Andersen NH, Hore PJ (2007) Nature 447:106
97. Schäublin S, Wokaun A, Ernst RR (1977) J Magn Reson 27:273
98. Barbieri R, Hore PJ, Luchinat C, Pierattelli R (2002) J Biomol NMR 23:303
99. Lyon CE, Suh ES, Dobson CM, Hore PJ (2002) J Am Chem Soc 124:13018
100. Hore PJ (1983) J Magn Reson 55:283
101. Hwang TL, Shaka AJ (1995) J Magn Reson A 112:275
102. Piotto M, Saudek V, Sklenář V (1992) J Biomol NMR 2:661
103. Liu M, Mao XA, Ye C, Huang H, Nicholson JK, Lindon JC (1998) J Magn Reson 132:125
104. Stob S, Scheek RM, Boelens R, Dijkstra K, Kaptein R (1988) Isr J Chem 28:319
105. Aue WP, Bartholdi E, Ernst RR (1976) J Chem Phys 64:2229
106. Jeener J, Meier BH, Bachmann P, Ernst RR (1979) J Chem Phys 71:4546
107. Tsentalovich YP, Morozova OB, Yurkovskaya AV, Hore PJ (1999) J Phys Chem A 103:5362
108. de Marco A, Zetta L, Petros AM, Llinás M, Boelens R, Kaptein R (1986) Biochemistry 25:7918
109. Sekhar A, Cavagnero S (2009) J Magn Reson 200:207
110. Sekhar A, Cavagnero S (2009) J Phys Chem B 113:8310
111. Lee JH, Sekhar A, Cavagnero S (2011) J Am Chem Soc 133:8062
112. Muszkat KA, Wismontski-Knittel T (1985) Biochemistry 24:5416
113. Muszkat KA, Khait I, Hayashi K, Tamiya N (1984) Biochemistry 23:4913
114. Canioni P, Cozzone PJ, Kaptein R (1980) FEBS Lett 111:219
115. Broadhurst RW, Dobson CM, Hore PJ, Radford SE, Rees ML (1991) Biochemistry 30:405
116. Canet D, Lyon CE, Scheek RM, Robillard GT, Dobson CM, Hore PJ, van Nuland NAJ (2003) J Mol Bio 330:397
117. Klein-Seetharaman J, Oikawa M, Grimshaw SB, Wirmer J, Duchardt E, Ueda T, Imoto T, Smith LJ, Dobson CM, Schwalbe H (2002) Science 295:1719

118. Capaldi AP, Kleanthous C, Radford SE (2002) Nat Struct Biol 9:209
119. Lenstra JA, Bolscher BGJM, Stob S, Beintema JJ, Kaptein R (1979) Eur J Biochem 98:385
120. Bolscher BGJM, Lenstra JA, Kaptein R (1979) J Magn Reson 35:163
121. Shelling JG, Sykes BD, O'Neil JDJ, Hofmann T (1983) Biochemistry 22:2649
122. Williams TC, Corson DC, McCubbin WD, Oikawa K, Kay CM, Sykes BD (1986) Biochemistry 25:1826
123. Norton RS, Beress L, Stob S, Boelens R, Kaptein R (1986) Eur J Biochem 157:343
124. Moonen CTW, Hore PJ, Müller F, Kaptein R, Mayhew SG (1982) FEBS Lett 149:141
125. van Schagen, Müller F, Kaptein R (1982) Biochemistry 21:402
126. Müller F, van Schagen CG, Kaptein R (1980) Methods Enzymol 66E:385
127. Draper RD, Ingraham LL (1970) Arch Biochem Biophys 139:265
128. Isenberg I, Szent-Gyorgi A (1958) Proc Natl Acad Sci USA 44:857
129. Stob S, Kemmink J, Kaptein R (1989) J Am Chem Soc 111:7036
130. Vogel HJ, Sykes BD (1984) J Magn Reson 59:197
131. Hore PJ, Volbeda A, Dijkstra K, Kaptein R (1982) J Am Chem Soc104:6262
132. Feeney J, Roberts GCK, Kaptein R, Birdsall B, Gronenborn A, Burgen ASV (1980) Biochemistry 19:2466
133. Fife DJ, Moore WM (1979) Photochem Photobiol 29:43
134. Hori A, Hayashi F, Kyogoku Y, Akutsu H (1988) Eur J Biochem 174:503
135. Day IJ, Wain R, Tozawa K, Smith LJ, Hore PJ (2005) J Magn Reson 175:330
136. Tsentalovich YP, Morozova OB (2000) J Photochem Photobiol A 131:33
137. Tsentalovich YP, Morozova OB, Yurkovskaya AV, Hore PJ, Sagdeev RZ (2000) J Phys Chem A 104:6912
138. Kuprov I, Goez MM, Abbott PA, Hore PJ (2005) Rev Sci Instrum 76:084103
139. Morozova OB, Yurkovskaya AV (2010) Angew Chem Int Ed Engl 49:7996
140. Miller RJ, Closs GL (1981) Rev Sci Instrum 52:1876
141. Blank B, Henne A, Fischer H (1974) Helv Chim Acta 57:920
142. Lee B, Richards FM (1971) J Mol Biol 55:379
143. Khan F, Kuprov I, Craggs TD, Hore PJ, Jackson SE (2006) J Am Chem Soc 128:10729
144. Dobson CM, Hore PJ (1998) Nat Struct Biol 5:504
145. van Nuland NAJ, Forge V, Balbach J, Dobson CM (1998) Acc Chem Res 31:773
146. Zeeb M, Balbach J (2004) Methods 34:65
147. Wenter P, Fürtig B, Hainard A, Schwalbe H, Pitsch S (2006) Chembiochem 7:417
148. Fürtig B, Wenter P, Reymond L, Richter C, Pitsch S, Schwalbe H (2007) J Am Chem Soc 129:16222
149. Fürtig B, Buck J, Manoharan V, Bermel W, Jäschke A, Wenter P, Pitsch S, Schwalbe H (2007) Biopolymers 86:360
150. van Nuland NAJ, Dobson CM, Regan L (2008) Protein Eng Des Sel 21:165
151. Balbach J, Forge V, Lau WS, van Nuland NA, Brew K, Dobson CM (1996) Science 274:1161
152. Schlörb C, Mensch S, Ritcher C, Schwalbe H (2006) J Am Chem Soc 128:1802
153. Day IJ, Maeda K, Paisley HJ, Mok KH, Hore PJ (2009) J Biomol NMR 44:77
154. Balbach J, Forge V, Lau WS, Jones JA, van Nuland NAJ, Dobson CM (1997) Proc Natl Acad Sci USA 94:7182
155. Mok KH, Nagashima T, Day IJ, Hore PJ, Dobson CM (2005) Proc Natl Acad Sci USA 102:8899
156. Lin YY, Lisitza N, Ahn S, Warren WS (2000) Science 290:118
157. Huang SY, Lin YY, Lisitza N, Warren WS (2002) J Chem Phys 116:10325
158. Neidigh JW, Fesinmeyer RM, Andersen NH (2002) Nat Struc Biol 9:425
159. Neidigh JW, Fesinmeyer RM, Prickett KS, Andersen NH (2001) Biochemistry 40:13188
160. Koradi R, Billeter M, Wüthrich K (1996) J Mol Graph 14:51
161. Copps J, Murphy RF, Lovas S (2007) Peptide Sci 88:427
162. Qiu L, Pabit SA, Roitberg AE, Hagen SJ (2002) J Am Chem Soc 124:12952
163. Eaton WA, Muñoz V, Thompson PA, Henry ER, Hofrichter J (1998) Acc Chem Res 31:745

164. Mayor U, Johnson CM, Daggett V, Fersht AR (2000) Proc Natl Acad Sci USA 97:13518
165. Solomon I (1955) Phys Rev 99:559
166. Neuweiler H, Doose S, Sauer M (2005) Proc Natl Acad Sci USA 102:16650
167. Ahmed Z, Beta IA, Mikhonin AV, Asher SA (2005) J Am Chem Soc 127:10943
168. Rogne P, Ozdowy P, Richter C, Saxena K, Schwalbe H, Kuhn LT (2012) PLoS One 7:e41301
169. Dobson CM, Karplus M (1999) Curr Opin Struct Biol 9:92
170. Brockwell DJ, Smith DA, Radford SE (2000) Curr Opin Struct Biol 10:16
171. Dinner AR, Šali A, Smith LJ, Dobson CM, Karplus M (2000) Trends Biochem Sci 25:331
172. Ferguson N, Fersht AR (2003) Curr Opin Struct Biol 13:75

Index

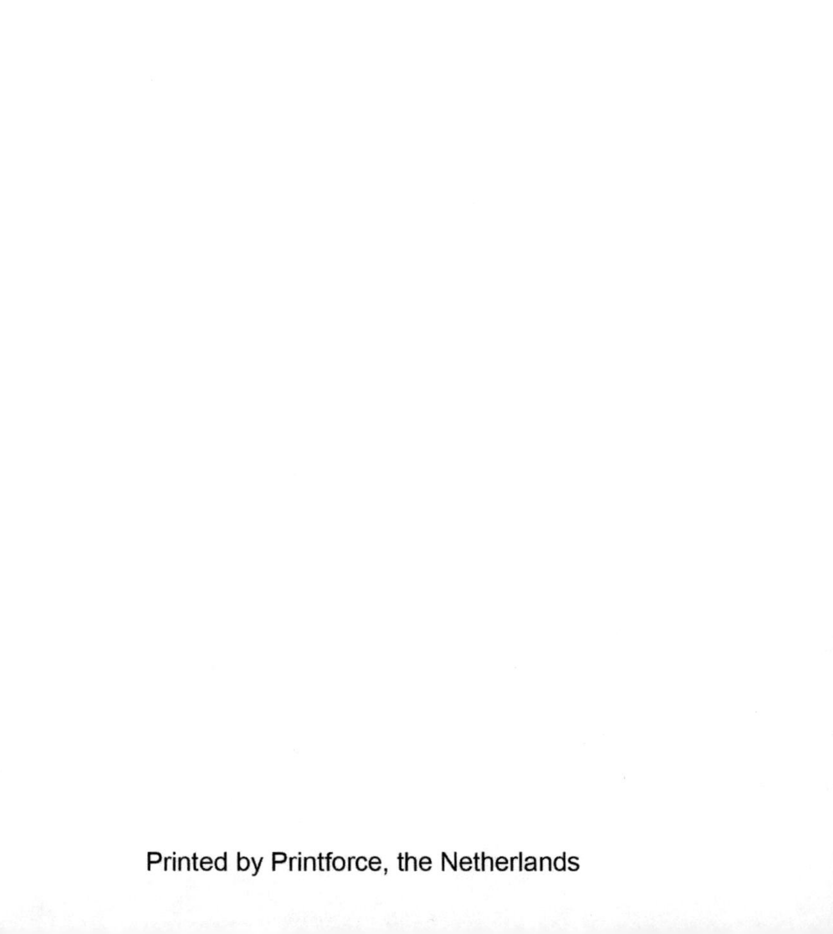

Printed by Printforce, the Netherlands